# MEI

## FOUNDATION
## MEI GCSE
## MATHEMATICS

### THIRD EDITION

PAT BRYDEN — JOHN SPENCER
DIANA COWEY — JULIAN THOMAS
DAVE FAULKNER — CHRISTINE WOOD

Hodder Murray
A MEMBER OF THE HODDER HEADLINE GROUP

**Acknowledgements**

The Publishers would like to thank the following for permission to reproduce copyright material:

**Photo credits**: p.16 © Hulton Archive/Getty Images; p.81 © Hodder; p.150 © Richard Klune/Corbis; p.218 (left) © SIPA PRESS/Rex Features; (right) © Reuters/Corbis; p.290 (top) M.C. Escher's Reptiles © 2006 The M.C. Escher Company-Holland. All rights reserved. www.mcescher.com; (bottom) M.C. Escher's Day and Night © 2006 The M.C. Escher Company-Holland. All rights reserved. www.mcescher.com; p.302 © Gary Merrin/Rex Features; p.390 © Purestock X; p.393 © Purestock X.

RAC, London's Transport Museum (London Underground logo is a registered trademark of Transport for London), Natwest, McDonald's, The Royal Bank of Scotland, Office for National Statistics and Ordnance Survey, Crown Copyright.

Every effort has been made to trace all copyright holders, but if any have been inadvertently overlooked the Publishers will be pleased to made the necessary arrangements at the first opportunity.

Although every effort has been make to ensure that website addresses are correct at time of going to press, Hodder Murray cannot be held responsible for the content of any website mentioned in this book. It is sometimes possible to find a relocated web page by typing in the address of the home page for a website in the URL window of your browser.

Hodder Headline's policy is to use papers that are natural, renewable and recyclable products and made from wood grown in sustainable forests. The logging and manufacturing processes are expected to conform to the environmental regulations of the country of origin.

Orders: please contact Bookpoint Ltd, 130 Milton Park, Abingdon, Oxon OX14 4SB. Telephone: (44) 01235 827720, Fax: (44) 01235 400454. Lines are open 9.00 – 5.00, Monday to Saturday, with a 24-hour message answering service. Visit our website at www.hoddereducation.co.uk

© Catherine Berry, Pat Bryden, Diana Cowey, Dave Faulkner, Geoff Rigby, John Spencer, Julian Thomas, Christine Wood, 1998, 2002, 2007

**First published in 1998 by**
Hodder Murray, an imprint of Hodder Education,
a member of the Hodder Headline Group,
An Hachette Livre UK Company
338 Euston Road
London NW1 3BH

**Second edition published 2002**

**This third edition first published 2007**

Impression number     10 9 8 7 6 5 4 3 2 1
Year                  2011 2010 2009 2008 2007

All rights reserved. Apart from any use permittted under UK copyright law, no part of this publication may be reproduced or transmitted in any form or by any means, electronic or mechanical, including photocopying and recording, or held within any information storage and retrieval system, without permission in writing from the publisher or under licence from the Copyright Licensing Agency Limited. Further details of such licences (for reprographic reproduction) may be obtained from the Copyright Licensing Agency Limited, Saffron House, 6–10 Kirby Street, London EC1N 8TS.

Cover photo © Michael Abbey/Science Photo Library
Illustrations were drawn by Ian Foulis and Associates, Maggie Brand, Tom Cross, Bill Donohoe, Tony Wilkins, Ann Kronheimer, Joseph McEwan and Barking Dog Art.
Typeset in 11.5pt New Baskerville by Pantek Arts Ltd, Maidstone, Kent
Printed in Dubai

A catalogue record for this title is available from the British Library.

ISBN: 978 0340 940 532

| YORK COLLEGE | |
|---|---|
| 150378 | |
| HJ | 19-Jun-2009 |
| 510 POR | £20.99 |
| | |

# Contents

**Information** .................. 2
   *How to use this book* .................. 2

## Chapter One: Co-ordinates ...... 4
   *Finding a square* .................. 4
   *Finding a point* .................. 6
   *Mathematical co-ordinates* ............. 8

## Chapter Two: Using numbers .. 12
   *Length* .................. 12
   *Weight and mass* .................. 14
   *Capacity* .................. 14
   *Time* .................. 16
   *Money* .................. 18

## Chapter Three: Types of number .................. 22
   *Multiples and factors* .................. 22
   *Prime numbers* .................. 24
   *Squares, square roots and cubes* ...... 26

## Chapter Four: Symmetry .......... 30
   *Reflection symmetry* .................. 30
   *Drawing reflections* .................. 32
   *Rotational symmetry* .................. 34
   *Angles round a point and on*
      *a line* .................. 36

## Chapter Five: Fractions ............ 40
   *Equivalent fractions* .................. 40
   *Adding and subtracting*
      *fractions (1)* .................. 42
   *Adding and subtracting*
      *fractions (2)* .................. 44
   *Improper fractions, mixed numbers*
      *and reciprocals* .................. 46
   *Mixed numbers* .................. 48
   *Fractions of a quantity* .................. 50

## Chapter Six: Maps and drawings .................. 54
   *Maps and scales* .................. 54
   *Scales* .................. 56
   *Angles* .................. 58

## Chapter Seven: Decimals ......... 62
   *Tenths and hundredths* .................. 62
   *Halves and quarters* .................. 64
   *Adding and subtracting decimals* .... 66
   *Multiples of 10* .................. 68
   *Multiplying decimals* .................. 70
   *Dividing decimals* .................. 72

## Chapter Eight: Shapes .............. 76
   *Sorting shapes* .................. 76
   *Triangles* .................. 78
   *Quadrilaterals* .................. 80
   *Other kinds of shapes* .................. 82
   *Circles* .................. 84

## Chapter Nine: Percentages ....... 88
   *25%, 50% and 75%* .................. 88
   *Finding percentages* .................. 90
   *Percentage calculations* .................. 92
   *From fractions to percentages* ......... 94

## Chapter Ten: Statistics .............. 98
   *Displaying data* .................. 98
   *Pie charts* .................. 100
   *Line graphs* .................. 102
   *Vertical line charts* .................. 102
   *Averages and spread* .................. 104
   *Grouping data* .................. 106
   *Displaying grouped data* .............. 108
   *Making comparisons* .................. 110

# Contents

## Chapter Eleven: Directed numbers ............114
- Negative numbers ............114
- Adding and subtracting ............116
- Positive and negative co-ordinates ...118

## Chapter Twelve: Ratio and proportion ............122
- Simple proportion ............122
- Ratio and proportion ............124
- Conversion graphs ............126

## Chapter Thirteen: Formulae ..130
- Using a formula ............130
- More formulae ............132
- Collecting like terms ............134
- Using brackets ............136
- Adding and subtracting with negative numbers ............138

## Chapter Fourteen: Surveys ......142
- Recording data ............142
- Stem-and-leaf diagrams ............144
- Planning a survey ............146
- The survey report ............148
- Social statistics ............150

## Chapter Fifteen: Using symbols ............152
- Being brief ............152
- Using negative numbers ............154
- Simplifying expressions with negative numbers ............156

## Chapter Sixteen: Spending money ............160
- Bills ............160
- Buy now, pay later ............162
- Value added tax (VAT) ............164

## Chapter Seventeen: Graphs ....168
- Gradients and intercepts ............170
- Obtaining information ............172
- Travel graphs: distance and time ..174
- Finding the speed from a travel graph ............176
- The gradient of a travel graph ......178

## Chapter Eighteen: Perimeter and area ............182
- Perimeter ............182
- Area ............184
- Triangles ............186
- Shapes made of rectangles and triangles ............188
- Circumference ............190
- Area of a circle ............192

## Chapter Nineteen: Three dimensions ............196
- Drawing solid objects ............196
- Using isometric paper ............198
- Nets ............200
- Volume ............202
- Surface area of a prism ............204

## Chapter Twenty: Earning money ............208
- Wages ............208
- Salaries ............210
- Simple interest ............212
- Tax ............214

## Chapter Twenty one: Estimation ............218
- Approximations ............218
- Decimal places ............220

# Contents

Significant figures......................222
Estimating costs .......................224
Using your calculator .................226

## Chapter Twenty two:
Equations ................................230
Solving equations (1) .................230
More equations .........................232
Using equations ........................234
Solving equations (2) .................236

## Chapter Twenty three:
Probability...............................240
Calculating probabilities ..............240
Working with probabilities ..........242
Estimating probabilities................244

## Chapter Twenty four: Using
indices.....................................248
Number patterns ......................248
Prime factorisation ....................250
Index notation..........................252
Rules of indices........................254
Calculators...............................256
Brackets ..................................258

## Chapter Twenty five:
Measuring and drawing............262
Triangles .................................262
More triangles .........................264
Using bearings.........................266

## Chapter Twenty six: Using
fractions ..................................270
Multiplying fractions..................270
Dividing fractions .....................272
Fractions to decimals ................274

Fractions to percentages................276
Making comparisons ...................278

## Chapter Twenty seven:
Angles and shapes ....................282
Parallel lines ............................282
Angles and triangles...................284
Quadrilaterals ..........................286
Interior angles of polygons ...........288
Exterior angles of polygons...........290
Tessellations ............................292

## Chapter Twenty eight: Using
decimals ..................................296
Tenths and hundredths................296
Multiplication and division .........298

## Chapter Twenty nine:
Sequences ................................302
Patterns ..................................302
More sequences .......................304
Finding n ................................306
More number sequences.............308

## Chapter Thirty: Using
percentages..............................312
Percentages, decimals and
    fractions .............................312
Percentage calculations ...............314
Proportions .............................316

## Chapter Thirty one:
Co-ordinates and graphs .........320
More co-ordinates .....................320
Using co-ordinates....................322
Equations and graphs ................324
Curved graphs.........................326

# Contents

**Chapter Thirty two: Using statistics** .................. 330
- *Pie charts* ............................. 332
- *Mean, mode, median and range* ... 334
- *Grouping data* ..................... 336
- *Scatter diagrams* ................. 340
- *Mean, median and mode of grouped data* .................. 340
- *Line of best fit* .................... 342

**Chapter Thirty three: Using formulae** .......................... 346
- *Using brackets* ..................... 346
- *Multiplying out brackets* ......... 348
- *More formulae* ..................... 350
- *Changing the subject of a formula* ......................... 352
- *Expanding two brackets* ......... 354

**Chapter Thirty four: Equations and inequalities** ...... 358
- *Solving equations* .................. 360
- *Trial and improvement* ............ 362
- *Using inequalities* .................. 364
- *Number lines* ........................ 366
- *Solving inequalities* ................ 368

**Chapter Thirty five: Ratio and proportion** ...................... 372
- *Simplifying ratios* .................. 372
- *Unitary method* .................... 374
- *Changing money* ................... 376
- *Distance, speed and time* ........ 378

**Chapter Thirty six: Area and volume** ............................. 372
- *Parallelograms and trapezia* ..... 382
- *Cuboids* ............................... 384
- *Volume of a prism* .................. 386
- *More about prisms* .................. 388

**Chapter Thirty seven: Transformations** .................. 392
- *Translations using column vectors* ........................ 394
- *Reflection* ........................... 396
- *Rotation* ............................. 398
- *Enlargement* ........................ 400
- *Scale factors less than 1* ........ 402
- *Similar shapes* ..................... 404

**Chapter Thirty eight: Locus** ... 408
- *Simple loci* ........................... 408
- *A point equidistant from two fixed points* .................. 410
- *A point equidistant from two lines* ......................... 412

**Chapter Thirty nine: Pythagoras' rule** .................. 416
- *Finding the hypotenuse* ........... 416
- *Finding one of the shorter sides* ..... 418

**Index** .................................. 422

# How to use this book

 This symbol next to a question means you are allowed to use your calculator. Don't use your calculator if you can't see the symbol!

 This symbol means you will need to think carefully about a point. Your teacher may ask you to join in a discussion about it.

## *Angles*

acute    right angle 90°    obtuse    reflex

## *Triangles*

equilateral    isosceles    right-angled    scalene

**Area of triangle = $\frac{1}{2}$ × base × height**

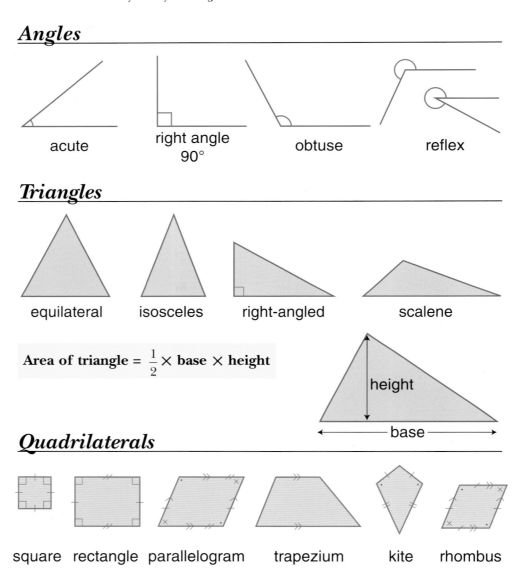

## *Quadrilaterals*

square    rectangle    parallelogram    trapezium    kite    rhombus

## Circles

Circumference of circle = π × diameter
                       = 2 × π × radius

Area of circle = π × (radius)$^2$

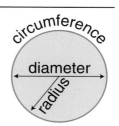

## Solid figures

**Volume of cuboid = length × width × height**

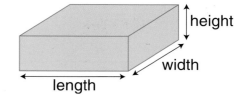

## Units

### Metric system

**Length**
1000 metres = 1 kilometre
1000 millimetres = 1 metre
100 centimetres = 1 metre
10 millimetres = 1 centimetre

**Mass**
1000 grams = 1 kilogram
1000 milligrams = 1 gram

**Capacity**
1000 litres = 1 kilolitre
1000 millilitres = 1 litre

kilo = 1000 times
centi = $\frac{1}{100}$ times
milli = $\frac{1}{1000}$ times

### Approximate conversions

1 km = $\frac{5}{8}$ miles

1 m = 39.37 inches

1 kg = 2.2 pounds (lb)

1 litre = $1\frac{3}{4}$ pints

1 foot = 30.5 cm

1 inch = 25.4 mm

1 pound = 454 g

1 gallon = 4.5 litres

## Long multiplication and long division examples

434 × 14

```
    434
     14
   ----
   4340
   1736
   ----
   6076
```

Answer: 6076

434 ÷ 14

```
      31
    _____
14 ) 434
      42
      --
       14
       14
       --
       ..
```

Answer: 31

# One

# Co-ordinates

## Finding a square

You are visiting the city of York.

Here is a street map to help you find your way around.

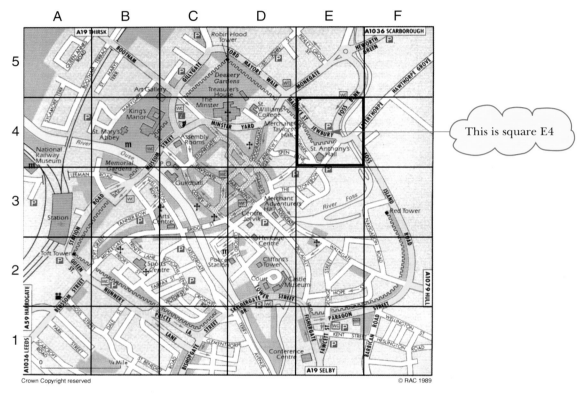

This is square E4

*Find Centre Jorvik.*

*Find the Police Station.*

Unless you already know York, it's like looking for a needle in a haystack, isn't it?

The square grid is there to help you. Each square has a number and a letter.

*Find these places.*
- *The Sports Centre in square* B2
- *Tower Street in square* D2
- *St Mary's Abbey in square* B4

You can see that it is easier to find things when you know which square to look in.

# 1: Co-ordinates

**1** Look at the map of York city centre.
In which square is
a) Clifford's Tower?
b) the Railway Station?
c) Red Tower?

**2** Look at this plan of Castle Bromwich Gardens.
In which square (or squares) is
a) the South gate?
b) the West pond?
c) the New orchard?
d) the Maze?

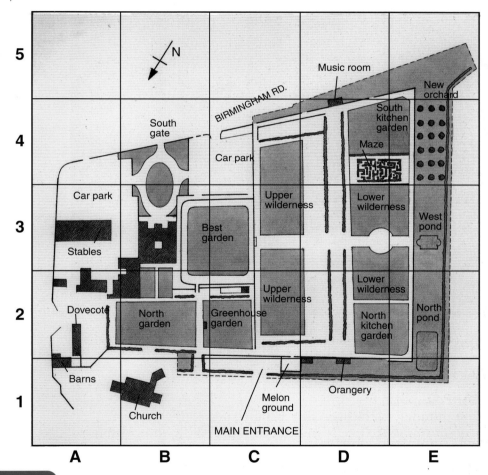

Airline seats often have a number (to tell you the row) and a letter (to tell you the seat). So 4H means row 4, seat H.

Make a list of other places in which a letter and a number are used to locate something.

How does this kind of system help?

## 1: Co-ordinates

# Finding a point

Sue is a landscape gardener.
She is planning the Smiths' garden.
She draws this plan. The distances are in metres.

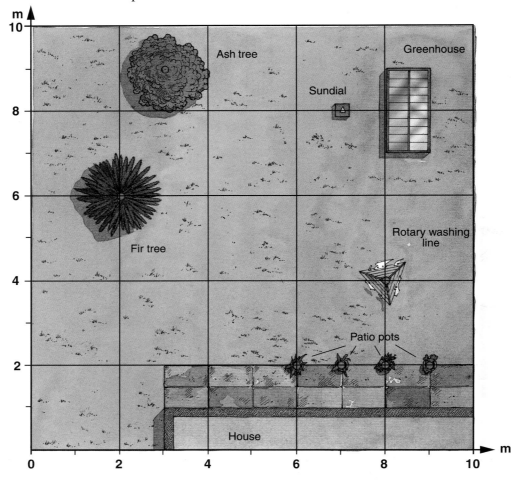

Sue has marked the positions of the trees.
She has used **co-ordinates**.

The fir tree is at (2, 6).
The first number, 2, tells you the distance across the garden.
The second number, 6, tells you the distance along the garden.

Using co-ordinates you can describe exactly where things are.

*The ash tree is at (3, 9).*
*How far is it from the left-hand side of the garden?*
*What are the co-ordinates of the washing line?*
*Where is the point (0, 0)?*
*Sue's assistant plants the fir tree at (6, 2) instead of (2, 6).*
*Where does she plant it?*
*The owner puts a garden gnome at $(3, 1\frac{1}{2})$.*
*Describe its position.*

# 1: Co-ordinates

Look again at Sue's plan.

**1** What are the co-ordinates of the sundial?

**2** What are the co-ordinates of the patio pots?
(There are 4 pots so you need 4 sets of co-ordinates.)

**3** How far is the sundial from the side of the greenhouse?

**4** How wide and how long are the patio slabs?

Look at this map of the Causeway Coast in Northern Ireland.
The distances are in kilometres.

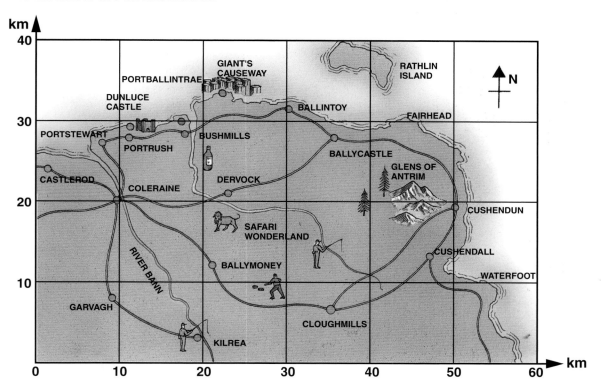

**5** Write down the co-ordinates of
a) Coleraine    b) Cushendun
c) Ballintoy    d) Ballycastle.

**6** Estimate the distance, as the crow flies
a) from Ballymoney to Dervock
b) from Dunluce Castle to Garvagh
c) from the Giant's Causeway to Kilrea.

*Remember that the first number is the distance across, and the second is the distance up*

Ordnance Survey maps use grid references, such as 251716.
These are just co-ordinates written in a different way.
Find an Ordnance Survey map of your area.
Write down the grid references of 5 interesting places.
Explain how grid references work.

## 1: Co-ordinates

# Mathematical co-ordinates

You have seen how a square grid can be used to locate an area or a point on a map or plan.

You can also use a square grid in mathematics.

You draw and label a pair of **axes**. (It is easiest to do this on graph paper.)

You can identify any point on the grid using co-ordinates.

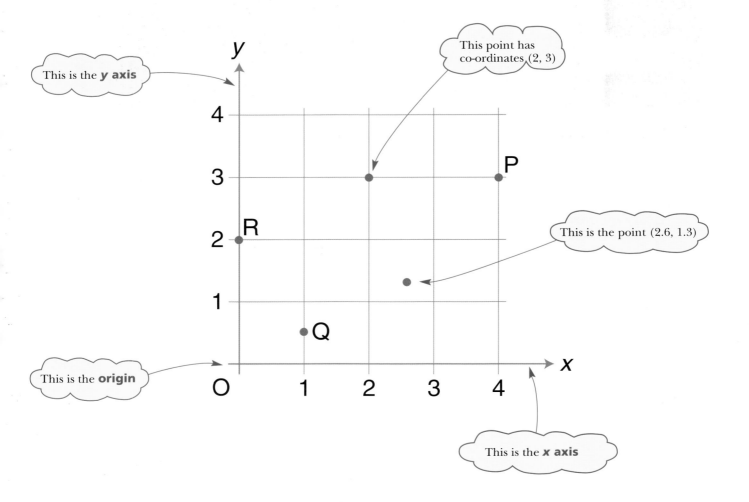

The distance along the $x$ axis is called the **x co-ordinate**.

The distance along the $y$ axis is called the **y co-ordinate**.

The $x$ (or horizontal) co-ordinate always comes first.

*What are the co-ordinates of the origin, O?*

*What are the co-ordinates of points P, Q and R?*

# 1: Co-ordinates

**1** These diagrams show the positions of the stars in two well-known constellations.
For each constellation, list the co-ordinates of all its stars.

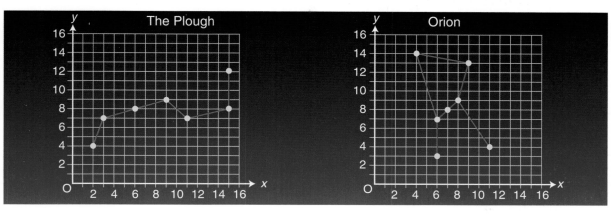

**2** On squared paper draw a pair of axes with 0 to 10 on each axis.
Mark on your grid in red the constellation of Leo:
(1, 1), (0, 3), (4, 4), (2, 5), (2, 6), (2, 7), (3, 8), (4, 8).

Mark in blue the constellation of Cancer:
(10, 3), (7, 3), (8, 5), (8, 6), (8, 7).

**3** Draw a grid with 0 to 4 on each axis.
Join (1, 1) to (1, 3) then (1, 3) to (3, 3), then (3, 3) to (3, 1), then (3, 1) to (1, 1).
What shape have you drawn?

**4** Write instructions for drawing this kite using co-ordinates.

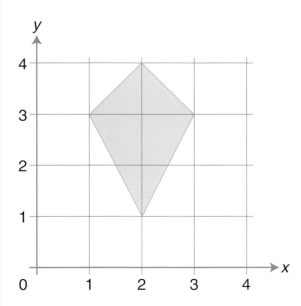

You are asked to mark out 3 badminton courts in a sports hall.

A badminton court is 13.4 m long and 6.1 m wide.

The sports hall is 18 m by 26 m.

On graph paper, draw a pair of axes going from 0 to 30 m ($y$ axis) and from 0 to 20 m ($x$ axis).

Draw a plan of the hall with one corner at the origin. Give the co-ordinates of all the corners of each court.

# 1: Co-ordinates

## Finishing off

**Now that you have finished this chapter you should**

★ understand why grids are used on maps and diagrams

★ understand what is meant by *x* axis, *y* axis and origin

★ be able to write down the *x* and *y* co-ordinates of a point on a grid

★ be able to plot a point using its co-ordinates

**Use the questions in the next exercise to check that you understand everything.**

### Mixed exercise

**1**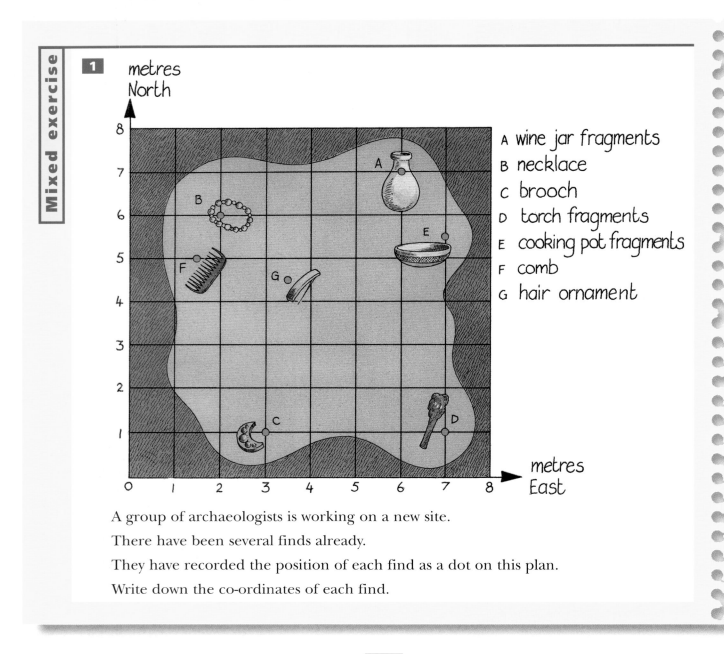

A group of archaeologists is working on a new site.

There have been several finds already.

They have recorded the position of each find as a dot on this plan.

Write down the co-ordinates of each find.

# 1: Co-ordinates

**2** On a piece of squared paper draw a 10 × 10 grid.
Join these points:
(2, 2), (2, 9), (3, 9), (3, 6), (6, 6), (6, 9), (7, 9), (7, 2), (6, 2), (6, 5), (3, 5), (3, 2) and back to (2, 2).
Shade in the shape you have made.
What letter of the alphabet is it?

**3** Find a partner. You are going to play a game called battleships.

This version of the game uses co-ordinates.

You will both need a 10 × 10 grid, with *x* and *y* axes on it.

Draw on your grid one 'submarine', one 'frigate' and one 'battleship' (see diagram).

Choose your own positions, and don't let your partner see.

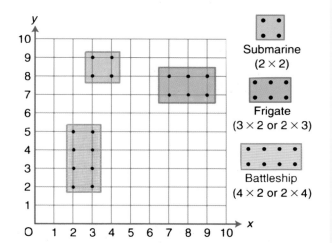

*Mixed exercise*

### The rules

Take it in turns to choose a pair of whole number co-ordinates.

You score a 'hit' when you pick a point **inside** one of your opponent's ships.

When you score a 'hit' your partner must cross out that point, and say what type of ship you have hit.

You get an extra turn when you score a 'hit'.

A ship is sunk when all the points inside it have been 'hit'.

The winner is the one to sink all the opponent's ships.

This grid is used in calculator displays.

Each line can be lit up.

The diagram shows how the calculator displays the number 3.

Draw diagrams for as many other numbers and letters as you can.

Which letters and numbers can you not draw? Which are you unable to tell apart?

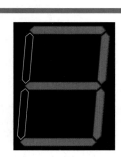

# Two

# Using numbers

## Length

*What is your handspan?*

*How far away is the nearest airport?*

*How thick is a pencil lead?*

*How tall are you?*

What would you use to measure these lengths? Here are some ideas.

You can measure lengths in inches, feet, yards or miles.

These units are from the **Imperial system**.

> 12 inches = 1 foot
> 3 feet = 1 yard
> 1760 yards = 1 mile

You can measure lengths in millimetres (mm), centimetres (cm), metres (m) or kilometres (km).

These units are from the **metric system**.

> 10 mm = 1 cm
> 100 cm = 1 m
> 1000 m = 1 km

You sometimes need to change lengths from Imperial to metric units (or the other way). When you just need a rough answer, you can use these conversions:

> 39 inches is about 1 m
> 1 foot is about 30 cm
> $\frac{5}{8}$ mile is about 1 km.

When you need to estimate lengths, you will find it helpful to remember things like

> the end of my thumb is about 1 inch long
> the span of my hand is about 20 cm
> 1 pace is about 1 m.

*Measure the widths of all your finger nails. Which one is closest to 1 cm?*

*Measure the distance from your nose to the end of your outstretched arm. How near is this to 1 m?*

# 2: Using numbers

**1** How many centimetres are there in  a) 2 m?  b) 4 m?

**2** How many metres are there in  a) 5 km?  b) 8 km?

**3** How many millimetres are there in  a) 3 cm?  b) 9 cm?

**4** How many centimetres are there in  a) 20 mm?  b) 35 mm?

**5** Greg drives from Bristol to Brighton. Here is a map of his journey.

a) How far has he driven when he gets to Southampton?
b) How far has he driven when he gets to Brighton?
c) Which is the longest part of the journey?

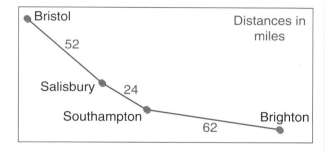

**6** A railway coach is 20 m long. How long is each of these trains?

a)

b)

**7** Hugh's office has a shelf 90 cm long. Hugh has 12 box files. Each box file is 7 cm thick. Will they all fit on the shelf?

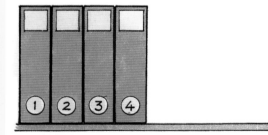

**8** Ella designs the tickets for a concert. She decides to use cards like this one.

(Not to scale)

She designs the decorated border which is 1 cm wide. The printing goes in the middle part. What are the width and depth of the middle part?

Find 8 examples of lengths written in catalogues or on packaging.

For each one, say whether the units are from the Imperial system or the metric system.

Write your answers in a table like this.

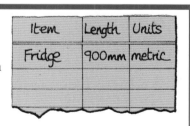

## 2: Using numbers

# Weight and mass

This cat is 3 kilograms, or just over $6\frac{1}{2}$ pounds.

In everyday English this is called the cat's **weight**. In Science the word **mass** is used instead.

*What is the mass of a leaf?*

*What is the mass of a person?*

*What is the mass of a car?*

*How would you measure these?*

You can measure masses in ounces (oz), pounds (lb), stones or tons.
These units are from the Imperial system.

You can measure masses in grams (g), kilograms (kg) or tonnes.
These units are from the metric system.

```
16 oz = 1 lb
14 lb = 1 stone
160 stone = 1 ton
```

```
1000 g = 1 kg
1000 kg = 1 tonne
```

You sometimes need to change masses from Imperial to metric units (or the other way). When you just need a rough answer you can use these conversions:

    1 oz is about 28 g
    2.2 lb is about 1 kg
    1 ton is about 1 tonne.

*An airline gives a baggage allowance of 30 kg.*
*How many pounds is this?*

# Capacity

This petrol can holds 5 litres, or just over 1 gallon.
This is the **capacity** of the can.

It is also called the **volume** of the can

*What is the capacity of a teacup?*

*What is the capacity of a bucket?*

You can measure capacity in pints or gallons.
These units are from the Imperial system.

```
8 pints = 1 gallon
```

```
1000 mL = 1 L
```

You can measure capacity in millilitres (ml) or litres (l).
These units are from the metric system.

When you need to convert from Imperial to metric units (or the other way), these rough conversions are helpful:

    $1\frac{3}{4}$ pints is about 1 litre      1 gallon is about $4\frac{1}{2}$ litres.

# 2: Using numbers

**1** Write
   a) 1 kg in grams
   b) 500 g in kilograms
   c) 2 litres in millilitres
   d) $\frac{1}{2}$ gallon in pints
   e) this weight in grams
   f) this volume in litres.

**2** Tony is shopping for his large family.
   a) He buys 4 bags of potatoes, each containing 5 kg. What is the total weight of the potatoes?
   b) He needs 1 kg of margarine. The biggest tubs in the shop contain 500 g each. How many of these tubs should Tony buy?
   c) Tony buys 2 kg of rice. How many grams is this?
   d) He buys 5 kg of onions. How many grams is this?

**3** Guy weighs 8 stone 10 pounds. How many pounds is this?

**4** Tom (105 kg), Fay (65 kg), Jenny (55 kg), Richard (80 kg) and Jill (70 kg) get into a lift. They see this notice in the lift.

   a) Can they all ride together in the lift safely?
   b) Jill offers to use the stairs. Can the other 4 go together in the lift?
   c) Richard offers to use the stairs instead. Can the other 4 go together in the lift?

**5** Mandy's baby needs 200 ml of milk at each feed.
   How many feeds can Mandy get out of
   a) a 1-litre carton of ready-to-drink baby milk?
   b) a tin of baby milk powder that makes 5 litres?

**6** Helga's doctor has given her a bottle of medicine containing 140 ml. She is to take a 5 ml dose 4 times a day.
   a) How many 5 ml doses are in the bottle?
   b) How many days will the bottle last?

---

Find 8 examples of weights (or masses) written in catalogues or on packaging.

Find 8 examples of capacity.

For each one, say whether the units are from the metric system or the Imperial system. Make a table like this.

| Item | Weight or Capacity | Units |
|------|--------------------|-------|
| Coffee | 200 g | Metric |

## 2: Using numbers

# Time

*How long does it take to run a mile?*
*How long does it take to boil an egg?*

*How would you measure these times?*

> 60 seconds = 1 minute
>
> 60 minutes = 1 hour

Jesse Owens ran 100 m in 10.2 seconds in 1936.

## Clock times

There are 24 hours in a day.

This clock shows only 12 hours. You have to decide whether it is before noon (am) or after noon (pm) when you read the time from it

This digital clock shows times from 00:00 to 23:59. It is a **24-hour clock**

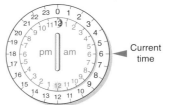

◄ Current time

This central heating timer shows both systems. You can see how to convert 24-hour clock times to 12-hour clock times.

## Timetables

| Hertford Bus Station | 0820 | 0845 | 0920 | 0945 |
| --- | --- | --- | --- | --- |
| Ware Railway Station | 0830 | 0855 | 0930 | 0955 |
| Hundred Acre Estate | — | 0905 | — | 1005 |
| Hoddesdon | 0840 | 0915 | 0940 | 1015 |
| Broxbourne | 0848 | 0923 | 0948 | 1023 |
| Cheshunt Railway Sta. | 0912 | 0947 | 1012 | 1047 |

These buses don't stop at Hundred Acre Estate

Most timetables use the 24-hour clock, like this one.

Tina lives on Hundred Acre Estate.
She wants to go by bus to Broxbourne in time for a job interview at 10.15 am.

*Which bus should Tina catch?*

## 2: Using numbers

**1** a) Change 120 minutes into hours.

b) Change 90 seconds into minutes and seconds.

**2** a) Change 3 minutes into seconds.

b) Change 1 hour 15 minutes into minutes.

**3** An athletics team has 4 runners who take 51 seconds, 52 seconds, 53 seconds and 49 seconds for the legs of a 4 × 400 metre relay. How long, in minutes and seconds, do they take altogether?

**4** Write these times using the 24-hour clock.

a) 4.35 am   b) 5.40 pm   c) 9.20 pm   d) 1.50 pm

**5** Write these 24-hour clock times using am or pm.

a) 0825   b) 1430   c) 2345   d) 1815

**6** a) For how many hours in the week is this shop open?

b) Is it open at 4 pm on Saturday?

c) Is it open at 7 pm on Monday?

Opening hours
Monday-Friday 0900-1800
Saturday 0900-1700
Sunday Closed

**7** a) How long does the News last?

b) How long does the Local news last?

c) What programme is on at quarter to seven?

d) How long does Tennis highlights last?

Evening TV
1800 The News
1830 Local news
1840 Tennis highlights
1915 EastEnders

**8** Use the bus timetable opposite for this question.

a) Lindsey wants to be at Ware Railway Station by 9 am. What is the latest bus that she can catch from Hertford?

b) Liam lives in Hoddesdon and wants to be at Cheshunt Railway Station by 1010. What is the latest bus that he can catch?

---

Find a real timetable.

Rewrite the timetable (or part of it), changing the times to 24-hour clock times (or vice versa).

Why is the 24-hour clock useful?

## 2: Using numbers

# Money

**Example**

Denise is a businesswoman. She books in at a hotel on Monday at £50 per night and she leaves on Friday morning.

a) How many nights does she stay at the hotel?

b) How much does it cost?

**Solution**

a) Mon    Tue    Wed    Thur    Fri
       1      2      3      4

*Notice that there are 5 days but only 4 nights*

b) The cost of the stay is 4 × £50 = £200.

## Booking a holiday

This chart shows the price (in pounds) of a holiday in Minorca in July for one person sharing a twin room.

| Hotel | Blue Water | | Golden Sand | |
|---|---|---|---|---|
| Room | Twin | | Twin | |
| Board | Half Board | | Full Board | |
| Departure | 7 Nights | 14 Nights | 7 Nights | 14 Nights |
| 6Jul-12Jul | 415 | 639 | 465 | 729 |
| 13Jul-19Jul | 449 | 669 | 505 | 759 |
| 20Jul-9Aug | 489 | 729 | 525 | 789 |
| Supplements per person per night | Single room £3 | | Single room £4 | |
| | Sea view £3 | | Balcony £5 | |
| | Full board £10 | | | |

Andrew and Verity want 14 nights at Hotel Blue Water, starting on 8 July. They want a twin room with a sea view.

The travel agent works out the cost.

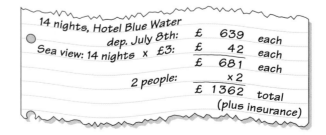

```
14 nights, Hotel Blue Water
     dep. July 8th:      £  639  each
Sea view: 14 nights × £3: £   42  each
                          £  681  each
         2 people:            ×2
                          £ 1362  total
                         (plus insurance)
```

*How much money do they save if they decide not to have a room with a sea view?*

## 2: Using numbers

**1** Paul and his 3 children decide to go on this boat trip.
How much does it cost in total?

**2** Vicky buys 2 T-shirts at £5 each and leggings at £9.
How much change does she get from a £20 note?

**3** Dave and Sophie go away for a long weekend.
They book in at a hotel on Friday at £26 per person per night.
How much will the bill be when they leave on Monday morning?

**4** Last Monday 700 cars and 20 lorries crossed this bridge.
How much money was collected?

**5** Jo organises a concert trip. She buys 12 seats at £8 each and 15 seats at £10 each.
What is the total cost of the tickets?

In questions 6 to 9 use the holiday price chart opposite.

**6** Sally and Mark want to go to Hotel Blue Water for 7 nights starting on 15 July. They want a twin room.
How much does this cost?

**7** Lee wants a single room at Hotel Golden Sand for 14 nights. His departure date is 30 July.
How much does this cost?

**8** Carla and Paul want a twin room with a balcony at Hotel Golden Sand from 19 July for a week.
How much does this cost?

**9** Mr Abbott wants a single room for a week's holiday full board. His departure date is 21 July.
Which hotel offers the cheaper deal, and by how much?

Find a 2-week holiday of your choice in a travel brochure.

How much will it cost?

Write down any choices you have made about when to go, what type of room to book and so on.

What are the cheapest times of year to go on your holiday?

Why do the prices vary through the year?

## 2: Using numbers

# Finishing off

**Now that you have finished this chapter you should be able to**

★ find length, mass, capacity and time

★ use the 24-hour clock and timetables

★ change a quantity from one unit to another

★ do money calculations

Use the questions in the next exercise to check that you understand everything.

## Mixed exercise

**1** a) How many millimetres are there in 4 cm?
   b) How many grams are there in 4 kg?
   c) How many litres is 3000 ml?
   d) How many seconds are there in 5 minutes?

**2** Kelly weighed 49 kg last year. She now weighs 52 kg. How much weight has she gained?

**3** Louise is stocking the chill-cabinets of a supermarket. She has a trolley with 4 shelves. There are 12 containers on each shelf. Each container holds 6 pints of milk.

   a) How many containers are there on the trolley?
   b) How many pints of milk are there on the trolley?

**4** Greg's Monday morning timetable is shown here.

|       | MONDAY     |
|-------|------------|
| 0850  | ASSEMBLY   |
| 0900  | MATHS      |
| 0940  | SCIENCE    |
| 1020  | BREAK      |
| 1035  | FRENCH     |
| 1115  | TECHNOLOGY |
| 1155  | LUNCH      |

   a) For how long, in hours and minutes, is he in lessons?
   b) How long, in hours and minutes, does the morning last?

**5** Look at this train timetable.

| London Kings Cross | 1938 | 2008 | 2038 | 2108 | 2138 |
|---|---|---|---|---|---|
| Finsbury Park | 1943 | 2013 | 2043 | 2113 | 2143 |
| Potters Bar | 1954 | 2024 | 2054 | 2124 | 2154 |
| Hatfield | 2000 | 2030 | 2100 | 2130 | 2200 |
| Welwyn Garden City | 2004 | 2034 | 2104 | 2134 | 2204 |

   a) How long does the 2038 train take to travel from Kings Cross to Welwyn Garden City?
   b) Carys gets to Kings Cross at 9.30 pm. She catches the next train to Potters Bar. What time does she arrive?

## 2: Using numbers

**Mixed exercise**

**6** Charlotte arrives at this guest house on Sunday evening and books to stay until Friday morning. She is too late for an evening meal on Sunday but orders it for the other nights.

How much is her bill?

**Bed and Breakfast £18**
**Evening Meal £7**

**7** This table shows the price (in pounds) of a holiday in Crete for one person sharing a twin room.

| Apartment | Cliffedge | | Beachside | |
|---|---|---|---|---|
| Departure date | 7 Nights | 14 Nights | 7 Nights | 14 Nights |
| 21Jul-11Aug | 399 | 475 | 429 | 499 |
| 12Aug-19Aug | 395 | 469 | 425 | 495 |
| 20Aug-28Aug | 375 | 445 | 399 | 475 |
| Supplements (per person per night) | Single room £4 | | Single room £3 | |
| | Sea view £3 | | Sea view £2 | |

a) Andrea and Tim want to start their 7-night holiday on 13 August. They would like to book 7 nights at Beachside with a sea view.

How much does this cost?

b) Jodie wants a single at Cliffedge for 14 nights, departing on 3 August.

How much does this cost?

---

This chart from a road atlas shows the distances in miles between 6 cities.

a) You can see that the distance from Southampton to Newcastle is 323 miles, but you need to travel from Southampton to Newcastle passing through all the other cities on the way.

*Cambridge is 61 miles from London.*

```
         Birmingham
    101  Cambridge
    108   207   Cardiff
    120   (61)   155   London
    202   230   318   285   Newcastle
    129   132   141    80   323   Southampton
```

Look at a map and decide which orders are sensible.

Find the one with the shortest total distance.

b) Make your own distance chart for 6 towns or villages near you.

# Three

# Types of number

## Multiples and factors

Debbie works for a food and drink company. She decides how items are packaged.

### Multiples

Debbie decides that tins of tea should be packaged in pairs.

*How many tins are there in 4 packages?*

*How many tins are there in 5 packages?*

1 package    2 packages    3 packages

The numbers of tins, 2, 4, 6, 8, 10, … are called the **multiples** of 2.

They are the answers to
the 2× table:   $1 \times 2 = 2$
                $2 \times 2 = 4$
                $3 \times 2 = 6$
                and so on.

> The multiples of 2 are the **even** numbers.
> The other numbers, 1, 3, 5, 7, …, are **odd**

### Factors

Debbie thinks that small bottles of Supa Juice will sell well in packages of 12.

Debbie designs this package.

It has 3 rows of 4.

$3 \times 4 = 12$. We say that 3 and 4 are **factors** of 12.

There are other ways of arranging 12 bottles in a rectangular package.

*Draw 2 other ways of arranging 12 bottles in a rectangular package.*

The different arrangements tell you all the factors of 12.

You may have thought of 6 rows of 2.    $6 \times 2 = 12$.

You may have thought of 1 row of 12.    $1 \times 12 = 12$.

The factors of 12 are    1, 2, 3, 4, 6 and 12.

> It is often useful to list the factors in order, like this

Another way of saying this is that 12 is **divisible** by 1, 2, 3, 4, 6 and 12.

*What numbers are factors of both 24 and 36?*

> These numbers are called **common factors**

# 3: Types of number

**1** Prem is delivering post to houses in Aspen Road.

He has post for numbers 8, 13, 21, 36, 44 and 47.

a) Write down the odd numbers.
b) Write down the even numbers.
c) How do you know which numbers are odd and which are even?
d) Prem delivers to the houses in order. How many times does he cross the road?

**2** Here is a box of chocolate eclairs.

How many eclairs are there in

a) 2 boxes?   b) 3 boxes?   c) 4 boxes?

**3** Sue is catering for a party.

She needs 50 eggs.

How many boxes does she need to buy?

**4** Lucy has 18 cork tiles.

She wants to use them to make a rectangular noticeboard in her room.

She sketches this arrangement.

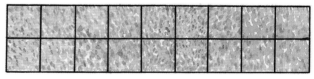

a) Draw all the other ways of making a rectangular noticeboard using exactly 18 tiles.
b) Now list all the factors of 18.

**5** Kay works in a nursery.

She is looking for a storage box for the building blocks.

There are 50 blocks.

She finds a box that can hold 16 blocks in each layer (4 rows of 4).

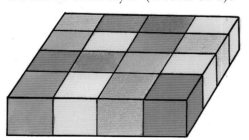

a) How many complete layers like this can Kay make?
b) How many blocks are left over?

**6** a) List all the factors of

(i) 10   (ii) 16   (iii) 7   (iv) 20
(v) 15   (vi) 30   (vii) 14   (viii) 21
(ix) 48   (x) 54   (xi) 28   (xii) 60

b) List the common factors of 28 and 60.
c) Which two numbers in a) have the most common factors?

Draw all the different ways of arranging 12 yoghurts in a rectangular package. Remember that you can use more than one layer.

Draw all the ways of arranging 18 yoghurts in a rectangular package.

Choose another number of yoghurts, and draw all the ways of arranging them in a rectangular package.

# 3: Types of number

## Prime numbers

Lauren goes on holiday to Portugal.

She likes the traditional tiles that she sees.

She wants to make 2 rows of 4 tiles above her hand basin at home, so she buys 8 identical square tiles.

Sadly, one tile gets broken on the way home. Only 7 are left.

How can Lauren arrange these in a rectangle?

There are only two ways.

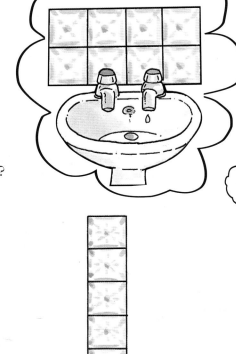

The different rectangles you can make tell you the factors of a number. 7 has only two factors, 1 and 7.

7 is an example of a **prime number**.

A prime number has just two different factors, 1 and itself.

 *How can you tell that 8 is not a prime number?*

The first three prime numbers are

2    (factors 1 and 2)

3   (factors 1 and 3)

5   (factors 1 and 5)

 *What is the next prime number?*

*What are the next four after that?*

*Can you find a quick way of identifying prime numbers?*

## 3: Types of number

**1** Work out whether each of these is a prime number.

a) 24   b) 23   c) 75   d) 31
e) 39   f) 82   g) 101  h) 236

**Investigation**

Eratosthenes lived from 276–194 BC (approximately).

One of the things for which he is famous is this method of finding prime numbers. It is called the 'Sieve of Eratosthenes'.

Get (or draw) a 10 × 10 number square like this, and follow these steps.

- Cross out 1
- Leave 2, but cross out all other multiples of 2 (4, 6, 8, …, 100)
- Leave 3, but cross out all other multiples of 3 (6, 9, 12, …, 99)
- 4 has already been crossed out so move on to 5
- Leave 5, but cross out all other multiples of 5 (10, 15, 20, …)
- Continue in this way for 6, 7, 8, 9 and 10.

| 1 | 2 | 3 | 4 | 5 | 6 | 7 | 8 | 9 | 10 |
|---|---|---|---|---|---|---|---|---|---|
| 11 | 12 | 13 | 14 | 15 | 16 | 17 | 18 | 19 | 20 |
| 21 | 22 | 23 | 24 | 25 | 26 | 27 | 28 | 29 | 30 |
| 31 | 32 | 33 | 34 | 35 | 36 | 37 | 38 | 39 | 40 |
| 41 | 42 | 43 | 44 | 45 | 46 | 47 | 48 | 49 | 50 |
| 51 | 52 | 53 | 54 | 55 | 56 | 57 | 58 | 59 | 60 |
| 61 | 62 | 63 | 64 | 65 | 66 | 67 | 68 | 69 | 70 |
| 71 | 72 | 73 | 74 | 75 | 76 | 77 | 78 | 79 | 80 |
| 81 | 82 | 83 | 84 | 85 | 86 | 87 | 88 | 89 | 90 |
| 91 | 92 | 93 | 94 | 95 | 96 | 97 | 98 | 99 | 100 |

How many numbers are left uncrossed? These are all the prime numbers less than 100.

You stopped following the steps at 10. Why didn't you have to follow the steps all the way up to 100? (You might like to discuss this with your group. The reason isn't easy to see.)

---

You are writing a quiz to raise money for your youth group. You decide to do a 'fill in the blanks' game about numbers. Here is the first question.

11 p_____ in a f_____ t_____

The correct answer is

11 p<u>layers</u>   in a f<u>ootball</u>   t<u>eam</u>

Make up 10 questions like this, using only prime numbers. (You can use the same number in more than one question.)

## 3: Types of number

# Squares, square roots and cubes

Ramesh designs products for a chocolate manufacturer.

He has designed a range of chocolate squares.

Here are the first three sizes:

 1 × 1 = 1     2 × 2 = 4     3 × 3 = 9

 *How many pieces would there be in a 4 × 4 square?*
*How many in a 5 × 5 square?*

The numbers 1, 4, 9, 16, 25, … are called **squares**, or **square numbers**.

A quick way of writing 4 × 4 is $4^2$

Similarly, 5 × 5 is $5^2$, and so on.

(Say this as '4 squared')

Ramesh's largest chocolate square has 36 pieces.

You know that
$$6^2 = 6 \times 6 = 36$$

A square with 36 pieces must have 6 pieces along each side.

You can write $\sqrt{36} = 6$.

The **square root** of 36 is 6.

Similarly, a square with 25 pieces has 5 pieces along each side.
You can write $\sqrt{25} = 5$.

Ramesh is working on a 'chocolate cube' for the Christmas market.

Each cube is made up of smaller cubes.

Here are the first three sizes:

 *How many chocolate cubes are there in the 3 × 3 × 3 cube?*

The numbers 1, 8, 27, … are called **cubes**, or **cube numbers**.

A quick way of writing 2 × 2 × 2 is $2^3$.

(Say this as '2 cubed' or '2 to the power 3')

# 3: Types of number

**1** Work out  a) $5^2$  b) $8^2$  c) $9^2$  d) $12^2$  e) $20^2$.

**2** Find the square root of

  a) 25    b) 49    c) 100    d) 121    e) 400.

(You will have to guess the square root, then check by squaring it.)

**3** A chessboard is a large square divided into 64 smaller squares.

How many squares has it along each side?

**4** During an earthquake appeal, Siba asks her friends to knit small squares. She collects the squares and sews them together to make square blankets.

  a) How many squares does she need for a blanket with 8 squares along each side?

  b) How many squares does she need for a blanket with 9 squares along each side?

  c) One of Siba's blankets has 100 squares. How many squares does it have along each side?

  d) Siba has 121 squares left. Can she make a square blanket with these? If so, how many squares will it have along each side?

**5** How many small cubes are there in each of these large cubes?

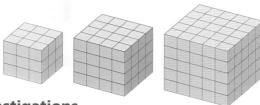

**6** We call 10 'ten' and $10^2$ 'a hundred'.

What do we call

  a) $10 \times 10 \times 10$, or $10^3$?
  b) $10 \times 10 \times 10 \times 10 \times 10 \times 10$, or $10^6$?
  c) $10^9$?

## Investigations

**1** Find out how to square a number on your calculator.

Check that you can use it to work out $4^2$.

Now use your calculator to work out

  a) $13^2$  b) $21^2$  c) $30^2$  d) $100^2$.

**2** Find out how to use the 'square root' button on your calculator.

Check that you can use it to work out $\sqrt{36}$. Now use your calculator to work out

  a) $\sqrt{121}$  b) $\sqrt{225}$  c) $\sqrt{256}$

  d) $\sqrt{625}$.

---

Make a list of 6 everyday objects that are squares or cubes.

For each one, state as accurately as you can the length of one side.

## 3: Types of number

# Finishing off

**Now that you have finished this chapter you should be able to**

★ recognise odd and even numbers
★ find factors
★ work out multiples
★ recognise prime numbers
★ work out square numbers and cube numbers
★ find square roots

Use the questions in the next exercise to check that you understand everything.

**Mixed exercise**

**1** The first and last houses on each side of Trinity Road are numbered on this map.

a) How many houses are there on the odd numbered side?

b) How many houses are there on the even numbered side?

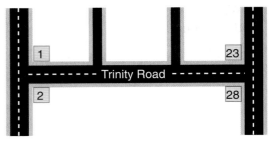

**2** Look at this number pattern:  6, 12, 18, …, …, …, …, …, …, ….

a) Copy and complete the pattern.

b) Describe the numbers.

**3** Phil has 24 square slabs to make a rectangular patio. He could arrange them like this:

a) Draw all the other ways of arranging 24 slabs in a rectangle.

b) List all the factors of 24.

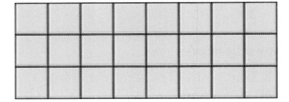

**4** A newspaper is made up of large sheets of paper. Each sheet is folded in half, to make 4 pages of newspaper.

a) How many pages are there in a newspaper made up of 8 sheets?

b) How many pages are there in a newspaper made up of 15 sheets?

c) Can a newspaper made up in this way have 42 pages?

**5** a) Write down the first 10 multiples of 5.

b) Is 95 a multiple of 5?

c) Is 107 a multiple of 5?

d) Explain how you can tell whether or not a number is a multiple of 5.

# 3: Types of number

**6** List all the factors of   a) 8   b) 35   c) 42   d) 100.

**7** Which number in the following list is a prime?

   18   95   70   41   99

**8** How many squares are needed for each of these patchwork designs?

 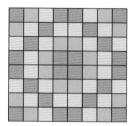

**9** Without using a calculator, write down the value of

   a) $5^2$   b) $8^2$   c) $3^3$
   d) $10^3$   e) $\sqrt{4}$   f) $\sqrt{36}$.

**10** Write down the value of

   a) $20^2$   b) $9^3$   c) $13^2$
   d) $\sqrt{225}$   e) $\sqrt{196}$   f) $\sqrt{625}$.

**Investigation**

Find the squares of 1, 2, 3, 4, ... up to 20.

Write them in a table like this, so that you can look at the last digits of the square numbers. (A few have been done for you.)

| Number | Square of number | Last digit of square |
|---|---|---|
| 1 | 1 | 1 |
| 2 | 4 | 4 |
| ... | | |
| 8 | 64 | 4 |

Look at the numbers in the right hand column.

Do you think that a square number could end with the digit 2?

Could 213 643 be a square number?

Buy yourself a chocolate orange.

How many people can share it equally with none left over? (There are several answers.)

You eat one piece while no-one is looking. How many people can share it equally now?

*Mixed exercise*

# Four

## Symmetry

### Reflection symmetry

Look at this picture of the Canadian flag.

If you stand a mirror on the dotted line, the flag will look exactly the same in the mirror as it does without the mirror.

The dotted line is a mirror line. It is also called a **line of reflection symmetry**.

*The flag is symmetrical about this line*

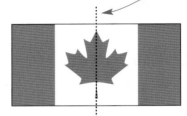

The Israeli flag has 2 lines of reflection symmetry. They are shown as dotted lines in the picture.

 *Check that the flag looks the same when you stand a mirror on each dotted line.*

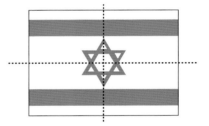

The English flag (the St George's Cross) also has 2 lines of reflection symmetry.

 *Do you think its diagonals are lines of reflection symmetry? Stand a mirror on the diagonal to find out.*

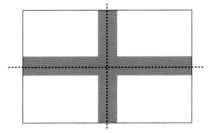

The Swiss flag has 4 lines of reflection symmetry. One is vertical, one is horizontal, and the other 2 are diagonal. The lines of symmetry are shown on the diagram.

 *Check them using a mirror.*

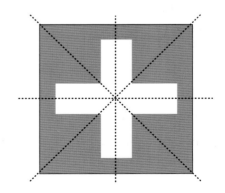

# 4: Symmetry

**1** Which of these signs have reflection symmetry?

A        B        C        D

  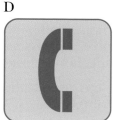

**2** Write down how many lines of symmetry each of these shapes has.

Copy each diagram and draw on the lines of symmetry.

a                b                c

 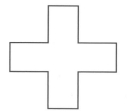 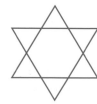

d                e                f

 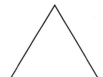 

**3** Make 3 copies of this diagram.

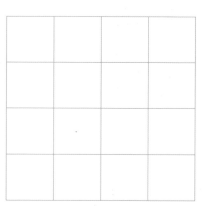

a) On the first copy, shade 8 of the squares to make a pattern with 1 line of symmetry.

b) On the second copy, shade 8 of the squares to make a pattern with 2 lines of symmetry.

c) On the third copy, shade 8 of the squares to make a pattern with 4 lines of symmetry.

Design a symmetrical pattern for a tile for a bathroom wall.

Show how the tiles would fit together to make an overall pattern.

## 4: Symmetry

# Drawing reflections

Sally is making a Christmas card. She is drawing a Christmas tree.

She wants both sides of the tree to be exactly the same, so that it has a line of reflection symmetry.

Sally has already drawn the left hand side of the tree.

*How can she draw the other side so that it is exactly the same?*

One way is to use tracing paper.

Sally traces her drawing and then turns the tracing over.

She uses the turned over tracing to draw the other half of the tree.

Another way to draw symmetrical shapes is to use squared paper or square dotted paper. You do not need to use tracing paper to draw the reflection. Instead, you use the squares (or dots) to help you, like this:

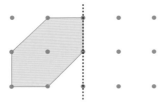 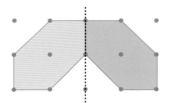

If there are 2 lines of reflection symmetry as in the next picture, reflect the shape in one of the lines first, then reflect the shape and its reflection in the other line.

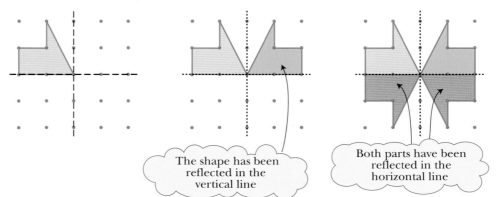

The shape has been reflected in the vertical line

Both parts have been reflected in the horizontal line

# 4: Symmetry

**1** The dotted lines in the patterns below are lines of reflection symmetry. Copy the patterns on squared paper and draw the rest of each pattern.

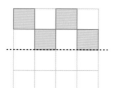 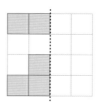 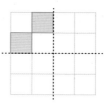

**2** Copy each of these patterns on square dotted paper, and draw the rest of it. (Again the dotted lines show the lines of reflection symmetry.)

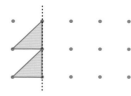 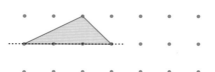

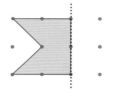 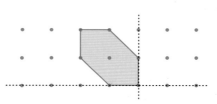

**3** Copy each of these shapes on square dotted paper. Draw its reflection in the line of symmetry shown.

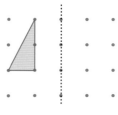 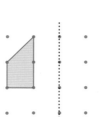  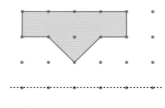

Design a pattern with reflection symmetry to go on a tea or coffee cup. The pattern should be symmetrical about the cup's handle.

## 4: Symmetry

# Rotational symmetry

The US dollar symbol has no lines of reflection symmetry. (You can check this using a mirror.)

However, it does have another kind of symmetry, called turn symmetry, or rotational symmetry.

 *Trace the dollar symbol. Now turn your tracing round through half a turn, so that it is upside down. You should find that you can fit the tracing exactly over the dollar symbol.*

The dollar symbol looks exactly the same in 2 different positions – the right way up and upside down.
It has **rotational symmetry with order 2**.

This is the symbol for recycling.

It can be turned so that it looks the same in 3 different positions. It has **rotational symmetry with order 3**.

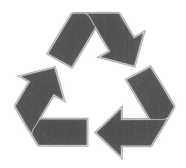

 *Trace the symbol and check that the tracing will fit exactly over the symbol in 3 different positions.*

Some patterns can have both reflection symmetry and rotational symmetry.

The symbol for First Aid has 4 lines of reflection symmetry. These are shown with dotted lines.

It looks the same if it is turned through a quarter turn, so it has
**rotational symmetry with order 4**.

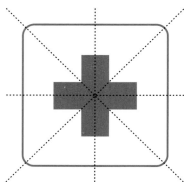

# 4: Symmetry

**1** Which of these playing cards have rotational symmetry?

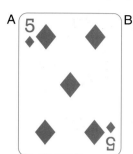 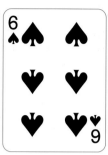  

**2** Write down the order of rotational symmetry for each of these patterns.

a) 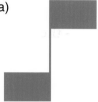   b)    c)

d) 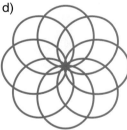   e) 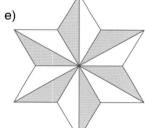   f)

**3** Copy the patterns below and shade in more squares so that each pattern has rotational symmetry of order 4.

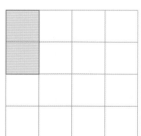   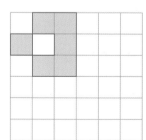

Design a pattern with rotational symmetry for a circular dinner plate. Write down the order of rotational symmetry of your plate.

## 4: Symmetry

# Angles round a point and on a line

Measure the angle at the centre of each piece of pizza.

For each pizza, add up the angles you have measured.

*What do you notice?*

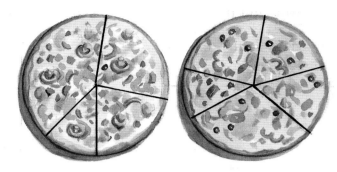

This works because there are 360° in a whole turn.

The number of degrees in half a turn is 180° (half of 360°).

An angle which is half a turn makes a straight line.

If several angles fit together to make a straight line, the angles must add up to 180°.

Half a turn or 180°

You can use these rules when you are working out angles in a diagram, without measuring them.

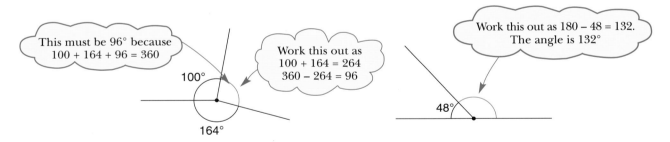

This must be 96° because 100 + 164 + 96 = 360

Work this out as
100 + 164 = 264
360 − 264 = 96

Work this out as 180 − 48 = 132.
The angle is 132°

100°
164°
48°

The two lines in this diagram meet at **right angles**.

They are **perpendicular**. The angles are all 90°.

*What fraction of a turn is a right angle?*

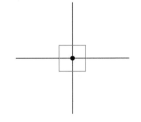

---

**Angles that fit round a point add up to 360°.**

**Angles that fit on a line add up to 180°.**

**A right angle is 90°.**

# 4: Symmetry

**1** Work out the angle marked with a letter in each of the diagrams. The diagrams are not drawn the correct size, so you must use the rules on the opposite page.

If you measure them, you will get the wrong answers!

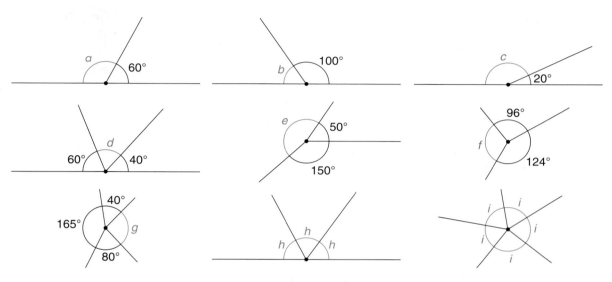

**2** I dropped three identical plates and they each broke into several pieces. Work out which pieces fit together to make each plate.

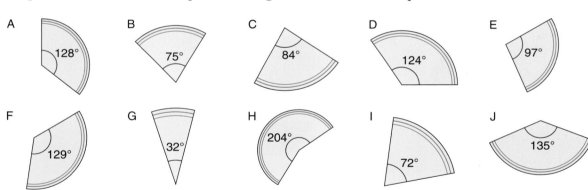

---

Try out this puzzle on your friends.

Draw 3 identical circles and cut each one into 3 pieces.

Make sure you cut the pieces from the centre of the circle, like the plates in question 2.

Now mix all the pieces up, and ask your friends to put the circles back together.

How long do they take?

# 4: Symmetry

# Finishing off

### Now that you have finished this chapter you should be able to

- ★ say how many lines of reflection symmetry a shape or pattern has
- ★ complete a pattern so that it has 1 or 2 lines of symmetry
- ★ draw the reflection of a shape in a horizontal or vertical line
- ★ say whether a shape or pattern has rotational symmetry, and if it has, give the order of rotational symmetry
- ★ know that angles round a point add up to 360°
- ★ know that angles on a straight line add up to 180°

Use the questions in the next exercise to check that you understand everything.

**Mixed exercise**

**1** For each of the logos below

a) say how many lines of reflection symmetry it has

b) say whether it has rotational symmetry, and if it has, what the order of rotational symmetry is.

(i) London Underground     (ii) NatWest     (iii) McDonald's

(iv) The Royal Bank of Scotland     (v) Train station

# 4: Symmetry

**2** Copy the patterns below.

Complete them so that the dotted lines are lines of symmetry.

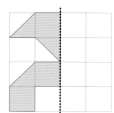 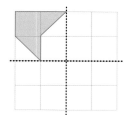

**3** Copy the letters of the alphabet shown below on squared paper and reflect them in the lines shown.

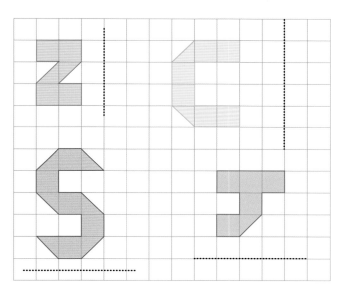

**4** Work out the marked angles.

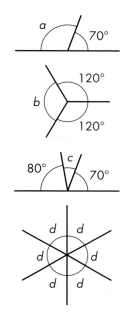

## Investigation

a) How many patterns with just one line of symmetry can you make by shading squares in a 3 × 3 grid? (Shade complete squares only.)

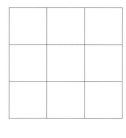

b) How many patterns with 2 lines of symmetry can you make?

c) How many patterns with 4 lines of symmetry can you make?

Using a 4 × 4 square grid, design a company logo and state what kinds of symmetry it has.

**Example**

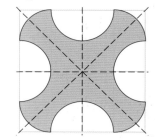

4 lines of symmetry and rotational symmetry of order 4.

# Five

# Fractions

## Equivalent fractions

Alf and Mo have allotments of the same size.
Alf divides his allotment into 4 equal parts.
He plants potatoes in 1 of the 4 parts.
You can write 1 out of 4 as $\frac{1}{4}$.
We call it one quarter. $\frac{1}{4}$ is a **fraction**.
The top, sometimes called the **numerator**, is 1.
The bottom, sometimes called the **denominator**, is 4.

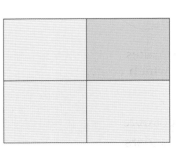

Mo divides her allotment into 8 equal parts.
She plants potatoes in 2 of the 8 parts.
You can write 2 out of 8 as $\frac{2}{8}$ (two eighths).
Look at the pictures of the allotments.
Who has the larger potato patch?
You can see that the patches are the same size.
$\frac{2}{8}$ is the same as $\frac{1}{4}$.

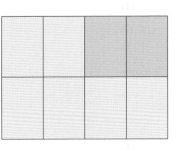

Elsa has an allotment too.
She has divided it up into 12 equal parts.

Elsa's potato patch is the same size as Alf's.
You can see that $\frac{3}{12}$, $\frac{2}{8}$ and $\frac{1}{4}$ all mean the same thing.
They are **equivalent fractions**. They can all be written as $\frac{1}{4}$.

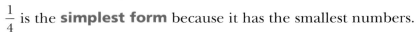

$\frac{1}{4}$ is the **simplest form** because it has the smallest numbers.

To find an equivalent fraction, you multiply (or divide) the top and bottom by the same number

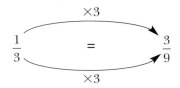
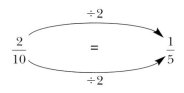

# 5: Fractions

**1** What fraction of this flag is green?

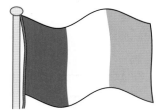

**2** What fraction of these eggs is brown?

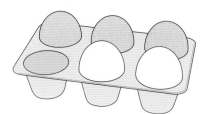

**3** Mel makes a chart showing which of her friends prefer Neighbours and which prefer Eastenders.

| Name | Abi | Rob | Gill | Ed | Jan | Prem | Guy | Jess |
|---|---|---|---|---|---|---|---|---|
| Neighbours | ✔ | | ✔ | ✔ | | ✔ | ✔ | |
| Eastenders | | ✔ | | | ✔ | | | ✔ |

a) What fraction of her friends prefer Neighbours?
b) What fraction of her friends prefer Eastenders?

**4** This chart shows when the sports hall is booked (◸).

a) What fraction of the sessions on Saturday are booked?
b) What fraction of the sessions on Sunday are booked?
c) For what fraction of the whole weekend is the hall booked?

**5** a) Find the missing number in each of these.

(i) $\dfrac{1}{2} = \dfrac{\square}{16}$    (ii) $\dfrac{3}{4} = \dfrac{\square}{16}$    (iii) $\dfrac{3}{8} = \dfrac{\square}{16}$

b) Write the fractions $\dfrac{1}{2}, \dfrac{3}{4}, \dfrac{3}{8}, \dfrac{7}{16}$ in order, smallest first.

**6** a) Find the missing number in each of these.

(i) $\dfrac{5}{6} = \dfrac{\square}{12}$    (ii) $\dfrac{2}{3} = \dfrac{\square}{12}$    (iii) $\dfrac{3}{4} = \dfrac{\square}{12}$

b) Write the fractions $\dfrac{5}{6}, \dfrac{2}{3}, \dfrac{3}{4}$ in order, smallest first.

---

This 4 × 4 grid has 1 out of 16 squares ($\dfrac{1}{16}$) shaded.

Draw similar grids for $\dfrac{2}{16}, \dfrac{3}{16}, \dfrac{4}{16}, \ldots$ up to $\dfrac{16}{16}$.

For each grid, say what fraction is shaded, and write the fraction in its simplest form.

# 5: Fractions

## Adding and subtracting fractions (1)

Martina and Gary have ordered a pizza. It has been cut into 8 equal parts.

Martina eats 3 of the 8 pieces. She eats $\frac{3}{8}$.

Gary eats 4 pieces. He eats $\frac{4}{8}$.

*How much have they eaten altogether?*

They have eaten 7 of the 8 pieces altogether.

$$\frac{3}{8} + \frac{4}{8} = \frac{7}{8}$$

Say this as '3 eighths plus 4 eighths makes 7 eighths'

Martina and Gary had 1 pizza to start with.

You can write this as $\frac{8}{8}$.

They have eaten $\frac{7}{8}$.

*How much is left?*

You can see that 1 of the 8 pieces is left. That is $\frac{1}{8}$.

$$\frac{8}{8} - \frac{7}{8} = \frac{1}{8}$$

Say this as '8 eighths minus 7 eighths makes 1 eighth'

Rabina fixes these sheets of hardboard together. How thick is the finished board?

5 sixteenths plus 7 sixteenths makes 12 sixteenths.

$$\frac{5}{16} + \frac{7}{16} = \frac{12}{16}$$

$\frac{12}{16}$ in its simplest form is $\frac{3}{4}$

**The finished board will be $\frac{12}{16}$ inch thick.**

Adding and subtracting fractions is easy when the denominators are the same.

Practise it by doing the questions opposite.

# 5: Fractions

**Mixed exercise**

**1** Work these out. Give your answers in their simplest form.

a) $\frac{1}{4} + \frac{1}{4}$  b) $\frac{5}{8} + \frac{1}{8}$  c) $\frac{3}{8} - \frac{1}{8}$  d) $\frac{3}{16} + \frac{7}{16}$

e) $\frac{1}{2} + \frac{1}{2}$  f) $\frac{5}{16} - \frac{3}{16}$  g) $\frac{3}{4} + \frac{1}{4}$  h) $\frac{9}{16} - \frac{5}{16}$

**2** Carl buys a chocolate bar. There are 8 pieces of chocolate in the bar. Carl eats 3 pieces and his brother eats 2 pieces.

a) What fraction of the bar do the boys eat?
b) What fraction of the bar is left?

**3** In each of these, copy out the sum and fill in the missing number.

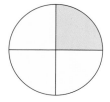 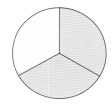 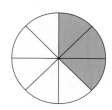

**4** This chart shows the times that Kim's family usually sleep (■) during a day (24 hours).

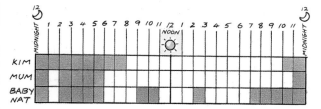

a) For how many hours is Kim asleep?
b) What fraction of the day is this?
c) For what fraction of the day is Kim's mum asleep?
d) For what fraction of the day is Kim's mum awake?
e) For what fraction of the day is Nat asleep?
f) For what fraction of the day is Nat awake?

**5** In these, write the 1 as a fraction then work out the answer.

a) $1 - \frac{1}{4}$  b) $1 - \frac{3}{8}$  c) $1 - \frac{1}{10}$  d) $1 - \frac{3}{4}$

e) $1 - \frac{3}{16}$  f) $1 - \frac{7}{10}$  g) $1 - \frac{1}{2}$  h) $1 - \frac{5}{8}$

---

Go to a DIY or hardware store.

How many different lengths of screw can you find, up to 2 inches? (Some will be in millimetres and some in inches.)

List them in order of size.

---

Find last season's results for a local football (or other) team.

What fraction of their games did they win, draw and lose?

Check that your fractions add up to 1.

# 5: Fractions

## Adding and subtracting fractions (2)

Glyn serves at the delicatessen.
There is a large pork pie on display.
The first customer buys $\frac{1}{4}$ of the pie, and the second customer buys $\frac{3}{8}$ of it.
How much of the pie do they buy altogether?

You need to add 1 quarter to 3 eighths.
But you cannot add quarters to eighths.
You need to change the quarters into eighths first:

$$\frac{1}{4} \xrightarrow{\times 2} \frac{2}{8} \qquad \frac{2}{8} \text{ is equivalent to } \frac{1}{4}.$$

**You can now write** $\quad \frac{1}{4} + \frac{3}{8} = \frac{2}{8} + \frac{3}{8} = \frac{5}{8}$

 +  =  +  =

 *Why can't we change the eighths into quarters instead?*
*What is $\frac{1}{4} + \frac{1}{8}$?   What is $\frac{1}{4} + \frac{5}{8}$?*

Glyn cuts large quiches into sixteenths before putting them on display.
There is $\frac{3}{4}$ of a quiche left, as shown.
The next customer buys $\frac{9}{16}$ of the quiche.
How much is left now?

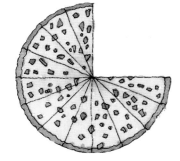

You need to subtract 9 sixteenths from 3 quarters. But you cannot subtract sixteenths from quarters. You have to change the quarters into sixteenths:

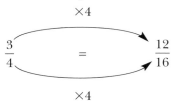

**You can now write** $\quad \frac{3}{4} - \frac{9}{16} = \frac{12}{16} - \frac{9}{16} = \frac{3}{16}$

 *Which is bigger, $\frac{7}{16}$ or $\frac{3}{4}$? How can you tell?*
*What is $\frac{3}{4} - \frac{7}{16}$?   What is $\frac{3}{4} - \frac{1}{16}$?*

# 5: Fractions

**Mixed exercise**

**1** Find the missing number in each of these.

a) $\frac{1}{2} = \frac{?}{4}$  b) $\frac{1}{3} = \frac{?}{6}$  c) $\frac{6}{8} = \frac{3}{?}$  d) $\frac{10}{15} = \frac{2}{?}$  e) $\frac{1}{4} = \frac{?}{12}$  f) $\frac{6}{9} = \frac{2}{?}$

g) $\frac{3}{4} = \frac{?}{16}$  h) $\frac{6}{10} = \frac{?}{5}$  i) $\frac{30}{40} = \frac{3}{?}$  j) $\frac{7}{8} = \frac{?}{16}$  k) $\frac{1}{6} = \frac{?}{30}$  l) $\frac{7}{10} = \frac{21}{?}$

**2** This map shows the distance, in miles, from the crossroads to each of 4 buildings.

Work out the distance by road from
a) Martha's house to the shop
b) Aron's house to the shop
c) Martha's house to Aron's house
d) Martha's house to the railway station.

**3** These boards are made up of 2 parts.

The total thickness and the thickness of the top layer are shown (in inches). Find the thickness of the bottom layer.

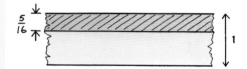

**4** Work these out and write your answers in their simplest form.

a) $\frac{1}{2} - \frac{3}{16}$  b) $\frac{7}{16} + \frac{3}{16}$  c) $\frac{5}{8} + \frac{1}{4}$

d) $\frac{3}{4} - \frac{1}{8}$  e) $\frac{1}{2} + \frac{5}{16}$  f) $\frac{7}{8} - \frac{1}{4}$

g) $\frac{5}{16} + \frac{3}{8}$  h) $\frac{3}{4} + \frac{5}{8}$

**5** Write down the larger of these fractions.

a) $\frac{2}{5}$ and $\frac{1}{3}$  b) $\frac{3}{10}$ and $\frac{1}{3}$  c) $\frac{4}{5}$ and $\frac{5}{7}$

In music, the symbol ♩ stands for a **crotchet**.

A crotchet takes a time of 1 unit.

Find out the symbols and names for other notes and how many units of time they take.

Written music is usually divided into short sections called **bars**.

Take a short piece of written music and add up the lengths of the notes in each bar.

What do you notice?

# 5: Fractions

## Improper fractions, mixed numbers and reciprocals

Clare is a waitress.
She is serving apple pie to 7 people.
Each serving is a quarter of a pie.
How many pies does Clare use?

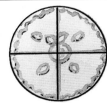

Clare uses 7 quarters ($\frac{7}{4}$).

$\frac{7}{4}$ is an **improper fraction**.
(Some people call it a **'top heavy'** fraction.)

The 7 quarters can be arranged to make 1 whole pie and 3 quarters of a pie.

**You can see that Clare uses $1\frac{3}{4}$ pies.**

$1\frac{3}{4}$ is a **mixed number**. You can write $\frac{7}{4} = 1\frac{3}{4}$.

How many pies does Clare need for 9 people?
She needs 9 quarters, $\frac{9}{4}$.

To see how many whole pies this is, we need to write it as a mixed number.
You divide the top (9) by the bottom (4):

$9 \div 4 = 2$ remainder 1.

So $\frac{9}{4} = 2\frac{1}{4}$.

(This is the number of whole pies)

(This is the number of extra quarters)

**Clare needs $2\frac{1}{4}$ pies.**

The **reciprocal** of a number is 1 divided by the number.

So the reciprocal of 4 is $\frac{1}{4}$; the reciprocal of $\frac{1}{4}$ is 4.

*What is the reciprocal of 8?*

*What happens if you multiply 8 and the reciprocal of 8 together?*

# 5: Fractions

**1** Change these improper fractions to mixed numbers.

a) $\frac{5}{4}$   b) $\frac{8}{3}$   c) $\frac{3}{2}$   d) $\frac{11}{4}$   e) $\frac{13}{8}$   f) $\frac{9}{2}$   g) $\frac{11}{6}$   h) $\frac{12}{3}$

**2** Change these mixed numbers to improper fractions.

a) $3\frac{1}{2}$   b) $1\frac{5}{8}$   c) $2\frac{3}{5}$   d) $3\frac{1}{3}$   e) $4\frac{3}{4}$   f) $1\frac{3}{16}$   g) $2\frac{5}{8}$   h) $6\frac{1}{4}$

**3** Brian works at a health centre. He sees 7 people for half an hour each. How many hours does it take him?

**4** Tess records 13 programmes each lasting quarter of an hour. How many hours does it take?

**5** The bottles of mineral water in this pack each contain $\frac{1}{2}$ litre.

How many litres of water does the pack contain?

**6** A doctor has a $2\frac{1}{2}$ hour clinic. How many $\frac{1}{2}$ hour appointments can be fitted in?

**7** Joanna has $2\frac{3}{4}$ hours left of a videotape.

How many $\frac{1}{4}$ hour programmes can she record?

**8** Parvez is a chef. He allows 1 kg of rice for 8 people.

How many people can he serve with $2\frac{1}{4}$ kg?

**9** Bottles of wine are packed in boxes of 6.

Diana has $3\frac{1}{2}$ boxes.

How many bottles of wine does Diana have?

Keep a record of how long you spend watching television for each of the seven days of a week.

Give each day's total to the nearest $\frac{1}{4}$ hour (for example, $1\frac{3}{4}$ hours).

How much television did you watch in the whole week?

**10** a) Write down the reciprocal of (i) 5 (ii) $\frac{1}{5}$. What is $5 \times \frac{1}{5}$?

b) Write down the reciprocal of (i) $\frac{1}{10}$ (ii) 10. What is $\frac{1}{10} \times 10$?

c) Write down the reciprocal of 1.

d) Why does 0 not have a reciprocal?

**11** a) Find the missing number in $\frac{2}{3} = \frac{?}{12}$.

b) Find the missing number in $\frac{3}{4} = \frac{?}{12}$.

c) Which is larger, $\frac{2}{3}$ or $\frac{3}{4}$?

d) What is the difference between $\frac{2}{3}$ and $\frac{3}{4}$?

# 5: Fractions

## Mixed numbers

Thomas wants to record two programmes on video.

The Match lasts $1\frac{3}{4}$ hours. Star Trek lasts $1\frac{1}{2}$ hours.

How much videotape time does he need?

You find the total time by adding $1\frac{3}{4}$ and $1\frac{1}{2}$.

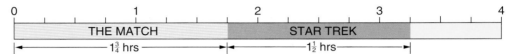

This is shown on the diagram above, but you can do it without drawing a diagram.

First add the whole numbers: $1 + 1 = 2$

Then add the fractions: $\frac{3}{4} + \frac{1}{2} = \frac{3}{4} + \frac{2}{4}$

*You have to change the half into quarters before you can add them*

$= \frac{5}{4} = 1\frac{1}{4}$

*The whole numbers*

So $1\frac{3}{4} + 1\frac{1}{2} = 2 + 1\frac{1}{4} = 3\frac{1}{4}$

*The fractions*

**Thomas needs $3\frac{1}{4}$ hours videotape time.**

Sam has $2\frac{3}{4}$ hours left unused on a videotape.

She records $1\frac{1}{2}$ hours of MTV.

How much time will be left?

You find this by taking $1\frac{1}{2}$ away from $2\frac{3}{4}$.

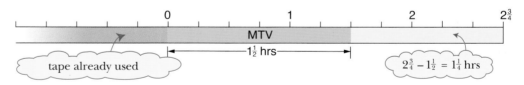

This is shown on the diagram above, but you can do it without drawing a diagram.

First subtract the whole numbers: $2 - 1 = 1$

Then subtract the fractions: $\frac{3}{4} - \frac{1}{2} = \frac{3}{4} - \frac{2}{4} = \frac{1}{4}$

*The whole numbers*

So $2\frac{3}{4} - 1\frac{1}{2} = 1 + \frac{1}{4} = 1\frac{1}{4}$

*The fractions*

**Sam has $1\frac{1}{4}$ hours left on her tape.**

# 5: Fractions

**1** 
a) $2\frac{1}{2} + 1\frac{3}{4}$
b) $1\frac{5}{8} + \frac{7}{8}$
c) $1\frac{5}{8} - \frac{1}{4}$
d) $1\frac{3}{4} + 2\frac{5}{8}$
e) $1\frac{1}{2} - \frac{3}{16}$
f) $\frac{3}{8} + 2$
g) $2 - \frac{1}{4}$
h) $3\frac{1}{2} - \frac{3}{4}$
i) $4\frac{1}{2} - 1\frac{7}{8}$

**2** This map shows the distances, in miles, between 5 villages.

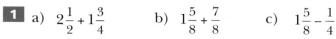

Ashley — $1\frac{1}{2}$ — Benton — $2\frac{1}{4}$ — Cowley — $\frac{3}{4}$ — Ditton — $2\frac{3}{4}$ — Elton

a) What is the distance from Benton to Ditton?
b) What is the distance from Ashley to Cowley?
c) What is the distance from Cowley to Elton?
d) Use your answers to b) and c) to find the distance from Ashley to Elton.

**3** Philip's train journey is expected to take $2\frac{3}{4}$ hours.

The train is delayed for $\frac{1}{2}$ hour on the way.

How long does the journey take?

**4** Brass is made from copper and zinc. In $3\frac{3}{4}$ kilograms of brass there are $2\frac{1}{4}$ kilograms of copper.

How much zinc is there?

**5** A coach journey takes $4\frac{3}{4}$ hours. The same journey by train takes $3\frac{1}{2}$ hours.

How much time is saved by going by train?

**6** The diagram shows a screw holding 2 pieces of wood together. All the measurements are in inches.

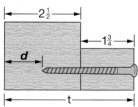

a) Find the total thickness, $t$ inches.

b) The screw is 3 inches long. Find the distance, $d$ inches, between the screw point and the side of the wood.

Blank videotapes are either 3 hours or 4 hours long.

Which would you use to record several episodes of

a) a $1\frac{1}{4}$ hour programme?

b) a $1\frac{1}{2}$ hour programme?

c) a programme of your choice (write down its name and the length of each episode, too)?

# 5: Fractions

## Fractions of a quantity

Look at this headline.

### Daily Globe
### One in four workers to lose jobs

There are 120 workers.

How many will lose their jobs?

1 in 4 is the same as $\frac{1}{4}$ (one quarter).

You need to find $\frac{1}{4}$ of 120.

You can write $\frac{1}{4} \times 120 = 120 \times 1 \div 4$

$= 120 \div 4$

$= 30$

*You can write 'of' as ×*

*First multiply by the top number…*

*…then divide by the bottom number*

**30 workers lose their jobs.**

The next week you see this headline.

### Daily Globe
### Two thirds of job losers offered new work

30 people lost their jobs: how many are offered new work?

Two thirds of 30 is $\frac{2}{3} \times 30$.

$\frac{2}{3} \times 30 = 30 \times 2 \div 3$

$= 60 \div 3$

$= 20$

*First multiply by the top number…*

*…then divide by the bottom number*

**20 people are offered new work.**

You may find it easier to find $\frac{2}{3}$ of 30 this way:

First find $\frac{1}{3}$ of 30:

$30 \div 3 = 10$

Then multiply by 2 to get $\frac{2}{3}$ of 30:

$10 \times 2 = 20$

*How many are not offered work?*

*What fraction of the job losers is this?*

# 5: Fractions

**1** Work these out.

a) $\frac{1}{3} \times 18$    b) $\frac{3}{4} \times 8$    c) $\frac{2}{5} \times 100$

d) $\frac{1}{4}$ of 36    e) $\frac{9}{10}$ of 80    f) $\frac{4}{3} \times 30$

**2** Holly buys this tent.
How much does she save by buying it in the sale?

**3** 24 people apply for a job. One quarter of them get an interview. How many get an interview?

**4** Donna takes her young son Samuel on this coach trip.

a) How much is Samuel's fare?

b) How much does it cost in total?

**5** Anneena's car cost £12 000 when new.

a) It is now worth two thirds of this. How much is it worth now?

b) Since Anneena bought her car, the price of the same model has gone up by a quarter. How much does a new one cost now?

**6** Hightown's annual rainfall last year was 96 cm.

a) One sixth ($\frac{1}{6}$) of the rain fell in January. How many cm is this?

b) One eighth ($\frac{1}{8}$) of the rain fell in April. How many cm is this?

**7** Last year the sales of a business were £600 000 a year.

In the current year the sales are expected to increase by one third.

What are the expected sales for the current year?

**8** There are 50 people at a party. Three fifths of them are female. How many are male?

---

Go to your local supermarket and find out whether bulk buys are always better value.

For example, is 10 packets of crisps for £2.40 better than 6 for £1.50?

(You need to work out the price of 1 packet in each case.)

# 5: Fractions

## Finishing off

**Now that you have finished this chapter you should be able to**

★ find equivalent fractions
★ find the simplest form of a fraction
★ add and subtract fractions
★ change between improper fractions and mixed numbers
★ find a fraction of a quantity
★ find the reciprocal of a whole number and of a fraction

Use the questions in the next exercise to check that you understand everything.

### Mixed exercise

**1** In each of these, write the fraction in its simplest form.

a) $\dfrac{12}{16}$   b) $\dfrac{10}{40}$   c) $\dfrac{15}{25}$   d) $\dfrac{18}{20}$   e) $\dfrac{24}{36}$   f) $\dfrac{25}{100}$

**2** In each of these, write the answer in its simplest form.

a) $\dfrac{3}{10} + \dfrac{1}{10}$   b) $\dfrac{3}{16} + \dfrac{11}{16}$   c) $1 - \dfrac{3}{5}$   d) $1 - \dfrac{5}{8}$

**3** Jo is making a fruit salad. She writes this list of the things she needs.

She finds these things in the fruit bowl.

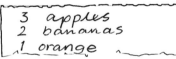

3 apples
2 bananas
1 orange

a) What fraction of the apples does Jo need?

b) What fraction of the bananas does she need?

c) What fraction of the oranges are left when she has made the fruit salad?

**4** Work these out.   a) $\dfrac{5}{8} + \dfrac{1}{4}$   b) $\dfrac{3}{4} - \dfrac{7}{16}$   c) $\dfrac{5}{6} - \dfrac{2}{3}$   d) $\dfrac{5}{8} + \dfrac{1}{2}$

**5** Hilary is the school nurse. She is doing health checks for all the students.

She spends about $\dfrac{1}{4}$ of an hour with each student.

a) How many hours does Hilary need to see 7 students?

b) How many students can Hilary see in $2\dfrac{1}{2}$ hours?

**6** Write these improper fractions as mixed numbers.

a) $\dfrac{5}{4}$   b) $\dfrac{11}{2}$   c) $\dfrac{17}{5}$   d) $\dfrac{15}{4}$   e) $\dfrac{13}{10}$

# 5: Fractions

**Mixed exercise**

**7** Write these mixed numbers as improper fractions.

a) $4\frac{1}{2}$   b) $3\frac{3}{4}$   c) $2\frac{7}{8}$   d) $1\frac{9}{10}$   e) $2\frac{1}{5}$

**8** Lyn is orienteering. Here is her map.

She has reached Checkpoint 2.

a) What distance has she travelled?

b) How far has she still to go?

c) What distance will she have travelled when she reaches the finish?

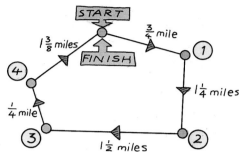

**9** Work these out and write your answers in their simplest forms.

a) $2\frac{1}{2} + \frac{3}{4}$   b) $2 - \frac{5}{8}$   c) $1\frac{5}{8} + 1\frac{1}{2}$   d) $1\frac{3}{4} + 3\frac{3}{8}$

e) $\frac{1}{4} + 2\frac{7}{8}$   f) $3\frac{5}{6} + 2\frac{1}{6}$   g) $2\frac{3}{16} - \frac{5}{8}$   h) $4\frac{1}{4} - 2\frac{3}{4}$

**10** Jack's water butt holds 180 litres.

The first picture shows the butt before Jack waters his plants.

The second picture shows the butt after he has watered the plants.

a) How much water is in the butt before he starts?

b) How much water is in the butt when he has finished?

c) How much water has he used?

(Write your answers in litres.)

**11** Here are the carriages of an Intercity train.

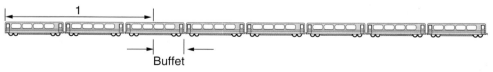

a) What fraction of the train is First class (1)?

b) What fraction of the train is buffet?

c) What fraction is Standard class (unmarked)?

d) Check that these fractions add up to 1.

**12** Write down the reciprocal of

a) 6   b) $\frac{1}{6}$   c) 7   d) $\frac{1}{7}$.

---

Take a standard 12-inch ruler and look at the scales for inches.

What lengths between 0 and 1 inch can you measure exactly?

(Give your answers as fractions of an inch, for example $\frac{7}{10}$ inch.)

# Six

# Maps and drawings

## Maps and scales

This is a map of the village of Greenbridge.

It shows some of the main buildings in the village, and the homes of three families, the Whites, the Greens and the Browns.

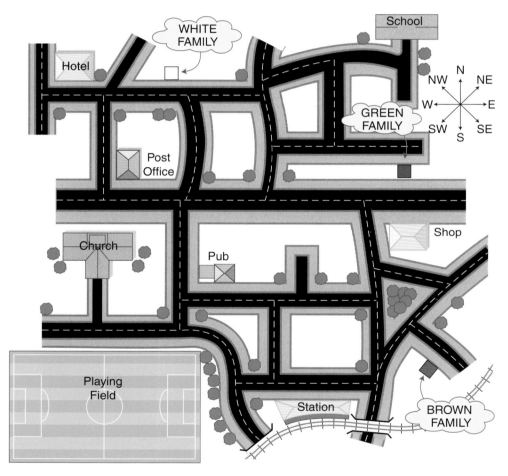

The **scale** of this map is 1 cm to 10 metres. This means that 1 cm on the map stands for 10 metres in real life.

Every map needs a scale.

The length of the playing field on the map is 5 cm.

As each centimetre stands for 10 metres in real life, the playing field must be 50 metres long.

 *Measure the width of the playing field on the map.*

*How wide is the field in real life?*

# 6: Maps and drawings

For questions 1 to 8, use the map of Greenbridge opposite.

**1**  a) Which family lives north of the shop?

b) Which building is south-east of the Post Office?

c) Which building is north-east of the pub?

d) Which family lives furthest west?

**2** What direction must a crow fly in to go directly from

a) the Whites' house to the hotel?

b) the Browns' house to the Whites' house?

c) the school to the playing field?

d) the Greens' house to the shop?

**3** The church is south-west of the school.

In which direction is the school from the church?

**4** Mrs White turns left out of her house. She takes the second right, then the second left. She stops just past the first turning on the right.

What building has she reached?

**5** Mr Green uses the train every morning to go to work.

Give directions for him to get from his home to the station.

**6** A visitor arrives at the station and wants to go to the hotel.

What directions would you give her?

**7**  a) Measure the distance on the map from the Post Office to the village shop.

b) How far would this distance be in real life?

**8** Jessica Brown, Mark Green and Ryan White all attend the village school.

a) Who do you think has the shortest walk to school?

b) Measure the distance that each has to walk. (You could use a piece of string or the edge of a piece of paper to make this easier.)

c) Work out how far each child has to walk in real life.

---

Measure a room at home using paces.

Draw a plan view of the room and its main contents using a scale of 1 cm to 1 pace.

# 6: Maps and drawings

## Scales

The map of Greenbridge had a scale of 1 cm to 10 metres.

This map is part of an Ordnance Survey map of South Devon.

Its scale is 1:50 000, 'one to fifty thousand'.

> When you write a scale like this it is called a **ratio**

This means that 1 unit on the map stands for 50 000 units in real life.

For example, 1 cm on the map is 50 000 cm in real life.

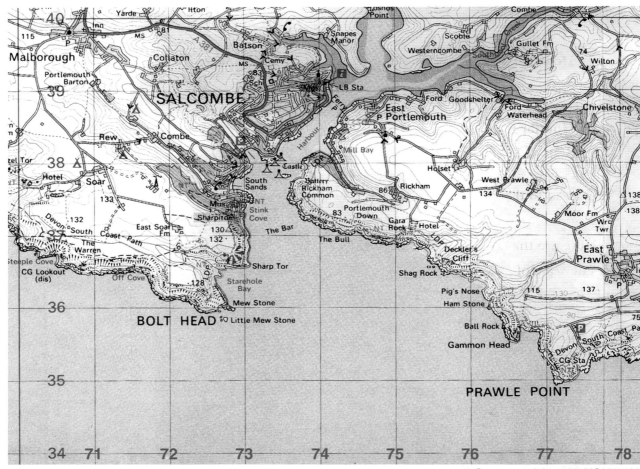

© Crown copyright 1997. MC88051M

Measure the distance from Bolt Head to Prawle Point.

You should find that it is about 9 cm.

9 cm on the map is 9 × 50 000 cm, or 450 000 cm.

> There are 100 cm in a metre, so divide by 100

450 000 cm = 4500 m
4500 m = 4.5 km

> There are 1000 m in a kilometre, so divide by 1000

So the distance from Bolt Head to Prawle Point is 4.5 km in real life.

It is easier to find things on the map if you use grid references.

The village of West Prawle is in grid square <u>76 37</u>

> This is the grid line at the left of the square

> This is the grid line underneath the square

56

# 6: Maps and drawings

**1** A map is drawn to a scale of 1:25 000. How far in real life are these distances on the map? (Give your answers in metres.)

   a) 1 cm   b) 6 cm   c) 8 mm   d) 3 mm   e) 9.5 mm

For questions 2 to 4, use the map of the Salcombe area opposite.

**2** a) Measure, in centimetres, the width of one grid square on the map.
   b) How many centimetres is this in real life?
   c) How many metres is it?
   d) How many kilometres is it?

**3** Find where the ferry crosses the estuary (square 7438).

   a) Measure, in millimetres, the distance that the ferry travels.
   b) How far is this in real life?

**4** Find Moor Farm in square 7737.

   Use string or the edge of a piece of paper to measure the distance by road to the church in East Portlemouth.

   How far is this in real life?

**5** Here are the plans of the ground floor and first floor of a house.

   The scale of the drawings is 1:150.

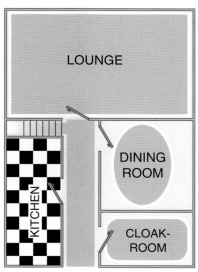
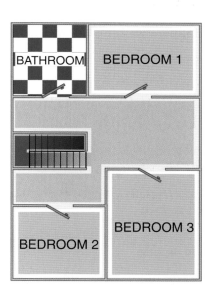

Write a description of the house, giving the length and width of each room in metres.

Using a scale of 1 cm to 1 m, make a plan of the ground floor of your home.

## 6: Maps and drawings

# Angles

A robot is being programmed to travel along this path.

The scale is 1 cm to 2 m.

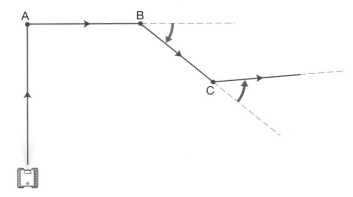

The first 3 instructions are   GO FORWARD 8 METRES
                               TURN RIGHT THROUGH A QUARTER TURN
                               GO FORWARD 6 METRES

The robot is now at point B. It needs to turn again.

This time it is not a quarter turn. It is less than that.

The robot needs to be told what **angle** to turn through.

Angles are measured in **degrees**. You need an **angle measurer** or a **protractor** to measure an angle.

*Measure the angle the robot needs to turn. It is shown by the red arrow.*

You should find that the angle is 40 degrees. This is written as 40°.

## Types of angle

A whole turn is 360°, so a half turn is 180° and a quarter turn is 90°.

An angle of 90° is called a **right angle**.

An angle less than 90° is called an **acute angle**.

An angle bigger than 90° but smaller than 180° (between a quarter turn and a half turn) is called an **obtuse angle**.

An angle bigger than 180° (between a half turn and a whole turn) is called a **reflex angle**.

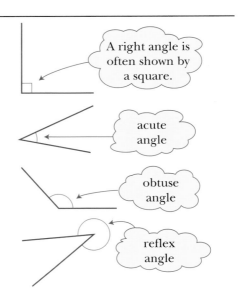

*Estimate the sizes of the acute, obtuse and reflex angles above.*

# 6: Maps and drawings

**1** For each of these angles, say (without measuring) whether it is an acute angle, a right angle, an obtuse angle or a reflex angle.

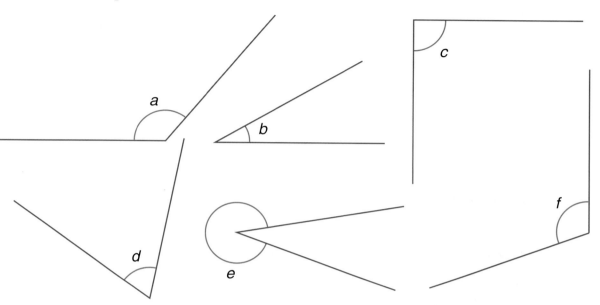

**2** Estimate each of the angles in question 1.

**3** Measure each of the angles in question 1 and see how good your estimates were.

**4** Look again at the robot path on page 58. The instruction for the corner at B was TURN RIGHT 40°.
Write the next 2 instructions for the robot.

---

You are designing a company logo.

Write a set of instructions to cut the logo from a large sheet of vinyl. Use the commands
FORWARD (and a distance)
        e.g. FORWARD 30 cm
RIGHT    (and an angle)
        e.g. RIGHT 90°
LEFT     (and an angle)
        e.g. LEFT 45°.

Your logo should be made up of straight lines.

Draw the logo accurately using the scale of your choice.

## 6: Maps and drawings

# Finishing off

**Now that you have finished this chapter you should**

- ★ be able to follow and give directions using a map
- ★ know how to use the scale on a map or scale drawing
- ★ be able to use the 8 compass directions
- ★ know what is meant by a right angle, an acute angle, a reflex angle and an obtuse angle
- ★ be able to use a protractor or angle measurer to measure angles in degrees

**Use the questions in the next exercise to check that you understand everything.**

## Mixed exercise

This map shows part of Edinburgh city centre.

The scale is 1 cm to 125 metres.

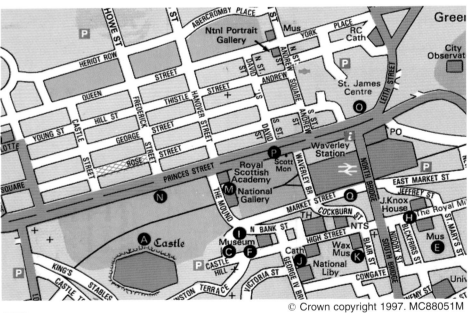

© Crown copyright 1997. MC88051M

**1** What is the distance along Princes Street between North Bridge and Castle Street?

**2** The Patel family have parked their car at the car park (marked P) by the castle.

  a) Give them directions to get to the National Portrait Gallery (north of Waverley station).

  b) How far do the Patels have to walk?

# 6: Maps and drawings

**Mixed exercise**

**3** The diagram shows the route of a sailing race.

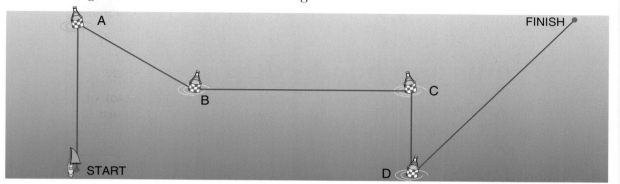

a) Copy and complete the table below.

| Stage of race | Distance on map | Distance in real life | Direction |
|---|---|---|---|
| START to A | 4 cm | 0.8 km | North |
| A to B | | | |
| B to C | | | |
| C to D | | | |
| D to FINISH | | | |

b) What is the total length of the race course?

**4** This is a doll's house.

The scale diagram below shows two of the pieces of wood used for the roof.

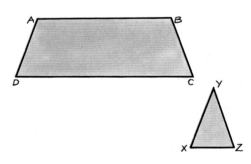

Measure the angles at each corner of the pieces of wood.

Write your answers like this.

Angle BAD = ........

---

Use a computer to draw the shape that you designed for your company logo on page 59.

Experiment with it to see if you can make it better.

Print out your final design.

# Seven

# Decimals

## Tenths and hundredths

Look at this **number line**.

It shows the whole numbers from 0 to 10.

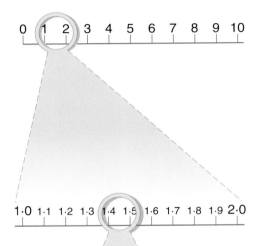

The part of the number line between 1 and 2 has been enlarged and divided into **tenths**.

1.6 is the same as one and six tenths, $1\frac{6}{10}$.

*How would you write 1.3 in mixed numbers?*

*How would you write $2\frac{5}{10}$ as a decimal?*

The part of the number line between 1.4 and 1.5 has been enlarged even more. It has been divided into 10 equal parts.

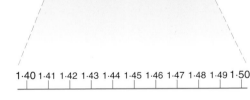

Each division is now one tenth of one tenth: it is **one hundredth**.

1.49 is the same as one and forty-nine hundredths, $1\frac{49}{100}$.

*How would you write 1.48 in mixed numbers?*

*How would you write $3\frac{14}{100}$ as a decimal?*

*What would each division be if we enlarged the part from 1.42 to 1.43 again, and divided it into 10 equal parts?*

# 7: Decimals

**1** Look at this number line.

Write down the decimal numbers at A, B and C.

**2** Write these decimals in order, smallest first.

10.2, 1.02, 100.1, 0.103, 0.110, 11.9

**3** For each of these, write down the water level.

 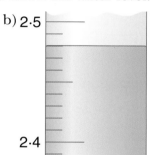 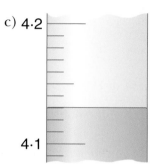

**4** Write each of these decimals as a fraction.

a) 0.7    b) 0.31    c) 0.4    d) 0.45

e) 2.9    f) 3.14    g) 4.641    h) 0.09

**5** Write each of these fractions as a decimal.

a) $\frac{9}{10}$    b) $\frac{71}{100}$    c) $\frac{3}{10}$    d) $\frac{28}{100}$

e) $2\frac{7}{10}$    f) $4\frac{23}{100}$    g) $3\frac{1}{10}$    h) $\frac{137}{100}$

**6** The graph shows the profits (in millions of pounds) made by companies A, B, C, D and E.

Write down the profit made by each company.

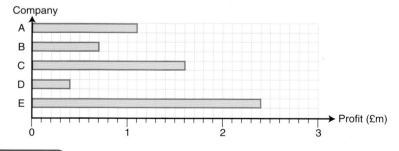

Find a supermarket till receipt with no more than 5 items on it.

Explain each number on the receipt.

# 7: Decimals

## Halves and quarters

Look at this number line from 0 to 1.

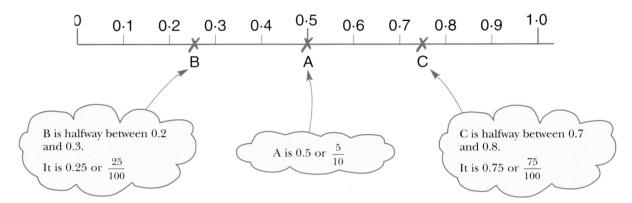

B is halfway between 0.2 and 0.3.
It is 0.25 or $\frac{25}{100}$

A is 0.5 or $\frac{5}{10}$

C is halfway between 0.7 and 0.8.
It is 0.75 or $\frac{75}{100}$

Write each of the fractions in its simplest form.
Check that you agree with the numbers in this table.
How would you write $2\frac{3}{4}$ in decimals?

| Fraction | Decimal |
|---|---|
| $\frac{1}{4}$ | 0.25 |
| $\frac{1}{2}$ | 0.50 |
| $\frac{3}{4}$ | 0.75 |

## Fifths

Look at the number line from 0 to 1 below. It is divided into tenths (but they are not all labelled).

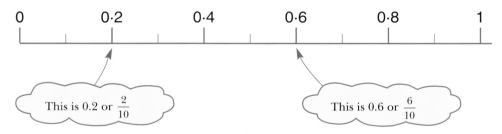

This is 0.2 or $\frac{2}{10}$

This is 0.6 or $\frac{6}{10}$

Write each of the fractions in its simplest form.
Check that you agree with the numbers in this table.
How would you write $3\frac{2}{5}$ in decimals?
Can you write any decimal as a fraction?
Can you write any fraction as a decimal?

| Fraction | Decimal |
|---|---|
| $\frac{1}{5}$ | 0.2 |
| $\frac{2}{5}$ | 0.4 |
| $\frac{3}{5}$ | 0.6 |
| $\frac{4}{5}$ | 0.8 |

# 7: Decimals

**1** Write these fractions as decimals.

a) $2\frac{1}{4}$  b) $3\frac{1}{2}$  c) $5\frac{3}{4}$  d) $1\frac{1}{2}$

e) $3\frac{3}{4}$  f) $7\frac{1}{4}$  g) $6\frac{1}{2}$  h) $1\frac{3}{4}$

i) $\frac{4}{5}$  j) $1\frac{3}{5}$  k) $3\frac{1}{5}$  l) $2\frac{2}{5}$

**2** Write these decimals as fractions or mixed numbers, in their simplest form.

a) 4.5   b) 3.25   c) 2.75   d) 8.5

e) 1.25   f) 5.5   g) 4.75   h) 6.25

i) 0.6   j) 1.4   k) 2.8   l) 3.2

**3** This number line from 4 to 5 has been divided into tenths.

Write down the decimal numbers at A, B and C.

**4** Draw a number line between 6 and 7 and divide it into tenths.

Mark the points   a) 6.5   b) 6.25   c) 6.75.

**5** This number line is divided into fifths.

Write down the decimal numbers at A, B and C.

**6** Draw a number line between 3 and 5 and divide it into fifths.

Mark the points   a) 3.6   b) 4.2   c) 3.1.

---

On lined paper, draw a line AB exactly 10 cm long, with A on the top line and B on the fifth line.

This diagram shows you how to do it, but it is not drawn to scale.

The lines on the paper cut AB into 4 equal parts, so each of them is $\frac{1}{4}$ of 10 cm long.

Measure them on your drawing. They should be 2.5 cm each.

Do similar drawings to find $\frac{1}{2}$, $\frac{1}{3}$, $\frac{1}{5}$, $\frac{1}{6}$, $\frac{1}{7}$, $\frac{1}{8}$ and $\frac{1}{9}$ of 10 cm.

What is $\frac{1}{7}$ of 1 cm?

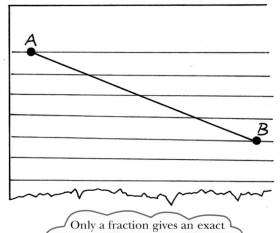

Only a fraction gives an exact answer. Your measured length is an approximation

# 7: Decimals

# Adding and subtracting decimals

Every time you add and subtract amounts of money you are working with decimals.

### Example

Jake buys these oranges and apples.

a) What is the total cost?
b) He pays with a £5 note. How much change does he get?

### Solution

a)
```
oranges    0.75
 apples   +1.05
total cost 1.80
```

*Remember to line up the decimal points*

**The total cost is £1.80.**

b)
```
 5.00
-1.80
 3.20
```

*When dealing with money you write £3.20, not £3.2*

**Jake gets £3.20 change.**

Here is a plan of Ann's lounge-diner.

Ann wants to know the length of the room.

She works it out like this:

```
  3·60
+ 2·25
  5·85
```

 *Why does Ann write 3.6 as 3.60?*

The lounge-diner is 5.85 m long.

Ann's mother has given her a carpet 8 m long for her lounge-diner. Ann wants to know how much will be left over.
She works it out like this:

```
  8·00
- 5·85
  2·15
```

 *Why does Ann write 8 as 8.00?*

**Ann has 2.15 m of carpet left.**

# 7: Decimals

**1** Work out the answers to these.

a) £1.10    b) £7.60    c) £3.55
   + £2.45      + £1.30      + £2.60

d) £10.20   e) £6.54   f) £7.15
   − £2.10      − £6.42      − £4.20

**2** Work out the answers to these.

a) 1.6 + 3.2       b) 4.8 + 3.12

c) 10.6 + 10.43    d) 4.9 − 3.01

e) 4.01 − 3.8       f) 5.51 − 3.8

**3** Kieran buys a T-shirt for £6.99.

How much change does he get from a £10 note?

**4** Sara is a waitress.

She takes this order from Table 3.

Work out the bill for Table 3.

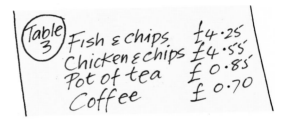

Table 3
Fish & chips £4.25
Chicken & chips £4.55
Pot of tea £0.85
Coffee £0.70

**5**

Kate walks to the woods. She passes the sign on the left.

Sometime later she passes the sign on the right.

How far has she walked in this time?

**6** Phil's lorry is 4.6 m high.

How much clearance will there be if he drives under this bridge?

(The clearance is the space between the bridge and the lorry.)

---

Use the price list from your canteen or a local take-away.

Write down a lunch order for yourself and a friend.

List each item and its price.

Work out the total cost.

How much change would you get if you paid with a £20 note?

# 7: Decimals

## Multiples of 10

Leo has 10 glasses to fill with water.
Each glass has a capacity of 0.25 litres.
How much water does Leo need?

You need to work out $0.25 \times 10$.
To multiply a decimal by 10, you move the decimal point 1 place to the right:

$0.25 \times 10 = 02.5$

*We usually leave out the zero and write this as 2.5*

*What is $0.75 \times 10$?*
*What is $0.4 \times 10$?*
*What is $4.3 \times 10$?*

How many litres would Leo need to fill 100 of these glasses?

You need to work out $0.25 \times 100$.
To multiply a decimal by 100, you move the decimal point 2 places to the right:

$0.25 \times 100 = 025.$

*We usually leave out the zero and the decimal point, and write this as 25*

**Leo needs 25 litres of water.**

*What is $0.75 \times 100$?*
*What is $0.4 \times 100$?*

*What is the rule for multiplying by 1000?*

Leo has 2.5 litres of water to divide equally between 10 glasses.
How much water should he put in each glass?

You have already seen that to put 0.25 litres in each of 10 glasses you need 2.5 litres.

**So Leo should put 0.25 litres in each glass.**

You can write
   $2.5 \div 10 = 0.25$

*÷10*
*2.5 becomes .25*

You can see that to divide by 10 you move the decimal point 1 place to the left.

*What is $7.5 \div 10$?*
*What is $4 \div 10$?*

To divide by 100 you move the decimal point 2 places to the left.

*How do you divide by 1000?*

# 7: Decimals

**1** Work out

a) 1.6 × 10
b) 3.01 × 10
c) 21.0 × 10
d) 0.35 × 100
e) 7.6 × 100
f) 0.032 × 100

**2** Ayla orders 15 boxes of computer disks. Each box contains 10 disks. How many disks does she order?

**3** a) How much do 10 of these pens cost?

b) How much do 100 cost?

Pens £0·24 each

**4** Tim and Joanne started their business with £5000. It is now worth 10 times that amount. How much is it worth now?

**5** Bernice is sending out 1000 brochures to advertise her business. Each brochure costs 35 pence to send. What is the total cost of the mailing?

**6** Work these out.

a) 70.1 ÷ 10
b) 7.3 ÷ 10
c) 0.6 ÷ 10
d) 632 ÷ 100
e) 82.6 ÷ 100
f) 0.611 ÷ 100

**7** A group of 10 friends hires this minibus.

They share the cost equally between them.

How much does each person pay?

MINI-BUS HIRE
10 seater £70

**8** 10 chairs cost £245. How much does one chair cost?

**9** A mirror is 150 cm wide. Divide this by 100 to get the width in metres.

**10** Sam and 9 friends want to play 5-a-side. They plan to share the cost equally.

**11** Elizabeth's car has a 1600 cc engine. Divide this by 1000 to get the engine size in litres.

5-a-side pitch
£35 per hour (Weekdays)
£45 per hour (Weekends)

How much will each pay if they play for an hour

a) at the weekend?

b) during the week?

Ask someone to measure your height, correct to the nearest mm.

Write your height in

a) millimetres (e.g. 1712 mm)
b) centimetres (e.g. 171.2 cm)
c) metres (e.g. 1.712 m)
d) kilometres (e.g. 0.001 712 km)

# 7: Decimals

## Multiplying decimals

Martine works in catering.

Her purchases and her recipes are not always in the same units.

She needs to work out how many pounds there are in 5 kg of potatoes. She knows that 1 kg is 2.2 pounds.

This is what Martine writes down.

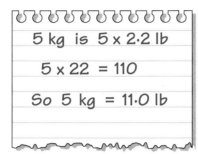

5 kg is 5 × 2·2 lb
5 × 22 = 110
So 5 kg = 11·0 lb

She makes the 2.2 into a whole number by moving the decimal point 1 place to the right…

…then she multiplies the whole numbers

She moves the decimal point 1 place to the left to get her final answer

 *Why does this method work?*

Martine needs to work out how many litres there are in 4.5 pints of milk. She knows that 1 pint is 0.57 litres.

She writes it like this:

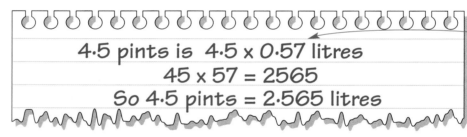

4·5 pints is 4·5 × 0·57 litres
45 × 57 = 2565
So 4·5 pints = 2·565 litres

Notice that 0.57 is between 0 and 1. When you multiply by a number between 0 and 1 the answer is smaller

You met squares in Chapter 3. You know that 4 squared is written $4^2$ and means $4 \times 4$.

You square decimals in the same way.

For example, 3.2 squared is written $3.2^2$ and means $3.2 \times 3.2$.

 *Use the $x^2$ key on your calculator to work out $3.2^2$. (You should get* 10.24.*)*

When you have to multiply a decimal by a nice round number (such as 50 or 400) you can do it quickly by hand.

For example,    £12.50 × 50 = £12.50 × 10 × 5

× 50 is the same as × 10 then × 5

= £125.00 × 5 = £625.00

 *How much does it cost a company to give its 400 employees a £10.50 bottle of champagne each at Christmas?*

# 7: Decimals

**1** Work out the answers to these.

a) $2 \times 3.6$  b) $4 \times 1.5$  c) $8 \times 1.11$  d) $5 \times 0.16$

e) $3 \times 0.04$  f) $7 \times 1.04$  g) $2 \times 0.031$  h) $9 \times 10.6$

**2** The table shows the heights of 4 children on their second birthdays. To estimate the height that each child will reach when fully grown, you double the height at age 2.

Copy this table and complete it by working out the estimated fully grown heights.

| Name | Height at age 2 (m) | Fully grown height (m) |
|---|---|---|
| Amy | 0.85 | |
| Jack | 0.83 | |
| Ryan | 0.96 | |
| Laura | 0.89 | |

**3** The map shows the route of a cycle race. Distances are in kilometres.

Using the fact that 1 kilometre is about 0.6 miles, write the distances in miles.

Set out your answers like this.

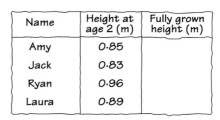

Guiseley ⟶ Bingley : 8 km = 8 × 0.6 miles = .......

**4** a) How much would 2 sessions on this sunbed cost?

b) How much would 5 sessions on the sunbed cost?

c) Is the price for 10 sessions good value?

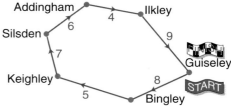

SUNBED
1 session   £1.75
10 sessions  £15.75

**5** Work out the square of each of these.

a) 1.5  b) 2.2  c) 0.5  d) 4.8

**6** A rock concert is sold out: 20 000 people have bought tickets at £13.50 each. How much have they paid altogether?

**7** Work these out.

a) $200 \times 4.5$  b) $8.3 \times 30$

c) $0.5 \times 700$  d) $4000 \times 2.5$

e) $400 \times 25$  f) $0.06 \times 300$

g) $0.01 \times 2000$  h) $500 \times 10.1$

**8** a) Work these out. (i) $25 \times 0.1$ (ii) $36 \times 0.5$ (iii) $200 \times 0.9$

b) In part a) you multiplied by 0.1, 0.5, and 0.9. These decimals are all between 0 and 1. Does multiplying a number by these decimals make it bigger or smaller?

In the UK, road atlases and signs have distances in miles.

A mile is about 1.6 kilometres, so to convert miles to km you multiply by 1.6.

This chart shows the distances between 5 major UK cities.

Copy the chart, but write all the distances in km.

Glasgow is 291 miles from Birmingham

| Birmingham | | | | |
|---|---|---|---|---|
| 108 | Cardiff | | | |
| 284 | 367 | Edinburgh | | |
| 291 | 393 | 46 | Glasgow | |
| 120 | 155 | 372 | 403 | London |

# 7: Decimals

## Dividing decimals

Dan is ordering headed notepaper at £31.50 for 5 reams.

Dan needs to know the price per ream.

This is what he writes.

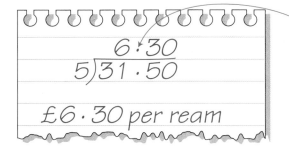

He is careful to put this decimal point in line with the one below it

$6.30$
$5)\overline{31.50}$

£6.30 per ream

Another supplier charges £28.80 for 4 reams.

Is this a better buy?

Gloria is ordering fruit juice for a training course.

One supplier sells 1.5-litre bottles at £1.20 each.

Gloria needs to work out the price per litre.

This is what she writes.

First she writes the division as a fraction

Then she multiplies the top and bottom by 10 to get a whole number on the bottom

$$1 \cdot 20 \div 1 \cdot 5 = \frac{1 \cdot 20}{1 \cdot 5} = \frac{12}{15} = \frac{4}{5} = 0 \cdot 8$$

Then she writes the fraction in its simplest form before doing the division

Price per litre = £0.80 = 80p

Another company sells 2.5-litre bottles of juice at £1.90.

Is this a better buy?

## Square roots

You met square roots in Chapter 3. You know, for example, that the square root of 36 (written $\sqrt{36}$) is 6.

$6^2 = 36$ so $\sqrt{36} = 6$

For most numbers, you need a calculator to work out the square root.

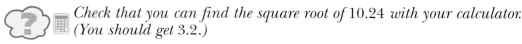

Check that you can find the square root of 10.24 with your calculator. (You should get 3.2.)

# 7: Decimals

**1** Work these out.
   a) 40.5 ÷ 5    b) 210.3 ÷ 3    c) 2.644 ÷ 4
   d) 0.42 ÷ 1.5  e) 460 ÷ 0.2    f) 1.43 ÷ 1.1

**2** Work out the price per litre of each of these products.

   a)    b)    c)

**3** Imran earns £164.85 for 5 days' work.
   How much does he earn per day?

**4** Claire earns £43.20 for working 12 hours.
   How much does she earn per hour?

**5** A swimming pool is 14 m wide. It is divided into
   8 lanes for a competition. How wide is each lane?

**6** Julian changes 120 Swiss francs into pounds.
   He gets £1 for every 2.5 Swiss francs.
   Work out 120 ÷ 2.5 to find the number of pounds he gets.

**7** A gallon of 4 star petrol costs £3.24.
   Work out the price per litre. (1 gallon is about 4.5 litres.)

**8** Mel buys a roll of elastic 10 metres long.
   She needs 15 pieces 0.7 metres long.
   Has she bought enough elastic?

**9** Work these out.
   a) 12 000 ÷ 20      b) 300 ÷ 20      c) 5000 ÷ 200
   d) 500 000 ÷ 500    e) 600 ÷ 400     f) 48 000 ÷ 30

**10** a) Work these out. (i) 25 ÷ 0.1  (ii) 36 ÷ 0.5  (iii) 200 ÷ 0.9
   b) In part a) you divided by 0.1, 0.5, and 0.9. These decimals are all
      between 0 and 1. Does dividing a number by these decimals
      make it bigger or smaller?

**11** Write down the square root of each of these.
   a) 6.25    b) 3.24    c) 7.84    d) 18.49    e) 51.84

---

Find a road atlas of Europe.
Choose 4 major cities, such as Paris, Brussels, Amsterdam and Bonn.
Find out the distance in km from each city to each of the other cities.
Convert the distances to miles (by dividing them by 1.6).
Present the information in a chart like the one on page 71.

# 7: Decimals

## Finishing off

**Now that you have finished this chapter you should be able to**

- ★ write tenths and hundredths in decimal form
- ★ write decimals as tenths and hundredths
- ★ write a half, a quarter and three quarters in decimal form
- ★ add and subtract decimals
- ★ multiply decimals
- ★ divide decimals
- ★ work out squares and square roots

Use the questions in the next exercise to check that you understand everything.

### Mixed exercise

**1** Write down a fraction equal to each of these decimals.

a) 0.3    b) 0.43    c) 9.2    d) 7.29

**2** Write down a decimal equal to each of these fractions.

a) $\frac{7}{10}$    b) $2\frac{3}{10}$    c) $\frac{47}{100}$    d) $3\frac{17}{100}$

**3** Ria takes the temperature of 3 patients. The readings are shown below (in degrees Centigrade). Write down each reading.

a) 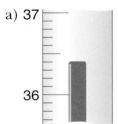   b) 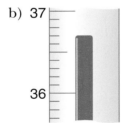   c)

**4** The diagrams show the height of a high jump bar (in metres). Write down the height of the top of the bar.

a) 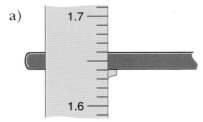   b)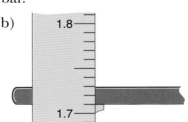

**5** Work these out.

a) 4.6 + 9.3    b) 18.3 + 5.26    c) 4.008 + 1.9
d) 10.9 − 5.4   e) 58.1 − 7.9    f) 21.3 − 4.26

# 7: Decimals

**6** Work out the amount each person spends at this snack bar.

a) Anna buys a can of cola and a jacket potato with beans.

b) Reena buys 2 cartons of apple juice, a vegeburger and a yoghurt.

c) Alex buys a bacon bap and a burger.

d) Kit buys 5 packs of biscuits.

e) Lu buys 3 cans and 2 cartons.

```
SNACK BAR MENU          £
Cans (cola, orange)     0.52
Cartons (apple, orange) 0.40
Jacket potatoes
   (cheese or beans)    1.10
Burger in a bun         1.29
Vegeburger in a bun     1.09
Bacon bap               1.29
Yoghurt                 0.32
Pack of biscuits        0.25
```

*Mixed exercise*

**7** Work these out.

a) $7.1 \times 100$   b) $2.16 \times 1000$   c) $3.32 \times 200$   d) $0.04 \times 40$

**8** An office desk is 180 cm wide. How many metres is this?

**9** Barbara has £120 to change to dollars. She gets $1.65 for each £1.

How many dollars does she get altogether?

**10** Wesley buys 0.6 kg of tomatoes and 1.2 kg of bananas at this stall.

a) How much does it cost?

b) How much change does he get from a £20 note?

Tomatoes £1.40 per kg   Bananas £0.95 per kg

**11** Hannah fits a wash-basin in the middle of a bathroom wall. The basin is 44.6 cm wide, and the wall is 87 cm wide.

How much space is there at each side of the basin?

**12** Work out

a) $42 \div 3.5$   b) $6 \div 0.25$   c) $25 \div 2.5$   d) $7.2 \div 1.5$

**13** A videotape box is 2.5 cm thick.

How many can be stored on a shelf 80 cm long?

**14** Find the value of

a) $7.2^2$   b) $\sqrt{7.84}$

c) $4.9^2$   d) $\sqrt{29.16}$

Find everyday objects (5 in all) with lengths as near as you can to 1 cm, 5 cm, 10 cm, 50 cm and 1 m.

Measure each of your objects and state its length.

# Eight

## Shapes

### Sorting shapes

Look at these shapes.

Can you organise them into groups? (There are lots of ways of doing this.)

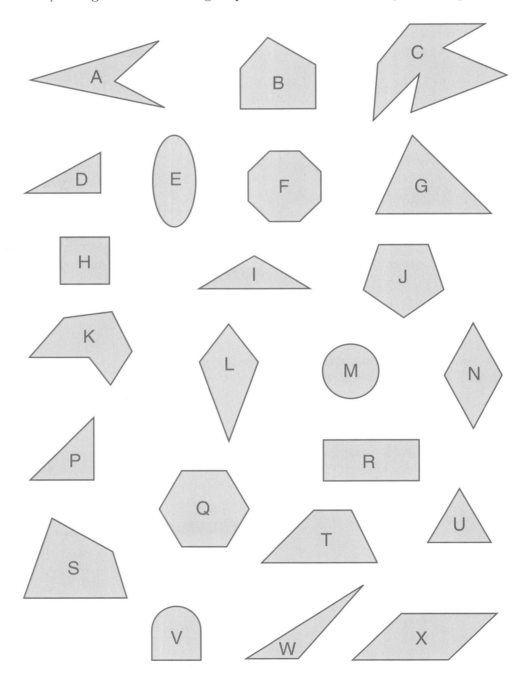

## 8: Shapes

**1** Here are some groups of shapes from the opposite page.

For each one, say what rule has been used to put shapes into the group.

You may know a special name for some of the groups.

a) D, G, I, P, U, W
b) A, H, L, N, R, S, T, X
c) B, J
d) K, Q
e) C, F
f) F, H, J, Q, U

**2** There are many other ways of putting the shapes into groups.

For each of the groups described below, write down the letters for all the shapes which would fit in the group.

a) shapes with curves in them
b) shapes with at least one right angle
c) shapes with at least one obtuse angle
d) shapes with at least one reflex angle
e) shapes with at least one pair of parallel lines
f) shapes with all sides the same length

**Remember:**
a right angle is a 90° angle,
an acute angle is an angle less than 90°,
an obtuse angle is an angle between 90° and 180°,
a reflex angle is an angle bigger than 180°

---

Look at these triangles. For each one, say how many sides are the same length, and what different types of angles it has (acute angles, right angles or obtuse angles).

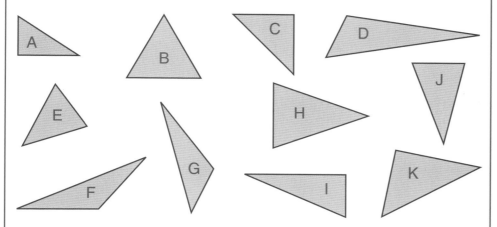

Group each triangle with others that have the same description.

How many different types of triangle are there altogether?

# Triangles

A shape with three sides is called a **triangle**.

Triangles can be described by how many sides with the same length they have.

A triangle whose three sides are all the same length is called an **equilateral triangle**.

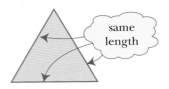

A triangle that has two sides the same length and the third side a different length is called an **isosceles triangle**.

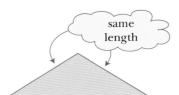

A triangle whose sides are all of different lengths is called a **scalene triangle**.

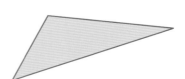

Another way of describing triangles is by what sort of angles they have.

An **acute-angled triangle** has three acute angles (all its angles are less than 90°).

An acute-angled triangle can be equilateral, isosceles or scalene.

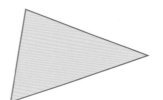

A **right-angled triangle** has one right angle (a 90° angle).

A right-angled triangle can be isosceles or scalene.

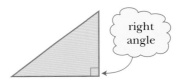

An **obtuse-angled triangle** has one obtuse angle (an angle greater than 90°).

An obtuse-angled triangle can be isosceles or scalene.

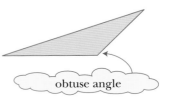

*Why is it impossible to have a triangle which is right-angled and equilateral?*

*Why it is impossible to have a triangle which is obtuse-angled and equilateral?*

# 8: Shapes

**1** Which of these triangles is
   a) a right-angled scalene triangle?
   b) an obtuse-angled isosceles triangle?
   c) an equilateral triangle?
   d) an acute-angled scalene triangle?
   e) a right-angled isosceles triangle?
   f) an obtuse-angled scalene triangle?
   g) an acute-angled isosceles triangle?

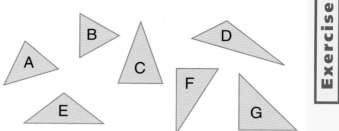

**2** In the diagram below there are lots of different triangles.

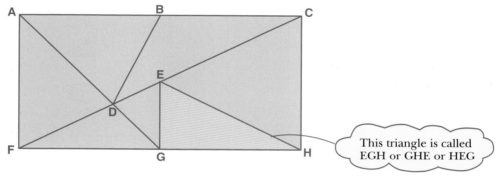

This triangle is called EGH or GHE or HEG

Find as many different triangles as you can. Describe each triangle that you find.

Write down your answers like this:

*EGH is a right-angled scalene triangle*

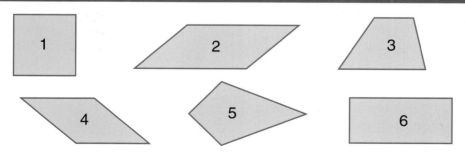

For each of these four-sided shapes, say which of the statements in the list below are true.

If you know the special name for the shape, write that down as well.

A  Four equal sides
C  No equal sides
E  Two pairs of equal angles
G  Two pairs of parallel sides

B  Two pairs of equal sides
D  Four right angles
F  One pair of equal angles
H  One pair of parallel sides

## 8: Shapes

# Quadrilaterals

Any shape with four sides is called a **quadrilateral**.

Some quadrilaterals have special names.

A **square** has four equal sides. All of its angles are right angles.

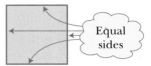

A **rectangle** has two pairs of equal sides. All of its angles are right angles.

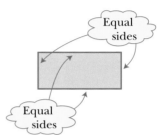

A **parallelogram** has its opposite sides equal and parallel.

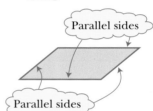

A **rhombus** has four equal sides. Its opposite sides are parallel.

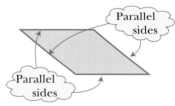

A **kite** has two pairs of equal sides. The equal sides are next to each other.

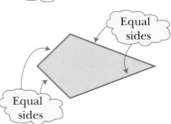

A **trapezium** has one pair of parallel sides.

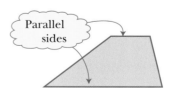

A shape with four sides which is none of these special shapes is called an **irregular quadrilateral**.

# 8: Shapes

**1** Draw each of these shapes.

For each one, show all of its lines of reflection symmetry.

a) square  b) rectangle  c) rhombus

d) parallelogram  e) kite  f) trapezium

**2** Here is a quadrilateral on a 3×3 pinboard.

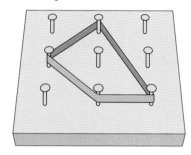

Find as many other different quadrilaterals as you can on a 3 × 3 pinboard.

Draw each one and write the name of the shape (e.g. parallelogram) beside your drawing.

**3** Write 'T' (for true) or 'F' (for false) for each of these statements. In each case, explain your choice.

a) A square is a special kind of rectangle.

b) A rectangle is a special kind of square.

c) A parallelogram is a special kind of rhombus.

d) A rhombus is a special kind of parallelogram.

e) A kite is a special kind of trapezium.

f) A square is a special kind of trapezium.

What other statements like these can you find? Which ones are true?

---

Trace this right-angled triangle 4 times.

Cut your triangles out so that you have 4 identical triangles.

Put them together to make as many other shapes as you can. Draw each shape and (where possible) give its name.

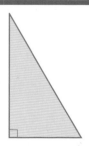

# 8: Shapes

# Other kinds of shapes

## Polygons

Shapes with three or more straight sides are called **polygons**.

You have already met two kinds of polygon: the triangle and the quadrilateral.

If all the sides of a polygon are the same length, and all its angles are the same, it is called a **regular polygon**.

Otherwise it is called an **irregular polygon**.

*What is the special name for a regular quadrilateral?*

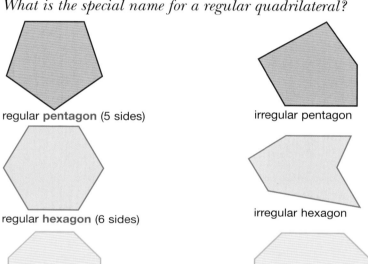

regular **pentagon** (5 sides)

irregular pentagon

regular **hexagon** (6 sides)

irregular hexagon

regular **octagon** (8 sides)

irregular octagon

This octagon has all of its angles equal, but it is not regular because the sides are not all the same

## Congruent shapes

*Trace triangle A.*

*Try to fit your tracing over each of the other triangles. You may need to turn it over for some of them. Which triangles fit the tracing exactly?*

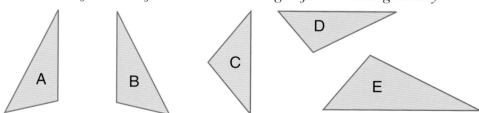

Shapes which are exactly the same shape and size are called **congruent**. Congruent shapes may be turned round or flipped over, but the tracing of a shape will always fit exactly over a shape which is congruent to it.

## 8: Shapes

**1** a) Copy or trace the polygons below and draw on the lines of symmetry. Label each shape with its correct name.

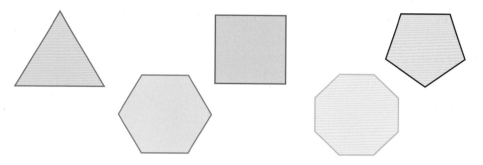

b) How many lines of symmetry do you think a regular polygon with 50 sides would have?

**2** Which of these shapes are congruent to shape A?

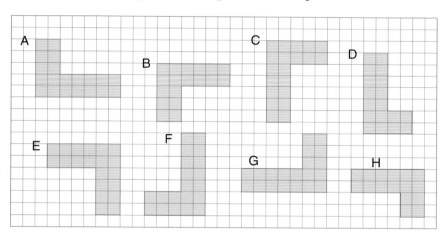

### Investigations

**1** Copy this pattern on to squared paper.

How many patterns, congruent to this one, can you draw on a 3 × 3 grid?

**2** Design your own pattern and draw some other patterns that are congruent to it.

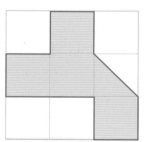

What shapes are used to make the cover of a football?

How many of each shape are needed?

How big is each shape?

## 8: Shapes

# Circles

The circle is a very familiar shape.

Car and bicycle wheels, dinner plates and CDs are all circular.

 *How can you describe a circle?*

There are several different ways of measuring circles to find out how big they are.

- The shaded region is called the **sector**. It is formed by two radii and an arc
- The **radius** is the distance from the centre to the edge of the circle
- Part of the circumference is called an **arc**
- The **diameter** is the distance all the way across the circle
- The **circumference** is the distance all the way round the edge of the circle

The radius is half the distance across the whole circle.

So the radius is half the diameter.

The circumference is difficult to measure because it curves.

It is easier to measure if you use a piece of string.

The tip of a windscreen wiper traces out part of a circle.
Part of a circle is called an **arc**.

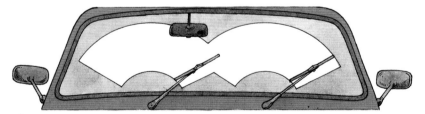

The bicycle wheel touches the line of the road. A line which just touches a circle is called a **tangent**.

A line which goes across the circle from one point on the circumference to another is called a **chord**.

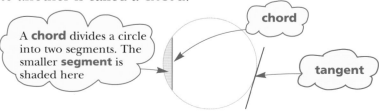

- A **chord** divides a circle into two segments. The smaller **segment** is shaded here
- chord
- tangent

# 8: Shapes

**1** Give 5 examples of circles in everyday life.

(Choose ones that have not been mentioned on the left-hand page!)

**2** Use compasses to draw

a) a circle with radius 4 cm
b) a circle with radius 5.4 cm
c) a circle with diameter 7 cm.

**3** Use a piece of string to measure roughly the circumference of each of the circles you drew in question 2.

**4** Draw a diagram showing the meaning of these words: circumference; diameter; radius; arc; sector; chord; segment; tangent.

## Investigations

**1** This is how to draw a pattern called a Mystic Rose.

   1. Draw a circle. Mark 5 points, roughly equally spaced, on the circumference.
   2. Draw chords from one point to each of the other points.
   3. Now do the same for each of the other points. You can colour the pattern if you like.

You can draw Mystic Roses with any number of points. Try drawing some more of your own.

**2** Here is another circle pattern for you to draw.

   1. Use compasses to draw a circle. Put the point of the compass on the circumference. Draw an arc inside the circle like this.
   2. Put the point of the compass on one end of the arc. Draw another arc.
   3. Repeat until the pattern is completed. Colour the pattern if you like.

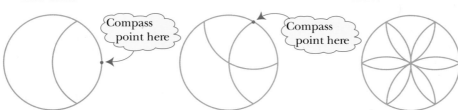

Make an accurate scale drawing of a darts board.
Write down the radius of each circle.
Explain how you score at darts.

# 8: Shapes

# Finishing off

**Now that you have finished this chapter you should recognise**

- ★ an equilateral triangle, an isosceles triangle and a scalene triangle
- ★ an acute-angled triangle, a right-angled triangle and an obtuse-angled triangle
- ★ a square, a rectangle, a parallelogram, a rhombus, a kite and a trapezium
- ★ a quadrilateral, a polygon, a pentagon, a hexagon and an octagon
- ★ the difference between a regular and an irregular shape
- ★ congruent shapes
- ★ a circumference, a radius, a diameter, a chord, an arc and a tangent of a circle

**Use the questions in the next exercise to check that you understand everything.**

## Mixed exercise

Look back at the shapes on page 76 to answer questions 1 – 4.

**1** Which of them are
  a) triangles?
  b) quadrilaterals?
  c) pentagons?
  d) hexagons?
  e) octagons?
  f) circles?

**2** Which of the triangles are
  a) isosceles?
  b) scalene?
  c) equilateral?
  d) right-angled?

**3** Which of the quadrilaterals are
  a) rectangles?
  b) parallelograms?
  c) kites?
  d) trapeziums ('trapezia')?

**4** Which of the pentagons, hexagons and octagons are regular?

**5** a) Sketch a circle as neatly as you can.
  b) On your circle, draw and label a diameter, a radius, a chord and a tangent.
  c) On your diagram, label and shade in different colours two segments and one sector.

**6** Look at Helen's design for a patchwork baby quilt. How many of each of these shapes does she need?
  a) right-angled triangles  b) squares
  c) parallelograms  d) trapezia  e) isosceles triangles

What kind(s) of symmetry does the design have?

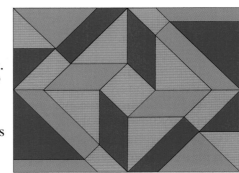

# 8: Shapes

**Investigation**

a) Find as many triangles as you can on a 3 × 3 pinboard.

  Draw the triangles on squared paper.

  Here are two examples.

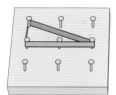 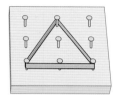

b) Describe each triangle that you have drawn.

  Use the words isosceles, scalene, right-angled, acute-angled and obtuse-angled.

c) Can you make an equilateral triangle?

---

Design a patchwork or mosaic of your own. Draw your design on squared paper.

Make a list of the shapes you have used.

What kind(s) of symmetry does your design have?

---

Collect 5 or 6 different-sized tins (such as baked beans tins, baby-milk tins, tomato purée tins and so on).

| Tin | Circumference (cm) | Diameter (cm) | Circumference ÷ Diameter |
|---|---|---|---|
| Baked Beans | | | |
| Tomato Purée | | | |

Make a table like this.

Use a piece of string to measure the circumference of each tin as accurately as you can.
Use a ruler or any other suitable method for measuring the diameter.
Write your results in the table.
Using a calculator, divide the circumference of each tin by its diameter.
Put the result in the last column.
When you have finished, look carefully at the figures in the last column.
Do you notice anything?
Write a sentence or two to explain your conclusions.

*Mixed exercise*

# Nine

# Percentages

## 25%, 50% and 75%

### Student Survey findings
**FACT** 50% of Year 10 have part-time jobs

The survey says that 50% of students have a part-time job.

That means 50 out of every 100.

You can show this in a 10 × 10 square like this:

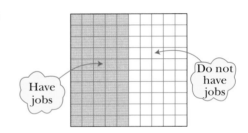

 *What fraction is shaded? Write the fraction in its simplest form.*

*Write this fraction as a decimal.*

You can see that 50% is the same as $\frac{1}{2}$ or 0.5. So half of Year 10 students have part-time jobs.

**FACT** 25% of Year 10 cannot swim

Again you can show this in a 10 × 10 square:

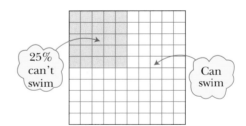

 *What fraction of the students can't swim? Write the fraction in its simplest form.*

You can see that 25% is the same as $\frac{1}{4}$ or 0.25.

You can also see from the diagram that 75% of the students can swim.

 *What fraction of the students can swim?*

75% is the same as $\frac{3}{4}$ or 0.75.

# 9: Percentages

**1** Copy and complete this table.

| Fraction | Decimal | Percentage |
|---|---|---|
| $\frac{1}{4}$ | | |
| | 0.5 | |
| | | 75% |

For questions 2 to 6 choose A, B or C. Use this pie chart to answer questions 2, 3 and 4. It shows the results of Ann's survey of the customers at her fitness centre. She asked them how they first heard of the centre.

**2** The percentage of customers who heard about it from friends is

A less than 25%   B 25%

C more than 25%.

**3** The percentage of customers who read about it in the local press is

A less than 50%   B 50%

C more than 50%.

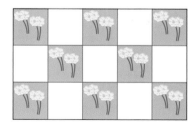

**4** The percentage of customers who are 'passing trade' is

A less than 25%   B 25%

C more than 25%.

**5** Ben scores 11 marks out of 20 in a test. Is this

A less than 50%?   B 50%?   or   C more than 50%?

**6** This is a small garden of flower beds and paving slabs.

Is the area covered by flower beds

A less than 50%?   B 50%?

or   C more than 50%?

---

You are asked to design a new flag for a small island.

It must be 50% blue and 50% green, and made of 4 rectangles as shown.

How many different flags are possible?

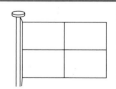

# 9: Percentages

## Finding percentages

Leo finds that 7 out of every 10 customers choose chicken.

What is this as a percentage?

You can draw this as 10 columns like this:

**LEO'S SNACK BAR**
*Lunch specials*
Chicken & chips £2.30
Fish & chips £2.30

7 out of 10 columns are chicken

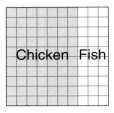
Dividing each column into 10 you get this

You can see that

7 out of 10 is the same as 70 out of 100, or 70%;

3 out of 10 is the same as 30 out of 100, or 30%.

Another way to do this is to use equivalent fractions:

$$\frac{7}{10} = \frac{70}{100} = 70\%, \quad \frac{3}{10} = \frac{30}{100} = 30\%$$

Using equivalent fractions in the same way,

$$\frac{1}{10} = \frac{10}{100} = 10\%, \quad \frac{1}{20} = \frac{5}{100} = 5\%$$

### Example

Write   a) $\frac{9}{10}$   b) $\frac{2}{5}$   as percentages.

### Solution

a) $\frac{9}{10} = \frac{90}{100} = 90\%$   b) $\frac{2}{5} = \frac{4}{10} = \frac{40}{100} = 40\%$

### Example

In a maternity hospital 520 out of 1000 births are boys.

What percentage are   a) boys?   b) girls?

### Solution

a)   520 out of 1000 is $\frac{520}{1000}$:

$$\frac{520}{1000} = \frac{52}{100} = 52\%$$

b)   The girls must be   $100\% - 52\% = 48\%$.

You can change decimals into percentages without doing any calculations.

0.17 is 17 hundredths, or 17%.

0.06 is 6 hundredths, or 6%.

# 9: Percentages

**1** Look at the diagram.

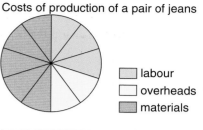

Costs of production of a pair of jeans

What percentage of the cost of making a pair of jeans is

a) labour?

b) overheads?

c) materials?

**2** Helen is a travel agent.

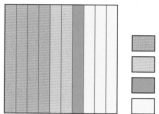

This diagram shows where her clients go on holiday.

a) What percentage go to Spain?

b) What percentage go to Greece?

c) What percentage go to Turkey?

**3** Anna does an experiment on the air someone breathes out.

Here are her results in litres.

| Total | Oxygen | Nitrogen | Carbon dioxide |
|---|---|---|---|
| 200 | 32 | 160 | 8 |

a) What percentage is oxygen?

b) What percentage is nitrogen?

c) What percentage is carbon dioxide?

**4** Here are three floor designs.

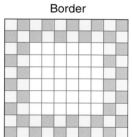 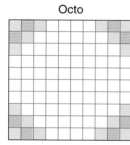 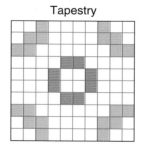

Border    Octo    Tapestry

Each is made up of 100 square tiles.

a) What percentage of each floor is coloured?

b) Write each percentage as a fraction in its simplest form.

c) Write each percentage as a decimal.

---

Find a newspaper article and count the number of letters in each of the first 50 words.

What percentage of the words have 1 letter, 2 letters, 3 letters,…?

Show your results in a table.

## 9: Percentages

# Percentage calculations

### 70% Success rate

A catering firm takes on 300 trainees. 70% of them are successful, completing their training and getting qualifications.

How many trainees is this?

The number who are successful is 70% of 300.

*70% means 70 out of 100*

*You could do this as*
$300 \times 70 \div 100 = 21000 \div 100 = 210$
*but it is easier to do the division first:*
$300 \div 100 \times 70 = 3 \times 70 = 210$

$$70\% \text{ of } 300 = \frac{70}{100} \times 300 = 210$$

*'of' means 'multiply'*

**210 trainees are successful.**

*What percentage of the trainees are not successful?*

*How many trainees are not successful?*

### 4% pay rise

Hasna earns £150 a week. She is given a 4% pay rise.

The pay rise is 4% of £150.

$$4\% \text{ of } 150 = \frac{4}{100} \times 150 = 6$$

*How much is the pay rise?*

*How much does Hasna earn after the pay rise?*

Carly buys this jacket.
The reduction is 40% of £70.

$$40\% \text{ of } 70 = \frac{40}{100} \times 70 = 28$$

*How much is the reduction?*

*How much does Carly pay for the jacket?*

# 9: Percentages

**1** Work out

a) 75% of 400  b) 20% of 600  c) 40% of 350
d) 50% of 144  e) 8% of 125  f) 60% of 225
g) 2% of 100  h) 1% of 200  i) 15% of 400
j) 10% of 220  k) 5% of 2000  l) 20% of 80

**2** Claire earns £9000 a year. She is given a pay rise of 10%.

How much extra money does she get?

**3** What are the savings on each of these cars when you buy in June?

**4** 6% of the cups made at a pottery are faulty.

How many faulty cups would you expect in a batch of 400?

**5** Sam's heating bills are £800 a year.

She insulates her loft and so her heating bills fall by 20%.

How much does she save each year?

**6** Rebecca gets 12% off a holiday priced at £400.

How much does she pay?

**7** Shamil's railway season ticket cost £200 last year.

The price has gone up by the rate of inflation, 6%.

How much does it cost this year?

**8** James receives a bill for £250.

He gets a 2% discount by paying within a week.

How much does he pay?

---

Get a copy of the Passenger's Charter from one of the railway companies.

Look through it and find the percentage refunds you can get if the trains run late.

Say exactly when you can get a refund.

## 9: Percentages

# From fractions to percentages

A new treatment for asthma is tested on 50 sufferers. Of these, 35 find that it is better than their old treatment.

You can write 35 out of 50 as a fraction, $\frac{35}{50}$. Its simplest form is $\frac{7}{10}$.

However, you often see results like this given as percentages.

There are two ways to change a fraction into a percentage. You need to be able to use both of them. Sometimes one is easier and sometimes the other: it depends on the numbers.

**Method 1: using equivalent fractions**

$$\frac{35}{50} = \frac{70}{100} = 70\%$$
(× 2 top and bottom)

Look at the bottom line, 50. You need to multiply it by something to make it 100. In this case you multiply by 2. Then multiply the top by the same number

**Method 2: multiplying by 100%**

$$\frac{35}{50} \times 100\% = 70\%$$

One way to work this out is
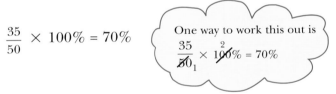

*Use both methods to change each of these fractions into a percentage.*

a) $\frac{9}{20}$   b) $\frac{3}{8}$

*Which is the easier method in each case?*

To convert from a percentage to a fraction, just remember that a percentage is a fraction with 100 on the bottom line.

### Example

Convert 48% to a fraction, giving the answer in its simplest form.

$$48\% = \frac{48}{100}$$
$$= \frac{12}{25}$$

Divide both top and bottom by 4 to get the simplest form

# 9: Percentages

**1** For each of these, write down a percentage and a decimal equal to the fraction.

a) $\frac{8}{25}$  b) $\frac{11}{20}$  c) $\frac{1}{8}$  d) $\frac{27}{40}$  e) $\frac{179}{250}$  f) $\frac{7}{15}$

**2** Look at these three patio designs.

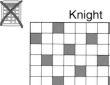
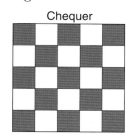

Knight  Chequer  Domino

For each design, write down

a) the fraction of the patio that is white

b) the percentage of the patio that is white.

**3** Sam's survey of 80 people found that 38 read Daily News, 25 read News Today and the rest read neither paper. Nobody reads both.

What percentage of people read

a) Daily News?  b) News Today?  c) neither paper?

**4** This bar chart shows the number of male and female employees at each of a firm's two sites.

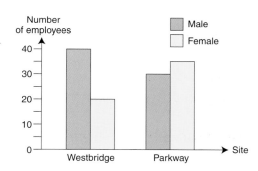

a) How many employees does the firm have altogether?

b) What percentage of the employees are male?

c) What percentage work at Parkway?

---

Collect five newspaper articles containing expressions like '1 in 3' or '20%'.

a) Work out each fraction as a percentage.

b) Work out each percentage as a fraction in its simplest form.

c) Explain whether you would use '44%' or '11 in 25' in a heading.

# 9: Percentages

# Finishing off

**Now that you have finished this chapter you should**

★ know the fraction and decimal equivalents of 25%, 50% and 75%

★ be able to change a percentage to a fraction or a decimal

★ be able to calculate a percentage of a number

★ be able to calculate the outcome of a percentage increase or decrease

Use the questions in the next exercise to check that you understand everything.

## Mixed exercise

**1** Work out

a) 10% of £750
b) 50% of £25
c) 40% of 400
d) 26% of 2000
e) 30% of 40
f) 50% of £4.50
g) 50% of £50 000
h) 75% of 16
i) 8% of 25

**2** Copy and complete this table.

| Fraction | Decimal | Percentage |
|---|---|---|
| 1/10 | | |
| | 0.2 | |
| | | 25% |
| 7/10 | | |
| | 0.8 | |
| | | 90% |

**3** Pat's food intake is 28% fat, 15% protein and the rest is carbohydrates.

What percentage is carbohydrates?

**4** What percentage of these fish are

a) angel fish (yellow)?
b) goldfish (orange)?
c) starfish (red)?

What percentage of the fish are

d) not angel fish?
e) not goldfish?
f) not starfish?

# 9: Percentages

**Mixed exercise**

**5** Ben usually buys a 250 ml can of orange at lunchtime.

One day the can is larger and is marked '20% extra'.

How much orange does this larger can contain?

**6** Crystal buys chocolates priced at £800 for her shop.

She gets a 15% trade discount.

How much does she pay?

**7** Jordan is doing a survey by post.

He sends out 250 questionnaires and expects to get 30% back.

How many replies does he expect?

**8** Sally wants to buy a washing machine.

She sees the model she wants in two different stores.

a) Which store offers the best deal if she wants it delivered?

b) Which store offers the best deal if she does not want it delivered?

## Investigations

**1**

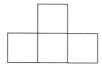

How many different ways are there of colouring 50% of this figure red and the rest white? (You may not colour part squares.)

**2**

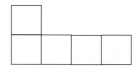

How many different ways are there of colouring 40% of this figure blue and the remainder white? (You may not colour part squares.)

---

Choose 5 kinds of packaged food.

On each one, find the label that tells you the amount of fat, protein and carbohydrate per 100 g.

Write the amounts as percentages in a table.

# Ten

# Statistics

## Displaying data

People often show their data in pictures. There are many ways of doing this. Some of these are shown here and others later in this chapter.

### *Pictograms*

This pictogram shows Karen's data about what people want to do for their evening out.

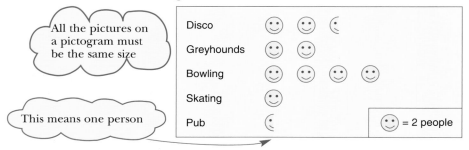

All the pictures on a pictogram must be the same size

This means one person

### *Bar charts*

Bar charts can be drawn vertically like the one below, or horizontally.

Notice that

- all the bars have the same thickness;
- the gaps between the bars are all the same;
- each bar is labelled in the middle.

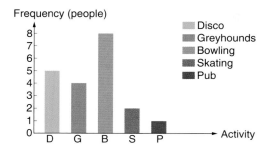

*Which do you prefer, the pictogram or the bar chart? They both show the same data.*

# 10: Statistics

**1** This pictogram shows the results of a netball team one season.

| Win | ● ● ● ● ● |
| Draw | ● ◗ |
| Lose | ● ● ◗ |

● means 4 matches

a) The symbol ◗ means 2 matches. Draw symbols for 1 match and 3 matches.

b) How many matches did they win, draw and lose?

c) How many matches did they play altogether?

**2** Neil draws this bar chart to show the sales of computers last week.

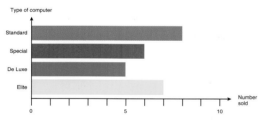

a) How many Standard are sold?

b) How many computers are sold in total?

The prices are Standard £550, Special £700, De Luxe £900 and Elite £1000.

c) Draw a bar chart showing the income from sales for these 4 computers.

d) Which type produces least income?

**3** A book on pets includes this pictogram showing the results of a survey into what happens when dogs meet other dogs.

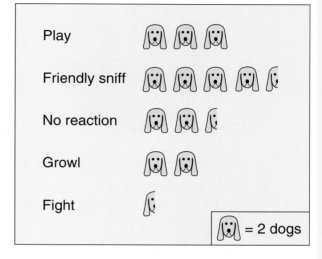

a) Make a frequency table showing how many dogs there are in each category.

b) What is the total number of dogs?

c) What percentage of dogs growl or fight?

d) Do you think these are good categories?

e) Draw a bar chart to illustrate these data.

Look through newspapers and magazines and find at least 2 examples of pictograms and 2 examples of bar charts or vertical line charts. Cut them out and paste them on a piece of paper.

## 10: Statistics

# Pie charts

Here is part of a report on an airline's business.

You can show this information on a pie chart.

It has a number of sectors. In the chart below, each sector is a different colour.

To draw a pie chart you need the angle for each sector.

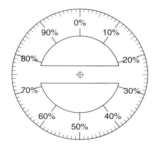

In this example an easy way is to use an angle measurer that is marked in percentages (sometimes called a pie chart scale).

But you should also know how to work out the angles in degrees, as follows.

The whole circle is 360°, so you work out the sector angles like this:

**Inside UK** $\quad \frac{25}{100} \times 360° = 90°$

**Rest of Europe** $\quad \frac{40}{100} \times 360° = 144°$

**North America** $\quad \frac{20}{100} \times 360° = ...$

**Rest of the World** $\quad \frac{15}{100} \times 360° = ...$

*Work out these two angles yourself to check that you can get the right answers*

Add up all the angles and check that they come to 360°.

Once you know the angles, you use a protractor or angle measurer to draw them.

# 10: Statistics

**1** In an election 40% of people vote for the Social Democrats, 30% for the Liberal Alliance, 20% for the Christian Democrats and the rest don't vote. You are going to show this on a pie chart.

   a) What percentage don't vote?
   b) Show that the angle for Social Democrats is 144°.
   c) Find the angles of the sectors for the other parties, and for those who don't vote.
   d) Draw the pie chart.

**2** In this question you are going to draw a pie chart for Karen's data from page 142.

| Activity | Disco | Greyhounds | Bowling | Skating | Pub |
|---|---|---|---|---|---|
| Frequency | 5 | 4 | 8 | 2 | 1 |

There are 20 people in the group so each person has 360° ÷ 20 = 18°.

   a) Show that the disco sector has an angle of 90°.
   b) Find the angles for the other activities.
   c) Draw the pie chart.

**3** This pie chart illustrates the results of a football team's matches one season.

The team played 60 matches.

   a) How many degrees is 1 match?
   b) Show that they lost 20 matches.
   c) How many matches did they win?
   d) How many matches did they draw?
   e) They score 3 points for a win, 1 for a draw and 0 for a loss.
      How many points did they get in the season?

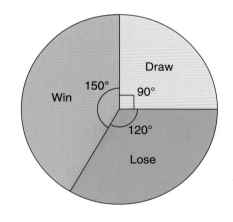

In the game of Shove-penny a coin is pushed along a board like this to score points.

Use a large sheet of paper to make a board and play the game 30 times.

Draw a pie chart to illustrate your results.

This scores 2

When it lies across a line it gets the lower score, 3

## 10: Statistics

# Line graphs

Nat is in hospital. Every 3 hours his temperature is taken and the points are plotted on a graph.

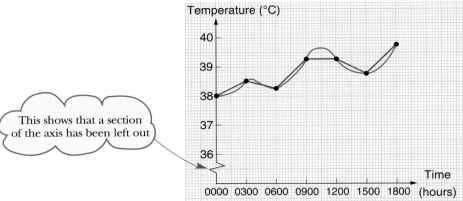

 *Why has the section of the temperature axis from 0 °C to 36 °C been left out?*

 *Describe what has happened to Nat's temperature during the day.*

In the graph the points have been joined by the blue straight lines. This is usual on a line graph, but be careful: it can be misleading.

You don't know what happened between the points: it could have been the red curve.

When data are collected at time intervals, they form a time series. It is quite usual to show them on a graph like this.

# Vertical line charts

A vertical line chart is often used for showing numerical data. This vertical line chart shows the number of people living in the 40 houses on one street.

It is like a bar chart with very thin bars.

In this diagram the zero on the horizontal axis has been offset to make the vertical line there clearer.

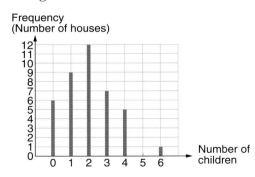

 *How many houses have no-one living in them?*

 *What percentage of houses have just two people living in them?*

# 10: Statistics

**1** Jake was born in January weighing 2.1 kg.

This is rather light for a baby so he was weighed every week for the next 10 weeks as a check.

| Week | 0 | 1 | 2 | 3 | 4 | 5 | 6 | 7 | 8 | 9 | 10 |
|---|---|---|---|---|---|---|---|---|---|---|---|
| Weight (kg) | 2.1 | 2.1 | 2.2 | 2.3 | 2.4 | 2.5 | 2.3 | 2.6 | 2.8 | 3.0 | 3.2 |

a) Plot these figures as a line graph.
b) Jake was ill one week.
   Which week do you think it was?
c) Estimate Jake's weight at $9\frac{1}{2}$ weeks.

**2** This table shows the population of voles each month on a river bank during 1997.

| Month | J | F | M | A | M | J | J | A | S | O | N | D |
|---|---|---|---|---|---|---|---|---|---|---|---|---|
| No. of voles | 40 | 30 | 28 | 16 | 16 | 20 | 14 | 28 | 56 | 100 | 78 | 60 |

a) Draw a line graph to illustrate the data.
b) Describe the annual pattern.
c) Over what months did the population increase most rapidly?

**3** One day in January, a jogging club asks all its members how many days they have been out running during the last week.

Here are their answers.

```
0  3  7  0  0     1  2  1  0  3
1  2  2  1  0     1  3  4  1  2
2  3  1  1  1     4  7  7  1  2
```

a) Record these figures on a tally chart.
b) Record the figures on a frequency table.
c) How many members does the club have?
d) Draw a vertical line chart to show the data.

The vertical line chart to the right shows the results when the members are asked the same question in June.

e) Look at the two vertical line charts, yours and the one above.
   What difference do you notice?
   Give a possible explanation for the difference.

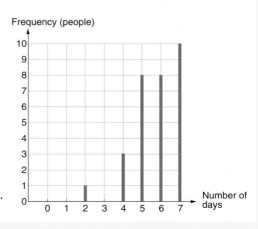

## 10: Statistics

# Averages and spread

Natasha and Paul are part of a group that go ten-pin bowling.

They decide to arrive early so that they can have some practice shots.

These are their scores.

| Natasha | 7 | 7 | 5 | 4 | 7 |
|---------|---|---|---|---|---|
| Paul | 10 | 10 | 3 | 1 | |

Who has done better?

Natasha's total score of 30 is greater than Paul's 24 but she had more turns.

You need to take an average. Here are three ways of doing it.

The **mode** is the most common score.

Natasha's mode is 7. What is Paul's mode?

The **mean** is the total divided by the number of turns played.

Natasha: $30 \div 5 = 6$   Paul: $24 \div 4 = 6$

They both have the same mean score.

The **median** is the middle score when the scores are put in order.

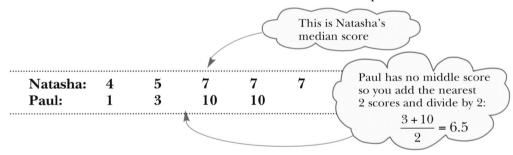

Natasha: 4   5   7   7   7
Paul:    1   3   10  10

This is Natasha's median score

Paul has no middle score so you add the nearest 2 scores and divide by 2:
$$\frac{3 + 10}{2} = 6.5$$

Natasha's median score of 7 is higher than Paul's 6.5.

*Who has done the better? It depends which calculation you do.*

*Which calculation would Natasha want to use?*

*Which calculation would Paul want to use?*

The **range** is the difference between the highest score and the lowest score.

Natasha: $7 - 4 = 3$   Paul: $10 - 1 = 9$

Paul has a larger range; his scores are more spread out.

Natasha has a smaller range: her scores are more consistent.

## 10: Statistics

**1** Find the mean, median and mode of

a) 1 1 2 3 8    b) 10 10 20 30 80

c) 11 11 12 13 18    d) 101 101 102 103 108

e) 
| Number | 20 | 21 | 22 | 23 | 24 |
|---|---|---|---|---|---|
| Frequency | 2 | 1 | 3 | 3 | 1 |

**2** Mr Doni is planning to start a company offering boat trips round the bay.

To find out what size of boats will be best he does 10 trial trips for one day. The numbers of people in them are

3  3  9  2  6    6  4  5  6  5

a) Work out the mean, median and mode of these figures.

b) Mr Doni has an 8-seater boat. How often will it be full?

**3** In a survey on TV-watching, a group of boys record how many hours TV they watch each week.

Boys:    16  23  21  5  12    0  5  13  14  11

a) Find the mean, median, mode and range of these figures.

A group of girls now do the same.

Girls:    9  11  4  16  15

b) From these results do the boys or the girls watch more TV?

**4** A hockey coach measures the times it takes members of the team to run 100 m (in seconds).

Hockey team:
| Time | 11 | 12 | 13 | 14 | 15 | 16 |
|---|---|---|---|---|---|---|
| Frequency | 4 | 2 | 3 | 3 | 2 | 1 |

a) Calculate the mean, median, mode and range.

The coach then asks the PE staff to run 100 m.

PE staff:    11  18  11  12  13

b) From these results, which group is the faster?

**5** The pay of 12 employees in a small company is:

£8 000    £8 000    £8 000    £8 000    £8 000

£11 000    £11 000    £11 000    £13 000    £18 000

£19 000    £65 000

a) Find the mean, median, mode and range.

b) What do you think the managing director earns?

---

Explain how batting and bowling averages are worked out for cricket. Give examples, using real players.

# 10: Statistics

## Grouping data

Sometimes it is easier to look at data when they are grouped. The figures below are the number of penalty points that some drivers have on their licences 2 years after passing their tests.

```
0 3 3 0 0      4 4 0 0 0      0 3 3 0 0
1 0 5 3 0 0    3 0 4 0 0      1 1 0 3 0 4
```

It is easier to see these as a grouped frequency table

| Penalty points | 0–2 | 3–5 | 6–8 | 9–11 |
|---|---|---|---|---|
| Drivers (frequency) | 16 | 12 | 0 | 2 |

*What happens if you get 12 penalty points?*

The number of points you get is always a whole number. You cannot get $4\frac{1}{2}$ or 5.8 points.

This is an example of **discrete** data: you can't have 'in-between' values.

Other data can have 'in-between' values, for example people's heights in centimetres. These data are called **continuous.**

*Think of some other examples of discrete and of continuous data.*

To display continuous data on a diagram you must group them.

The figures below are the heights in cm of the members of Avonford youth club.

```
152.1  150.0  160.3  140.7  128.0    134.1  135.7  151.5  163.4  181.0
182.8  180.8  154.6  161.5  171.4    129.2  151.8  138.4  153.4  165.0
```

These can be put into groups, for example, from 120 cm up to 130 cm, from 130 cm up to 140 cm, and so on.

*Jean is 130 cm tall. Which group does Jean belong in?*

Make sure that

- every person fits into a group
- every person belongs to only one group.

To make sure that the groups meet but do not overlap, use the ≤ sign.

> This can be written as $120 < x \leq 130$ where $x$ is the height

> $120 < x \leq 130$ is read as '120 is less than $x$, which is less than or equal to 130'

*Complete the grouped frequency table.*

| Height (cm) | $120<x\leq130$ | $130<x\leq140$ | $140<x\leq150$ | $150<x\leq160$ | $160<x\leq170$ | $170<x\leq180$ | $180<x\leq190$ |
|---|---|---|---|---|---|---|---|
| Frequency | 2 | 3 | | | | | |

# 10: Statistics

**1** Arif wants to know if more babies are born at some times of the year than others.

He asks 24 people what month they were born in.

| June | Feb | Jan | July | Dec | May | Sept | June | Nov | May |
| Oct | Jan | Sept | Aug | June | Oct | Aug | Nov | April | Feb |
| Aug | Sept | June | Sept | | | | | | |

Arif groups the data by season.

a) Copy and complete this frequency table.

| Season | Winter<br>Dec, Jan, Feb | Spring<br>March, April, May | Summer<br>June, July, Aug | Autumn<br>Sept, Oct, Nov |
|---|---|---|---|---|
| Frequency (people) | | | | |

b) Draw a bar chart to illustrate the frequency table.

c) Do you think Arif has collected enough data to draw any conclusions?

**2** The committee of a squash and tennis club are worried that not enough young people are coming to play.

They record the ages of people who come the next day.

| 76 | 69 | 68 | 70 | 12 | 15 | 16 | 18 | 19 | 41 |
| 36 | 27 | 49 | 56 | 61 | 45 | 39 | 33 | 26 | 44 |
| 52 | 53 | 36 | 24 | 31 | 41 | 43 | 44 | 42 | 19 |
| 18 | 18 | 42 | 53 | 21 | | | | | |

a) Make a tally chart, using groups 10–19, 20–29, …
b) Make a frequency table.
c) Do you think there should be more young people playing?

*Notice that age 19 goes from your 19th birthday to the day before your 20th birthday*

**3** A time trial for 30 students running 100 m is carried out.

Times are in seconds.

| 14.3 | 15.2 | 16.4 | 14.8 | 13.9 | 14.7 | 15.0 | 14.8 | 13.9 | 12.7 |
| 16.7 | 12.8 | 18.3 | 15.5 | 15.7 | 16.9 | 14.2 | 14.0 | 18.5 | 14.6 |
| 15.4 | 16.4 | 17.5 | 18.5 | 12.1 | 13.8 | 15.2 | 16.7 | 16.5 | 16.9 |

a) Make a tally chart, using groups $12.0 \leq x < 13.0$, $13.0 \leq x < 14.0$, …
b) Make a frequency table.
c) Draw a frequency chart.
d) Do you think the times are fast or slow? Give a possible explanation for them being so.

Keep a record of how long it takes you to get to school or college on 20 occasions.
Show this on a frequency chart and explain its main features.

## 10: Statistics

# Displaying grouped data

The table shows the heights of 20 members of Avonford youth club.

| Height (cm) | 120<x≤130 | 130<x≤140 | 140<x≤150 | 150<x≤160 | 160<x≤170 | 170<x≤180 | 180<x≤190 |
|---|---|---|---|---|---|---|---|
| Frequency | 2 | 3 | 1 | 6 | 4 | 1 | 3 |

The data are illustrated on the grouped frequency chart below.

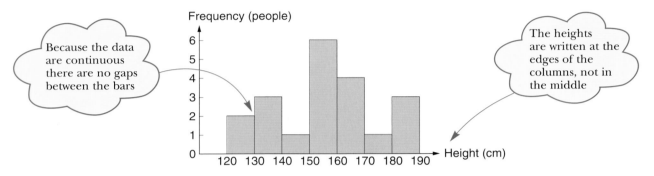

*Because the data are continuous there are no gaps between the bars*

*The heights are written at the edges of the columns, not in the middle*

Which group has the highest frequency?

Which class has the tallest column on the graph?

*This is called the **modal class***

You can compare two sets of data by drawing their **frequency polygons**.

The members of the youth club visit a youth club in Holland.

Look at the diagram below. It compares the heights of the clubs' members.

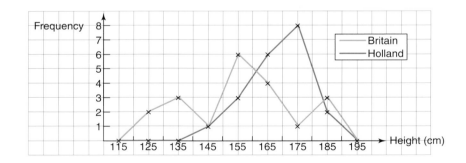

Make a grouped frequency table for the data from the youth club in Holland.

Use the same intervals as those in the table at the top of the page.

What does the highest peak on each graph show?

Which peak is further right?

Which youth club has the taller members?

Which youth club has the more variable heights?

# 10: Statistics

**1** Danielle takes a party of 48 people from York to Amsterdam.
She arrives at the coach station at 7 a.m.
She keeps a record of the length of time, $t$ minutes, she waits for people to arrive.

| Time $t$ (mins) | 5≤$t$<10 | 10≤$t$<15 | 15≤$t$<20 | 20≤$t$<25 | 25≤$t$<30 | 30≤$t$<35 | 35≤$t$<40 | 40≤$t$<45 |
|---|---|---|---|---|---|---|---|---|
| Frequency | 0 | 2 | 4 | 7 | 15 | 14 | 5 | 1 |

Draw a grouped frequency chart to display the data.

**2** Barbarella has two shops. One is for young people, the other for older people.
She asks customers to complete a questionnaire.
One question asked them to tick their age range (in confidence).

| Age range | 0≤$x$<10 | 10≤$x$<20 | 20≤$x$<30 | 30≤$x$<40 | 40≤$x$<50 | 50≤$x$<60 | 60≤$x$<70 | 70≤$x$<80 |
|---|---|---|---|---|---|---|---|---|
| Shop A | 0 | 1 | 6 | 12 | 20 | 10 | 1 | 0 |
| Shop B | 0 | 10 | 25 | 13 | 2 | 0 | 0 | 0 |

a) Using the same axes, draw two frequency polygons.
b) State the modal group for each shop.
c) Which shop has the younger customers?
d) In which shop is the age more variable?

**3** On a coach trip the courier knew the ages of his passengers from their passports.

| Age of passengers | 20≤$x$<24 | 24≤$x$<28 | 28≤$x$<32 | 32≤$x$<36 | 36≤$x$<40 | 40≤$x$<44 | 44≤$x$<48 | 48≤$x$<52 |
|---|---|---|---|---|---|---|---|---|
| Males | 0 | 2 | 6 | 8 | 12 | 9 | 2 | 1 |
| Females | 4 | 10 | 14 | 9 | 3 | 0 | 0 | 0 |

a) Using the same axes, draw two frequency polygons.
b) State the modal group for males and for females.
c) Are the males younger than the females?
d) Do the female ages vary more than the males ages?

---

Measure the heights of a number of boys and the same number of girls in your group.

Plot the two frequency polygons on a large diagram to put on the wall.

What would you expect to find?

Are you right?

## 10: Statistics

# Making comparisons

Many professional basketball players are very tall. Imran collects these data on the heights of 120 basketballers and 120 footballers (all to the nearest whole inch).

| Height (inches) | 70 | 71 | 72 | 73 | 74 | 75 | 76 | 77 | 78 | 79 | 80 | 81 |
|---|---|---|---|---|---|---|---|---|---|---|---|---|
| Footballers | 0 | 12 | 24 | 31 | 34 | 12 | 4 | 2 | 1 | 0 | 0 | 0 |
| Basketballers | 0 | 0 | 0 | 0 | 4 | 10 | 25 | 37 | 25 | 12 | 7 | 0 |

He draws these two frequency polygons showing the players' heights.

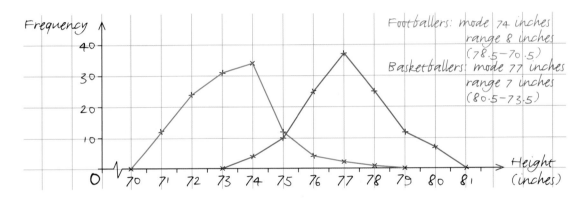

Footballers: mode 74 inches, range 8 inches (78.5 − 70.5)
Basketballers: mode 77 inches, range 7 inches (80.5 − 73.5)

 What does this graph tell you?

 Which team has the taller players?

 In which team do the players have the more variable heights?

# 10: Statistics

**1** Emma has done a survey of the number of items in people's shopping baskets at the checkouts of two different shops. Here are her results.

| Number of items | 1–5 | 6–10 | 11–15 | 16–20 | 21–25 | 26–30 | 31–35 |
|---|---|---|---|---|---|---|---|
| Frequency: shop A | 24 | 49 | 13 | 8 | 4 | 2 | 0 |
| Frequency: shop B | 5 | 8 | 15 | 33 | 29 | 10 | 0 |

a) Which is the modal group for each shop?

b) On the same set of axes, draw the frequency polygon for each shop.

c) Compare the frequency polygons. What do they tell you about the two shops?

d) Suggest what type of shop each might be.

**2** This table shows the results of a survey that was done to check the phasing of the traffic lights at the site of some major roadworks. The time spent waiting by 200 cars travelling in each direction was recorded one morning.

| Delay, $t$ (minutes) | $0 \leq t < 2$ | $2 \leq t < 4$ | $4 \leq t < 6$ | $6 \leq t < 8$ | $8 \leq t < 10$ | $10 \leq t < 12$ | $12 \leq t < 14$ | $14 \leq t < 16$ |
|---|---|---|---|---|---|---|---|---|
| North-bound | 50 | 80 | 60 | 10 | 0 | 0 | 0 | 0 |
| South-bound | 10 | 25 | 35 | 55 | 45 | 20 | 10 | 0 |

a) On the same set of axes, draw a frequency polygon for the delay times in each direction.

b) Do you think the traffic lights are sensibly phased?

c) Might the survey have produced different results at a different time of the day?

## 10: Statistics

# Finishing off

**Now that you have finished this chapter you should be able to**

- ★ draw pictograms, bar charts, vertical line charts, pie charts, frequency charts, line graphs and frequency polygons
- ★ work with grouped data
- ★ calculate the mean, median and mode of a set of data
- ★ calculate the range of a set of data

**Use the questions in the next exercise to check that you understand everything.**

## Mixed exercise

**1** Salma, Bob and Carole start a new business. Salma contributes £30 000, Bob contributes £40 000 and Carole contributes £50 000.

a) Draw a pie chart for their wall to show what proportion each owns of the company.

The first year they make a profit of £18 000 which is to be shared in the same proportion as their contributions.

b) Use your pie chart to find how much profit each receives.

One year Carole receives £30 000 profit share.

c) How many pounds are represented by 1° on the pie chart?

d) How much does Salma receive that year?

**2** Anita works in a small library. One week the library lent out 540 books in 4 categories. Anita drew a pie chart to show this information. The 'Thrillers' sector had an angle of 150°, the 'Biographies' sector had an angle of 70°, 45 'Reference' books were on loan and the other category was 'Romance'.

a) Calculate the number of 'Thrillers' on loan.

b) Find the angle of the 'Romance' sector.

# 10: Statistics

**3** The pie chart below shows the mixture of cement, sand and coarse aggregate used in concrete for paths.

a) Measure the angles of the 3 sectors of the pie chart.

b) Write down the fraction of the concrete that is cement, the fraction that is sand and the fraction that is coarse aggregate.

c) Rachel's path will need 3 m³ of concrete. How much cement, sand and coarse aggregate will she need?

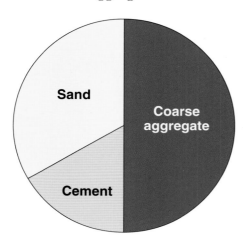

d) Draw a bar chart to illustrate the amounts of cement, sand and coarse aggregate.

e) Which do you think is clearer, the pie chart or the bar chart?

**4** Here are the numbers of goals scored by a football team in their matches this year.

6 0 1 0 4   5 5 1 2 0

a) Find the mean, median, mode and range of these data.

b) What can you say about their performance?

**5** A garden pond contains goldfish.

On 1 March each year the fish are counted.

| Date (1 March) | 1995 | 1996 | 1997 | 1998 | 1999 | 2000 |
|---|---|---|---|---|---|---|
| Number of goldfish | 12 | 16 | 17 | 20 | 10 | 18 |

a) Draw a graph to illustrate these data.

Use the horizontal axis for the date.

b) Can you answer the question 'How many goldfish were there on 1 September 1997?'?

**6** Here are the weights, in kg, of 20 male students, all aged 18.

61 98 72 84 63   77 77 81 85 72

90 83 76 82 77   81 80 83 75 68

a) Use a tally chart like the one below to group the data.

| Weight $w$ (kg) | Tally |
|---|---|
| $60 < x \leq 65$ | |
| $65 < x \leq 70$ | |
| $70 < x \leq 75$ | |
| $75 < x \leq 80$ | |
| $80 < x \leq 85$ | |
| $85 < x \leq 90$ | |
| $90 < x \leq 95$ | |
| $95 < x \leq 100$ | |

b) Make a frequency table.

c) A healthy weight for a male 18-year-old is about 70–80 kg. Comment on the weights of this group of students.

d) Draw a frequency polygon to display the data.

# Eleven

# Directed numbers

## Negative numbers

Peter has £40 in his bank account.
He has these bills to pay.

 *How much is left in his account if he pays the gas bill?*

*How much is left if he then pays the car repairs bill?*

He decides to pay both bills as he has a credit arrangement with the bank.

They allow him to owe them up to £100.

When he has paid the bills he is *overdrawn* by £50.

His balance is –£50.

This will be shown on his bank statement as £50.00 DR.

 *Peter pays another bill, this time for £30.*

*What does his statement say now?*

Here is the control panel for the lift in a large store.

Jo gets into the lift at the floor –1 and presses the button for floor 3.

How many floors does she go up?

You can see in the diagram that she goes up 4 floors.

She comes back to the lift at floor 3.

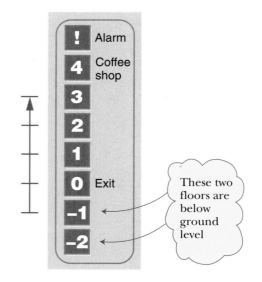

These two floors are below ground level

 *Where will she be if she goes down*

a) *2 floors?*  b) *3 floors?*  c) *5 floors?*

*What other situations can you think of in which negative numbers are used?*

# 11: Directed numbers

**1** Write the following numbers in order.

7, −2, −100, 1000, 25, 0, −3, 10

**2** Look at the lift buttons opposite.

For each of these trips say how many floors the lift goes up or down.

Example: 1 to 3 is up 2 floors.

a) 3 to 4   b) 4 to 3   c) 3 to 0
d) 0 to −2   e) −2 to 2   f) 2 to −1

**3** You have £200 in your bank account.

a) You pay a gas bill of £100. How much is left?
b) Now you pay £200 towards your holiday. How much is left?
c) You pay in your wages of £100. How much is in the account now?

**4** Kate buys some jackets to sell on her market stall. They cost her £50 each.

a) She prices the jackets at £70.

What is the profit on a jacket?

b) She sells the last jacket for £45.

What is her profit on this one?

**5**

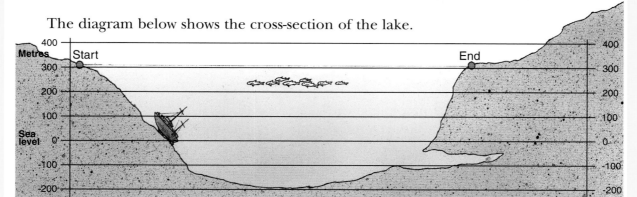

The diagram below shows the cross-section of the lake.

Work out how much the submarine goes up or down between

a) the start and the wreck
b) the wreck and the fish shoal
c) the fish shoal and the underwater cave
d) the underwater cave and the end.

Look at a bank statement that goes overdrawn.

Explain the entries in the 'balance' column.

## 11: Directed numbers

# Adding and subtracting

In winter, the temperature is often near zero.

It sometimes goes below zero.

The temperature one day is 2 °C.
At nightfall it drops by 5 °C.

The new temperature is

    2 °C – 5 °C = –3 °C

You can see this on the thermometer.

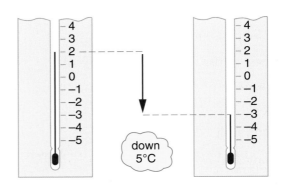

During the night, the temperature drops by another 1 °C.

The new temperature is

    –3 °C – 1 °C = –4 °C

Check this on the thermometer.

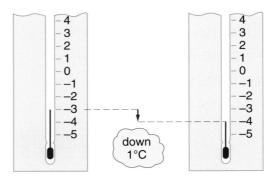

The next morning, the temperature rises by 7 °C.

The new temperature is

    –4 °C + 7 °C = 3 °C

Again you can check this on the thermometer.

 *Which is warmer, 2 °C or –5 °C ?*

Drawing a number line helps you to add and subtract positive and negative numbers.

The scale on the thermometer is a number line.

Usually we draw number lines across the page, like this:

 *What is –5 + 6 ?*

*What is +3 – 7 ?*

*How would you draw a number line to help you work out +10 – 25 ?*

# 11: Directed numbers

**1** Use the number line opposite to work out

a) 5 – 1    b) 0 – 4    c) 1 – 7    d) –1 + 2
e) 5 + 2    f) –5 – 2   g) 8 – 5    h) –8 + 5
i) 5 – 2    j) –5 + 2   k) 3 + 4    l) –3 – 4

**2** Work out

a) 20 – 21    b) –1 + 101    c) –50 + 30    d) 1000 – 2000

**3** Copy and complete this table, using the information below.

The temperature in London is 4 °C.

Manchester is 5 °C colder than London.

Leeds is 2 °C colder than Manchester.

Inverness is 3 °C colder than Leeds.

Accra is 36 °C hotter than Inverness.

| City | Temperature (°C) |
|---|---|
| London | |
| Manchester | |
| Leeds | |
| Inverness | |
| Accra | |

**4** This scale measures the water level of a river (in feet).

'Normal' river level is zero on the scale.

On Monday evening after a long spell of dry weather, there is a heavy rainstorm. This table show the water level at 3-hourly intervals on Tuesday.

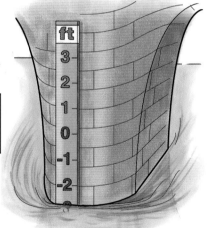

| Mid-night | 3 am | 6 am | 9 am | Noon | 3 pm | 6 pm | 9 pm | Mid-night |
|---|---|---|---|---|---|---|---|---|
| –2 | –1.5 | –1 | 0 | 1 | 1.5 | 1 | 0 | –1 |

a) What is the level at midnight on Monday?

b) How much has the river level risen by 3 am on Tuesday?

c) How much further has it risen by 3 pm?

d) How much does the level drop from 3 pm to 9 pm on Tuesday?

---

You have offered to make a special lunch for your family.

You are to serve it at 1 pm.

Choose the menu.

Write down all the stages in its preparation.

Work out how long before 1 pm each job has to be done.

Write a timetable for the morning to ensure that lunch is ready at exactly 1 pm.

## 11: Directed numbers

# Positive and negative co-ordinates

Snaefell is the highest point on the Isle of Man,
There is a telecommunications aerial on it.
On the map below, Snaefell is the origin.

Snaefell is (0, 0), Ramsey is (2, 2).

What about Peel, Port Erin and Douglas?

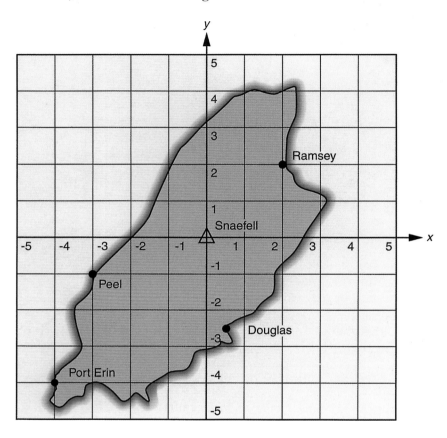

Points on the *x* axis to the left of the origin have negative *x* co-ordinates.

Points on the *y* axis below the origin have negative *y* co-ordinates.

```
x−     x+
y+     y+
-------+-------
x−     x+
y−     y−
```

Peel has co-ordinates (−3, −1).

Port Erin is at (−4, −4).

Douglas has co-ordinates (0.5, −2.5).

# 11: Directed numbers

**1** Write down the co-ordinates of points A to H.

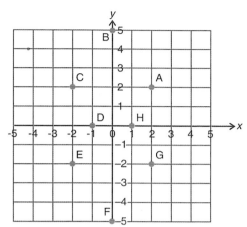

**2** Draw a grid with –5 to 5 on each axis.

Join these points.

(3, 4)   (5, 4)   (5, –3)   (3, –3)
(3, 1)   (1, –1)   (–1, 1)   (–1, –3)
(–3, –3)   (–3, 4)   (–1, 4)   (1, 2)

and back to (3, 4)

Shade in the shape you have made.

What letter of the alphabet is it?

**3** Pizza Palace in New York offers a delivery service.

A charge is made for each block travelled.

a) On a piece of squared paper draw a grid like the one in question 1, with Pizza Palace at the origin.

b) Using different coloured pens mark points that are 1 block, 2 blocks, 3 blocks, . . . etc., away from Pizza Palace.

Make lists of the points that you have marked in each group.

For example,

Points 3 blocks away are (–2, –1), . . . etc.

c) Write down a rule to decide if a point is in a particular group.

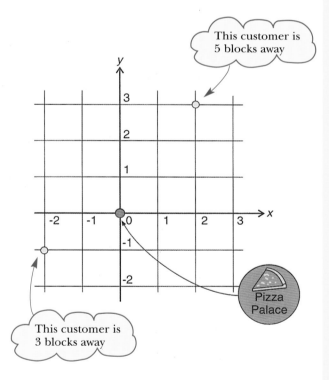

---

On graph paper, make a scale drawing of the stairs in your school or college.

Think carefully about where you want to have the origin and what scale you want to use.

Give the co-ordinates of the top of each stair.

# 11: Directed numbers

## Finishing off

**Now that you have finished this chapter you should**

★ know when to add and when to subtract

★ be able to add and subtract positive and negative numbers

★ be able to draw a number line with positive and negative numbers

★ be able to use positive and negative co-ordinates

Use the questions in the next exercise to check that you understand everything.

## Mixed exercise

**1** Work these out.

a) 6 – 8     b) –4 + 3     c) 5 – 8     d) –2 + 7

e) –5 – 3     f) –10 + 6     g) 20 – 25     h) –15 – 9

**2** Work these out.

a) £23.50 – £26.50     b) £4.25 – £5.00     c) £10.25 – £11.50

**3** Which of these temperatures is colder:

a) –2 °C or 3 °C?     b) 0 °C or –1 °C?     c) –2 °C or –4 °C?

**4** Oliver has £125 in his bank account.

He writes a cheque for £58 to pay his phone bill.

Then he writes a cheque for £104 to pay for car repairs.

Finally he pays in a cheque for £29.

What is the new balance of his account?

**5** When she is born, Rebecca weighs 4 kg.

The graph shows her weight gain after each of the next 3 weeks.

a) What does she weigh after 1 week?

b) What does she weigh after 2 weeks?

c) How much weight does she put on during the third week?

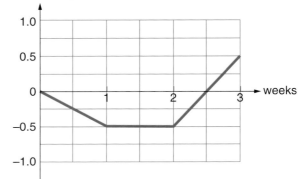

## 11: Directed numbers

**Mixed exercise**

**6** Local time in Calais is 1 hour ahead of local time in Dover.

The ferry crossing from one to the other takes 90 minutes.

a) A ferry leaves Dover at 0945 local time.

What will be the time in Calais when it arrives?

b) A ferry leaves Calais at 1635 local time.

What will be the time in Dover when it arrives?

**7** Kim and Tessa plan to start a computer games company. They have saved £20,000 to start the company.

They will need to do a lot of work designing the games before they can start to sell them.

They go to the bank with this plan. It shows what they expect their bank balance to be each year.

|              |           | LOSS    | PROFIT  |
|--------------|-----------|---------|---------|
| Start        | Equipment | £10,000 |         |
| Year 1       | Loss      | £25,000 |         |
| Year 2       | Loss      | £5,000  |         |
| Year 3       | Profit    |         | £15,000 |
| Year 4       | Profit    |         | £30,000 |

Bank balance

How much profit or loss do they expect to make in

a) Year 1?   b) Year 2?   c) Year 3?   d) Year 4?

---

Look at the people in your maths group.

Choose one person whose height you think is about average for the group.

Stick a piece of masking tape vertically on a wall.

Mark Mr (or Ms) Average's height on the tape.

Draw centimetre scales on the tape above (+) and below (−).

Now measure and record everyone else's height using your scales.

When you have finished, add up all the results.

Do you think your Mr (or Ms) Average really is close to the average height?

# Twelve

# Ratio and proportion

## Simple proportion

Kim is filling up her petrol tank.

She looks at the display on the petrol pump.

After a few seconds the display looks like this:

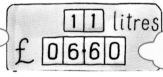

Kim has put 2 litres of petrol in her tank

The price so far is £1.20

A little while later it looks like this:

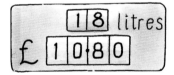

Now Kim has put 11 litres in her tank

The price is £6.60

When Kim has finished the display looks like this:

*How much petrol has Kim put in her tank?*

*How much will it cost her?*

While she fills her tank, Kim can watch the amount of petrol increasing.

As the amount increases, so does the price.

The connection between the amount and the price is shown in this arrow diagram.

*What is the price of 3 litres of petrol?*

*How many litres do you get for 360 pence?*

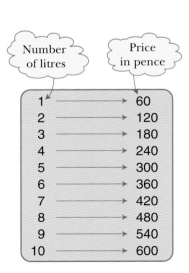

| Number of litres | Price in pence |
|---|---|
| 1 | 60 |
| 2 | 120 |
| 3 | 180 |
| 4 | 240 |
| 5 | 300 |
| 6 | 360 |
| 7 | 420 |
| 8 | 480 |
| 9 | 540 |
| 10 | 600 |

# 12: Ratio and proportion

**1** At the petrol pump on the opposite page, what is the price of a litre of petrol?

**2** Draw an arrow diagram for petrol priced at 70 pence per litre. Put the number of litres on the left, and the price in pence on the right. (Go up to 10 litres.)

**3** Draw an arrow diagram for this parking meter.

Put the number of hours on the left, and the price in pence on the right. (Go up to 4 hours.)

**4** These diagrams show the prices of different items at a fruit and vegetable stall. Copy and complete each diagram.

a)
```
    Potatoes
kg            £
1  ──────→  0.20
2  ──────→  0.40
3  ──────→  0.60
4  ──────→  .......
5  ──────→  .......
```

b)
```
    Apples
kg            £
1  ──────→  .......
2  ──────→  1.00
3  ──────→  .......
4  ──────→  2.00
5  ──────→  2.50
```

c)
```
    Bananas
kg            £
1  ──────→  .......
2  ──────→  .......
3  ──────→  1.20
4  ──────→  .......
5  ──────→  2.00
```

**5** Thomas is doing a sponsored swim. He has lots of sponsors.

He will raise £2.50 for every length that he swims.

a) Draw an arrow diagram with the number of lengths on the left and the amount of money raised on the right. (Go up to 20 lengths.)

b) Thomas swims 18 lengths. How much money does he raise?

---

Look in a car magazine.

Find a model of car that you like.

Write down its fuel consumption in miles per gallon (mpg).

Write down its fuel tank capacity in gallons.

Draw an arrow diagram showing how far the car can travel on 1 gallon, 2 gallons, 3 gallons … up to a full tank of fuel.

## 12: Ratio and proportion

# Ratio and proportion

'Pea green' paint is made by mixing yellow and blue.

**MIXING INSTRUCTIONS**
- 3 parts yellow to 1 part blue

For example, 3 tins of yellow are mixed with 1 tin of blue.

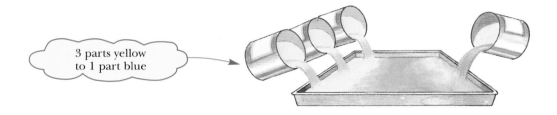

3 parts yellow to 1 part blue

 *How many tins of each colour do you need for twice as much pea green?*

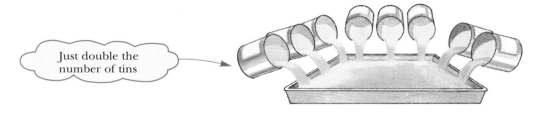

Just double the number of tins

3 parts to 1 part is the same ratio as 6 parts to 2 parts.

3 parts to 1 is written 3:1.

 *What other statements are the same as 3:1?*

*How many tins of yellow go with 5 tins of blue?*

*How many tins of blue go with 30 tins of yellow?*

The amount of yellow is $\frac{3}{4}$ or 75% or 0.75.

 *What fraction of the tins are blue?*

*What is this as a percentage and as a decimal?*

# 12: Ratio and proportion

**1** 'Purple Haze' paint is made by mixing red and blue.

> **MIXING INSTRUCTIONS**
> - 2 parts red to 1 part blue

  a) How many tins of blue do you mix with 4 tins of red?

  b) How many tins of red do you mix with 5 tins of blue?

**2** A recipe for scones uses 4 parts of flour to 1 part of butter.

  a) How much flour is mixed with 50 grams of butter?

  b) How much butter is mixed with 300 grams of flour?

**3** Linda and Jane agree to pay the phone bill in the ratio 2:1.

  a) Linda pays £120. How much does Jane pay?

  b) What fraction of the bill does Jane pay?

**4** This pie chart shows Mr Singh's sales of local newspapers on Tuesday. He sells 20 copies of The Weekly.

  a) What is the ratio of sales of The Weekly to The News?

  b) How many copies of The News does he sell?

On Friday Mr Singh sells 100 copies of The Weekly. The ratio of sales is the same as for Tuesday.

  c) How many copies of The News does he sell on Friday?

  d) Why do you think he sells five times as many local newspapers on a Friday?

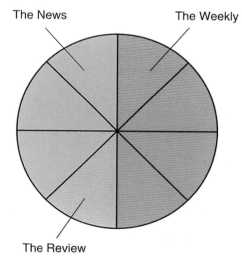

---

Go into a DIY store that has a machine for mixing paint.
How does it work?
Give examples of 3 colours that can be mixed.
State the proportions of the different colours in the mixtures.

## 12: Ratio and proportion

# Conversion graphs

The ferry's speed is 17 knots.

One knot is one nautical mile per hour.

A nautical mile is about 1.2 miles, so 1 knot is about 1.2 miles per hour.

You can use this table to convert between knots and miles per hour.

| knots | 5 | 10 | 15 | 20 | 25 | 30 | 35 | 40 |
|---|---|---|---|---|---|---|---|---|
| mph | 6 | 12 | 18 | 24 | 30 | 36 | 42 | 48 |

*What is* 15 knots *in miles per hour (mph)?*

*What is* 30 mph *in knots?*

*What is* 17 knots *in mph?*

It is not easy to use the table to convert 17 knots into miles per hour. 17 knots is not shown in the table.

A conversion graph like this is better.

You can use it to convert any speed (up to 40 knots).

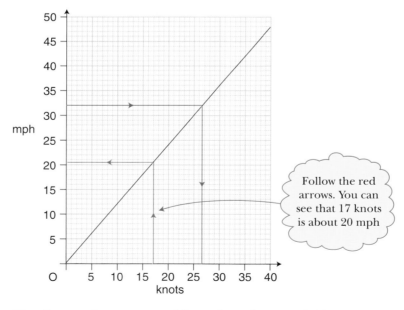

Follow the red arrows. You can see that 17 knots is about 20 mph

*Use the graph to convert* 32 mph *to knots. (Follow the green arrows.)*

*Use the graph to convert* 36 knots *into mph.*

*How would you convert* 45 knots *into miles per hour?*

## 12: Ratio and proportion

This graph is on display at a bureau de change. It tells you how many US dollars ($) you will get for any number of pounds sterling (£).

**1** Use the graph to work out how many dollars you will get for

    a) £10    b) £25    c) £50    d) £80

**2** Use the graph to work out how many pounds you will get for

    a) $32    b) $48    c) $144    d) $160

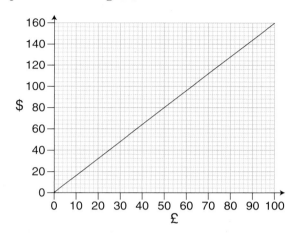

This graph tells you the cost of a phone call to Australia.

**3** a) How much does a 12-minute call cost?

    b) How many minutes do you get for £15?

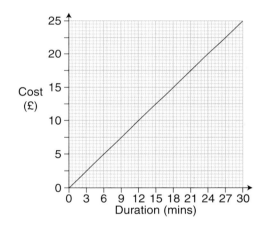

**4** Draw a graph to convert a measurement from metric to Imperial units, for example to convert a length from centimetres to inches. (You will find the information you need on page 3.)

---

Babies are sometimes weighed in pounds and sometimes in kilograms.

Draw a chart that health visitors and parents can use to convert babies' weights from one to the other.

Make sure your chart covers a sensible range of weights for babies up to 1 year old.

## 12: Ratio and proportion

# Finishing off

**Now that you have finished this chapter you should be able to**

★ draw arrow diagrams  ★ use conversion graphs
★ use ratio and proportion

Use the questions in the next exercise to check that you understand everything.

## Mixed exercise

**1** Look at this graph for converting litres into gallons.

   a) John buys 27 litres for his car.

      How many gallons is this?

   b) The petrol tank of John's car holds 8 gallons.

      How many litres is this?

   c) A garden centre sells 40-gallon water barrels.

      How many litres do they hold?

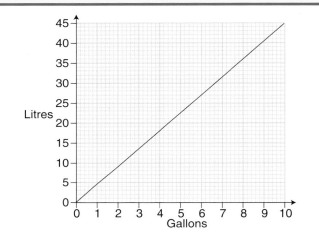

**2** 'Sunny orange' paint is made by mixing red with yellow.

   a) How many tins of yellow do you mix with 3 tins of red?

   b) How many tins of red do you mix with 8 tins of yellow?

**MIXING INSTRUCTIONS**

■ 2 parts yellow to 1 part red

**3** One unit of cleaning fluid is mixed with 25 units of water before use.

   a) How much water is mixed with 4 units of cleaning fluid?

   b) How much cleaning fluid is mixed with 150 units of water?

## 12: Ratio and proportion

**4** In each of these, write down the missing number.

a) 1 : 3 is the same as 5 : ☐
b) 25 : 2 is the same as 75 : ☐
c) 1 : 10 is the same as 5 : ☐
d) 5 : 20 is the same as 1 : ☐
e) 3 : 2 is the same as 15 : ☐
f) 5 : 2 is the same as 50 : ☐
g) 80 : 2 is the same as 320 : ☐
h) 4 : 200 is the same as 20 : ☐
i) 9 : 7 is the same as 45 : ☐
j) 7 : 2 is the same as ☐ : 10
k) 5 : 9 is the same as ☐ : 45
l) 11 : 2 is the same as ☐ : 22

**5** Write each ratio in its simplest form.

a) 1 hour : 30 minutes
b) 20 grams : 1 kilogram
c) 25 cl : 1 litre
d) £5 : 75p
e) 4 cm : 20 m
f) 3 cm : 1 km
g) 2 mm : 1 km
h) 2 hours : 45 seconds
i) 2 hours 30 minutes : 3 hours 45 minutes

*Mixed exercise*

---

A length of 38 mm is almost exactly the same as $1\frac{1}{2}$ inches.

Draw a graph for converting millimetres to inches.

What other lengths in millimetres convert into convenient numbers of inches?

---

On some kettles you can see the amount of water in them.

It is shown on a scale marked in numbers of cups.

By putting measured amounts of water into a kettle like this, find how much you need for 1, 2, 3, … cups.

Draw a conversion graph for the measured volume of water to the number of cups.

Find a formula connecting the number of cups, $N$, to the volume, $V$.

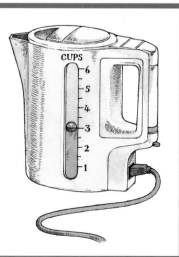

# Thirteen

# Formulae

## Using a formula

Here is the arrow diagram for the petrol pump again. It shows the connection between the number of litres and the price.

| Number of litres | Price in pence |
|---|---|
| 1 | 60 |
| 2 | 120 |
| 3 | 180 |
| 4 | 240 |
| 5 | 300 |
| 6 | 360 |
| 7 | 420 |
| 8 | 480 |
| 9 | 540 |
| 10 | 600 |

Each time you follow an arrow you multiply by 60.

$1 \xrightarrow{\times 60} 60$
$2 \xrightarrow{\times 60} 120$

You can see that to find the price (in pence) you multiply the number of litres by 60.

You can write this as

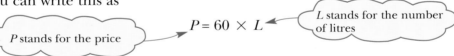

$P = 60 \times L$

*P* stands for the price

*L* stands for the number of litres

*P* and *L* are called **variables**. They can take different values

This is called a **formula** for *P*.

You can use the formula to find the price of any number of litres of petrol.

For example, if $L = 9$, $P = 60 \times 9 = 540$.

**The price of 9 litres is 540 pence.**

The formula is more useful than the arrow diagram, because you can use it for any number of litres.

What is the price of 12 litres of petrol?

Putting $L = 12$ in the formula gives $P = 60 \times 12 = 720$.

**The price of 12 litres is 720 pence.**

*How would you work out the price of*

a) $\frac{1}{2}$ litre *of petrol?*   b) $4\frac{1}{2}$ litres *of petrol?*

## 13: Formulae

**1** Here is the completed arrow diagram for the potatoes on page 123.

a) Copy and complete this sentence:

  'To find the price (in £) you multiply the weight by ...'

b) Copy and complete this formula for the price:

  $P = \ldots \times W$

c) Use the formula to work out the price of 12 kg of potatoes.

Potatoes

| kg | £ |
|---|---|
| 1 → | 0.20 |
| 2 → | 0.40 |
| 3 → | 0.60 |
| 4 → | 0.80 |
| 5 → | 1.00 |

**2** Draw an arrow diagram for each formula.

(They have been started for you.)

a) $y = 3 \times x$

| x | y |
|---|---|
| 1 → | 3 |
| 2 → | 6 |
| 3 → | ...... |
| 4 → | ...... |
| 5 → | ...... |
| 6 → | ...... |
| 7 → | ...... |
| 8 → | ...... |
| 9 → | ...... |
| 10 → | ...... |

b) $B = A + 3$

| A | B |
|---|---|
| 1 → | 4 |
| 2 → | 5 |
| 3 → | ...... |
| 4 → | ...... |
| 5 → | ...... |
| 6 → | ...... |
| 7 → | ...... |
| 8 → | ...... |
| 9 → | ...... |
| 10 → | ...... |

**3** You saw opposite that when petrol costs 60 pence a litre, the formula

  $P = 60 \times L$

gives the price, $P$ pence, of $L$ litres of petrol.

a) Find the price of 5 litres of petrol.

b) Find the price of 12 litres of petrol.

c) How much petrol can you buy for £6.00?

d) How much petrol can you buy for £18.00?

e) The price of petrol is increased to 70 pence a litre. Write down the new formula for $P$.

---

You are working on a market stall.

You decide to write out arrow diagrams to help you to work out customers' bills quickly.

Here is the one for grapefruits.

Grapefruits

| 1 → | 29p |
| 2 → | 58p |
| 3 → | 87p |

Find typical prices for oranges and for punnets of strawberries.

Draw arrow diagrams for each of these.

# 13: Formulae

## More formulae

Sonal is getting her holiday photos developed.

She gets 4 sets developed (so she can give some to her friends).

She works out the cost like this:

> £5 for my set and 3 lots of £1 for the rest
> £5 + 3 × £1 = £8

The formula she uses is

$C = 5 + (N \times 1)$

£$C$ is the cost

$N$ is the number of extra sets

 *How much do 10 sets cost?*

John is also in the shop collecting his photos.
He pays £10.

Sonal works out how many sets John has like this:

> £5 for 1 set + £1 for each extra set is £10
> 5 + N = 10
> N = 5
> John buys 6 sets altogether

 *How many sets cost £12?*

Remember: $N$ is the number of **extra** sets

# 13: Formulae

**1** At this car park, how much does it cost for

a) 10 days?   b) 2 weeks?

**Airport Car Park**

| 1 day | £4 | 4 days | £16 |
|---|---|---|---|
| 2 days | £8 | 5 days | £20 |
| 3 days | £12 | 6 days | £24 |

**2** John opens a bank account.

He puts £5 in the bank every week.

The formula for the money in the bank, £$M$, after $W$ weeks is

$M = 5 \times W$

a) How much is in the bank after

(i) 3 weeks?

(ii) 10 weeks?

b) How many weeks does John take to save £100?

**3** Use the formula $P = 5t - 2$ to find the value of $P$ when

a) $t = 2$   b) $t = 9$   c) $t = 13$.

**4** Gil spends £33 on T-shirts.

T-shirts only £5
SPECIAL OFFER
£2 off the total sale for every customer

How many does he buy?

**5** Find the value of $3x^2 + 4$ when

a) $x = 1$   b) $x = 2$   c) $x = 3$   d) $x = 5$.

**6** The Post Office works out how much to charge for parcels over 20 kg with this formula:

$C$ stands for the charge in £

$C = 1.5 \times W - 25$

$W$ stands for the weight in kg

How much does it cost to post each of these parcels?

a)  25 kg

b)  22 kg

c)  30 kg

---

Find 4 formulae that you use in other subjects.

Write them down and explain what the letters in them mean.

# 13: Formulae

## Collecting like terms

The entry gate of an adventure park has a computer. It works out the total cost of tickets.

This is what it looks like when it is ready to use.

Number of adults ____
Number of children ____

The computer works out

7 × number of adults + 4 × number of children

 *Where do the 7 and 4 come from?*

In algebra you could write this as $7a + 4c$

*c stands for the number of children*

*a stands for the number of adults*

 *Notice that 7 × a is written 7a. We can leave out the × sign in algebra. Why can't you leave out the × sign in arithmetic?*

*How much does it cost for 4 adults and 5 children to visit the adventure park?*

The **expression** $7a + 4c$ has 2 **terms**. These are **unlike** terms because they use different letters.

The terms in the expression $2x + 5x - 6x$ are **like** terms because they all have the same letter, $x$. This expression can be simplified to $(2 + 5 - 6)x = x$

*You write $1x$ as $x$*

The expression $7a + 4c$ cannot be simplified any more. It is in its **simplest form**.

Sometimes you need to simplify expressions with like and unlike terms.

### Example

Simplify $2x + 3y - 4x + 2y + 8x - 3y - x - y$

*Collect the $+x$ terms then the $-x$ terms, then the $+y$ terms and the $-y$ terms*

**Solution**

$2x + 8x \quad -4x - x \quad +3y + 2y \quad -3y - y$

(+x terms) (−x terms) (+y terms) (−y terms)

$= 10x \quad -5x \quad +5y \quad -4y$

$= 5x + y$

*This is the answer. It cannot be tidied up any more*

## 13: Formulae

**1** Copy each expression and write it as simply as possible.

State how many terms there are in each simplified expression.

a) $5 \times a + 7 \times a$
b) $3 \times p + 6 \times p$
c) $4 \times q - 2 \times q$
d) $4 \times y + 3 \times w + 5 \times z$
e) $1 \times r - 7 \times s + 3 \times t$
f) $3 \times a + 6 \times b + 3 \times a + 8 \times b$
g) $2 \times h + 7 \times k + 6 \times h - 3 \times k$
h) $2 \times m + 5 \times n + 8 \times p$
i) $4 \times d + 3 \times e + 5 \times d$
j) $4 \times y - 4 \times y$

**2** Copy each of these and write it more simply.

a) $4a + 5a$
b) $5b - 3b$
c) $8c - 5c$
d) $4d + 5d - 3d$
e) $12x - 9x + 3x$
f) $5y + 2y - 4y$
g) $4x + 11x - 6x$
h) $6y + 9y + 4y - 15y$

**3** For each of these, put the like terms into separate lists, then add the terms in each list.

a) $x, 5x, 12, -7, 9x$
b) $y, 8a, 5y, 3y, 9a$

**4** Tidy up the following expressions.

a) $12x + 3x + 10y - 5y$
b) $13a + 15 + 4a + 2$
c) $2x + 16y + 8x - 3y$
d) $17p - q + 3p - 1$
e) $8m + 16n - m + n$
f) $8m + 25n + 17m - 15n - 5n$
g) $15x + 8 + 2x - 7$
h) $13p + 7q + 5p - 3q$
i) $2x + 3y + 5x - 2y$
j) $6x + 7y + 13x - 6y$
k) $5a + 10b - 4a - 7b$
l) $5p + 3q + 6p - 2q$

**5** This question refers to the adventure park opposite.

a) Use the expression $7a + 4c$ to work out the cost of adventure park tickets for

(i) 2 adults and 3 children

(ii) a school party of 40 children and 3 teachers.

b) The following year the prices for the adventure park rose to £9 for an adult and £6 for a child.

Write down a new expression for working out the total cost.

---

A charity box contains a number of coins.

Write down expressions for the number of coins, their total value and their weight.

(Hint: let $a$ be the number of 1p coins, $b$ be the number of 2p coins, $c$ be the number of 5p coins, etc.)

# 13: Formulae

## Using brackets

Gus and Dougie are buying crisps for a party. They decide to get 3 of these special-offer bags.

How many packets of crisps will they have altogether?

Gus and Dougie have different ways of working it out.

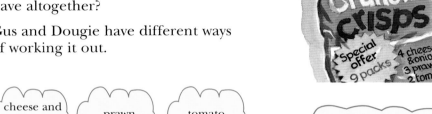

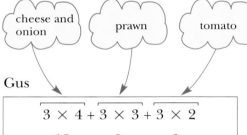

Gus

$$3 \times 4 + 3 \times 3 + 3 \times 2$$
$$= 12 + 9 + 6$$
$$= 27$$

Dougie

$$3 \times (4 + 3 + 2)$$
$$= 3 \times 9$$
$$= 27$$

You can see that they both get the same answer:

$$3 \times 4 + 3 \times 3 + 3 \times 2 = 3 \times (4 + 3 + 2)$$

If they buy 5 special offer bags, the number of packets will be

$$5 \times 4 + 5 \times 3 + 5 \times 2 = 5 \times (4 + 3 + 2)$$

*This is called factorising*

You can write the same for any number of bags, say $n$:

$$n \times 4 + n \times 3 + n \times 2 = n \times (4 + 3 + 2)$$

*or, more neatly*
$$4n + 3n + 2n = n(4 + 3 + 2)$$

 Check that this works for $n = 6$.

A different special-offer bag contains $c$ packets of cheese and onion, $p$ packets of prawn and $t$ packets of tomato.

 What does $c + p + t$ represent?

Describe the meaning of $6(c + p + t)$.

Describe what each term represents in $6c + 6p + 6t$.

 The total number of packets is $n(c + p + t)$.

How many packets are there when $n = 6$, $c = 5$, $p = 4$ and $t = 3$?

 Write an expression equivalent to $4(x + 2y - 3z)$.

*What you have done is called* **expanding the brackets**.

# 13: Formulae

**1** Simplify these expressions.

a) $x(5 + 2)$    b) $(6 + 4 + 3)y$
c) $n(22 + 77 + 1)$   d) $17x + 3x + 10x$

*Remember to write the number first in the answer*

**2** Copy each of these and expand the brackets.

a) $3(a + b)$    b) $5(c + d + e)$    c) $10(x + 2)$
d) $2(3 + y + z)$    e) $7(f + 3 + g)$    f) $2(x - 5)$
g) $4(x - y)$    h) $(p + q - r) \times 2$    i) $(c + d + 1) \times 8$

**3** a) What is the value of $2m + 3$ when $m = 5$?

b) What is the value of $4(2m + 3)$ when $m = 5$?

c) Expand $4(2m + 3)$.

d) Find the value of your expanded expression when $m = 5$. Check that your answer is the same as the one for b).

**4** a) What is the value of $(3x - 2y)$ when $x = 4$ and $y = 2$?

b) What is the value of $5(3x - 2y)$ when $x = 4$ and $y = 2$?

c) Expand $5(3x - 2y)$.

d) Find the value of your expanded expression when $x = 4$ and $y = 2$. Check that your answer is the same as the one for b).

**5** Simplify each of these by expanding the bracket then adding the like terms. Check each one by substituting $x = 4$ and $y = 3$ in both the question and your answer; both should work out to be the same.

a) $2(x + y) + 3y$    b) $3(x + 5) - 7$    c) $10(x + 3) + 2x$
d) $4(5 + x) + 8$    e) $2(y + 2x) + x$    f) $6(x + y) + 2x$
g) $4(x + 4y) + 2x$    h) $7(x + y + 7) + y + 7$

**6** Factorise the following.

a) $3x + 3y + 3z$    b) $3x + 6y + 9z$
c) $6a - 3b$    d) $8a - 4b + 16c$
e) $36p - 12q + 18r$    f) $100u - 25v - 75w$

**7** Find the value of $(x + y)(x + y)$ when

a) $x = 5, y = 2$    b) $x = 10, y = 1$
c) $x = 7, y = 3$    d) $x = 8, y = 8$

Write down the key strokes you need to enter $3 \times 5 + 3 \times 6$ into your calculator
a) as it is written and
b) using brackets.
How many key strokes are involved in each?
What about $3 \times 5 + 3 \times 6 + 3 \times 7$? How many key strokes do you save by using brackets in a long calculation like this?

# 13: Formulae

## Adding and subtracting with negative numbers

George works in the Hillside Hotel.

He takes a bottle of champagne in the lift from the cellar (floor –5) to the honeymoon suite (floor 4).

The diagram shows George's trip.

His journey can be written as

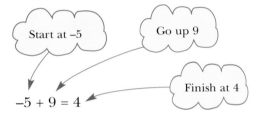

$-5 + 9 = 4$

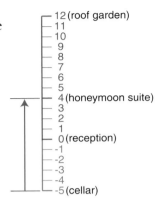

The honeymoon couple are in the roof garden at floor 12. They take the lift down to their suite. The lift stops on the way to let someone else get in. Here is an expression for the honeymoon couple's trip down from the rooftop garden:

 Draw a diagram to show their trip.

At what floor did the lift stop on the way?

You could show these trips on a horizontal number line to save space.

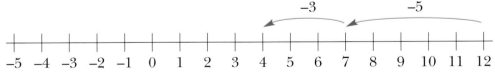

 What does $12h - 5h - 3h$ come to?

What does $-3h - 5h + 12h$ come to?

 If $h$ is the distance from one floor to the next, what does $4h$ represent?

 How do you key in –5 on your calculator?

Work out $-5 + 4$, $-3 + 12$ and $-4 - 6$ on your calculator.

Check your answers on a number line.

## 13: Formulae

**1** Draw a number line to illustrate each of these, and write down the answer to each.

a) 6 – 4
b) –4 + 6
c) –6 + 4
d) 4 – 6
e) 2 + 3 – 5
f) 2 – 4 – 1
g) –11 + 3 + 5
h) –18 – 21
i) –55 + 61

**2** There are many ways of making 7 using the numbers 12, 4 and 1.
For example, –4 – 1 + 12 = 7.
Write down 3 other ways of making 7 using these numbers.
Check each way using a calculator and a number line.

**3** Write these more simply.

a) $6x - 4x$
b) $4x - 6x$
c) $-4x - 6x$
d) $7y - 3y$
e) $-3y + 7y$
f) $3y - 7y$
g) $2x + 3x - 4x - x$
h) $2y - 4y + 3y$
i) $-7y - 3y - 15y$

**4** Substitute $x = 2$ and $y = 3$ in the expressions in question 3, and in each of your answers, and check that you get the same results.

**5** Write these as simply as possible.
(Remember that you can only add and subtract *like terms*.)

a) $2 + 3 - 5p$
b) $6q - 3q + 4 - 3$
c) $p + 5 - 11$
d) $6 - 4p - 3p$
e) $7 + 5q - 4 + 6q$
f) $8p + 5 - 9p - 2$
g) $q + 3p - 4q + p$
h) $-46p - 18q + 46p$
i) $-3 + q + 3 - q$

**6** Substitute $p = 1$ and $q = 2$ in the expressions in question 5, and in each of your answers, and check that you get the same results.

A lift sometimes goes past the floor you want, or stops at another floor on the way.

Write 4 different ways for George at the Hillside Hotel (see opposite page) to get to the kitchen on floor –3 after delivering the champagne on floor 4.

**7** Write a set of instructions to enable a friend to use your calculator for adding and subtracting negative numbers.

## 13: Formulae

# Finishing off

**Now that you have finished this chapter you should**

★ know that
 $1 \times n = n$, $2 \times n = 2n$,
 $3 \times 2n = 6n$

★ understand the words
 *expression* and *term*

★ be able to simplify an expression, doing the steps in the right order

★ be able to substitute into a formula

★ be able to work with brackets and expand them

★ be able to add and subtract like terms

★ be able to use a number line for negative numbers

Use the questions in the next exercise to check that you understand everything.

## Mixed exercise

**1** Work out the value of each of these.

a) $4 + 6 \times 2$

b) $(4 + 6) \times 2$

c) $12 - 6 + 1$

d) $12 - (6 + 1)$

e) $3 \times 2^2$

f) $(3 \times 2)^2$

g) $2n^2$ when $n = 5$

h) $(2n)^2$ when $n = 5$

i) $12 + 3 \times 4 - 2 \times 8 + 6 \div 2$

j) $8 - 2 \times 3 + 4 \div 2 - 1$

k) $2(x + 6)$ when $x = 3$

l) $2x + 12$ when $x = 3$

m) $3(9 - 2x)$ when $x = 2$

n) $27 - 6x$ when $x = 2$

**2** Copy the following expressions and write them as simply as possible.

State how many terms there are in each.

a) $3 \times a + 6 \times b$

b) $12 - 6 \times x$

c) $2 \times y - 5 \times z + 4$

d) $3 \times 5 \times c$

e) $1 \times p + 1 \times q + 1 \times r$

f) $99 \times f - g \times 12$

g) $3 \times x - 6 \times y + 5 \times 3$

h) $n \times 7 - 3 \times m$

i) $3 \times q + 7 \times p$

j) $5 \times r + 6 \times s - 2 \times r - 5 \times s$

## 13: Formulae

**Mixed exercise**

**3** Copy each expression and collect together the like terms. Write each expression as simply as possible.
   a) $4a + 2a$
   b) $4a + 1 + 2a$
   c) $2x - 8 + 5x$
   d) $2x + 8 - 5x$
   e) $5n + 8n - 6 + 2n$
   f) $7k - 2 - 4k - 2k + 5$
   g) $p + 4x + 7y - 6x + 13p$
   h) $9a + 5c - 3d - 2c - 3a$
   i) $101d - 52w + 97 - 7d + w$
   j) $-4a - 16 - 13a + 7$

**4** Alf has a decorating business and he can claim back the VAT that he spends on materials. When he spends £$P$ the VAT is given by the formula
$$V = \frac{7}{47} \times P$$
Find the amount of VAT he can claim for paint costing £23.50.

**5** Work out the time Mac should cook a turkey weighing 6.3 kg by:
   a) using the rule 15 *minutes per* 450 g *plus* 15 *minutes*;
   b) using the formula $T = 33W + 15$ where $T$ is the time in minutes for a turkey weighing $W$ kg. Is your answer close enough to a)?

**6** Copy and expand these expressions.
   a) $2(n - 5)$
   b) $2(7 + 3x)$
   c) $(3a + 6) \times 4$
   d) $3(4y + 2z)$
   e) $5(6 - a)$
   f) $(a + b + c) \times 3$

**7** Copy and simplify these expressions.
   a) $2(x + 8) + 1$
   b) $12 + (5 + y)$
   c) $3 + 4(2n - 1)$
   d) $(6d + 5) + d$
   e) $7 + 6(4a - 2)$
   f) $3x + 9(5 - x)$
   g) $2(x + 3y) + 3(x + y)$
   h) $2p + 3(p + q)$
   i) $12(2p - 5q) - 6(2p + p)$
   j) $3(2s + 4r + 11) - 6s - 2r - 33$

**8** a) Substitute $a = 4$, $b = 5$ and $c = 6$ into
   (i) $3a + 12b + 6c$
   (ii) $3(a + 4b + 2c)$.
   b) Why are your answers the same?
   c) Check that you get the same answers when you substitute $a = 2$, $b = 1$ and $c = -3$ into the two expressions.

### Investigation

You have some scales and a 1 g, a 3 g and a 9 g weight.

Explain how you can use them to weigh objects of 1 g, 2 g, ..., 13 g.

You are allowed a fourth weight. What do you choose?

# Fourteen

## Surveys

### Recording data

Karen's department at work decide to have an evening out together. They ask Karen to organise it.

Karen goes round asking everyone what they would most like to do.

She records their answers like this:

*This is called a **tally chart***

*This means 5*

Disco            ||||
Greyhound racing ||||
Ten-pin bowling  |||| |||
Ice-skating      ||
Pub              |

*Ten-pin bowling has the most tallies. It has 5 + 3 = 8. It is the most popular*

*These are the tallies*

Keeping a tally is an easy way to count things. It is easy to add them up afterwards too, especially if you can count in fives.

When numbers are collected like this they are called **data**.

After she has collected the data, Karen shows the results to everyone as a **frequency table**.

| Activity  | Disco | Greyhounds | Bowling | Skating | Pub |
|-----------|-------|------------|---------|---------|-----|
| Frequency | 5     | 4          | 8       | 2       | 1   |

*These are the frequencies*

When you add these frequencies up you get 20. This is the number of people who are going out for the evening.

In this case the data are **categorical** because they are grouped by categories such as *ice-skating* and *disco*.

If the data are grouped by numbers, for example, the ages of a class of students, they are called **numerical**.

# 14: Surveys

**1** The evening is to end with a meal, so Karen also asks everyone where they would like to eat.

Here is her tally chart.

Fish + chips  |
Indian  ||||| ||||| |
Chinese  ||
Pub meal  ||||| |

a) Show their answers as a frequency table.

b) Which was the most popular choice?

c) Has Karen asked everyone? Explain your answer.

**2** Samantha goes to a theme park with a party of friends. Afterwards she asks them which ride they liked best. Here are their answers.

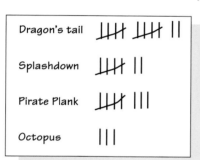

a) Show their answers as a frequency table.

b) How many are in the party?

c) Which is the most popular ride?

**3** For her own interest, Karen also records her data for the department's evening out as a 2-way table. She has different rows for men and women. Copy and complete the table.

| Activity | Disco | Greyhounds | Bowling | Skating | Pub |
|---|---|---|---|---|---|
| Frequency (women) | 4 | 1 | | | 0 |
| Frequency (men) | | | 5 | 1 | |

---

Find out the most popular music group among your friends or classmates.

Design a tally chart and use it to record their answers when you question them.

Show the results both as tallies and as a frequency table.

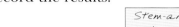

# 14: Surveys

## Stem-and-leaf diagrams

Shani and Lisa measure the heights of all the girls in their class.

Look at how they record the results.

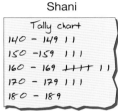
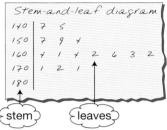

*What advantages does Lisa's stem-and-leaf diagram have over Shani's tally chart?*

*Are there any advantages in Shani's tally chart?*

Next Lisa orders her stem-and-leaf diagram. She adds a key to explain it.

```
Heights of girls
140 | 5 7
150 | 4 7 9
160 | 1 2 2 3 4 4 6
170 | 1 1 2
       120/5 represents 125
```
key

*How many girls are 164 cm?*

*Marion's height is 159 cm. Find the entry corresponding to Marion.*

*What is the smallest height?*

*How tall is the tallest girl?*

Shani and Lisa also measure the boys and Lisa draws a back-to-back diagram to compare the two sets of data. She puts the number of leaves beside the diagram.

```
              Heights of
    Boys     class 5A     Girls
0         | 140 | 5 7           2
1       9 | 150 | 4 7 9         3
4   7 6 6 5 | 160 | 1 2 2 3 4 4 6  7
6 8 8 6 5 5 4 | 170 | 1 1 2        3
4     7 5 4 3 | 180 |              0
              120/5 represents 125
```

*How tall is the tallest student in the class?*

*Jeff is 167 cm. How many girls are taller than Jeff?*

*How many boys are taller than Jeff?*

*Are the boys taller than the girls?*

*Which group (e.g. 140 ≤ height < 150) is the largest?*

This is called the **modal group**

# 14: Surveys

**1** A butcher keeps a record of the number of sausages he sells each day.

| 30 | 35 | 61 | 73 | 64 | 62 | 59 | 33 | 42 | 55 |
| 34 | 36 | 45 | 42 | 39 | 51 | 47 | 38 | 42 | 30 |
| 43 | 45 | 65 | 53 | 57 | 42 | 45 | 34 | 37 | 65 |

a) Construct a sorted stem-and-leaf diagram to show the data.

b) What is the smallest number sold?

c) What is the greatest number sold?

d) On how many days are 45 sausages sold?

**2** The numbers of miles travelled by students to a conference were as follows.

| 23 | 46 | 63 | 25 | 39 | 47 | 52 | 61 | 69 | 58 |
| 47 | 53 | 47 | 45 | 63 | 47 | 59 | 47 | 65 | 33 |
| 29 | 47 | 35 | 47 | 68 | 21 | 25 | 57 | 47 | 47 |

a) Construct a sorted stem-and-leaf diagram to show the data.

b) What was the smallest number of miles travelled?

c) What was the greatest number of miles travelled?

d) How many students travelled 47 miles?

**3** The ages of swimmers at a local swimming pool were recorded in a survey.

| 7 | 24 | 27 | 17 | 9 | 11 | 13 | 25 | 22 | 18 |
| 37 | 27 | 14 | 17 | 23 | 31 | 16 | 8 | 19 | 6 |
| 23 | 15 | 14 | 27 | 34 | 26 | 17 | 13 | 14 | 14 |

a) Construct a sorted stem-and-leaf diagram to show the data.

b) How old is the youngest swimmer?

c) How old is the oldest swimmer?

d) How many 17-year-olds are present?

e) What is the most common age?

**4** Mick and Bob live next door to each other. They both planted 20 hollyhocks. Mick used a fertiliser but Bob didn't.

The heights, in cm, of their hollyhocks were as follows.

| Mick: | 185 | 201 | 256 | 248 | 200 | 254 | 239 | 234 | 196 | 223 |
|  | 199 | 243 | 257 | 239 | 246 | 222 | 229 | 254 | 180 | 250 |
| Bob: | 208 | 245 | 150 | 228 | 230 | 215 | 217 | 228 | 215 | 158 |
|  | 164 | 179 | 208 | 212 | 226 | 230 | 188 | 196 | 214 | 226 |

a) Construct a sorted back-to-back stem-and-leaf diagram to show the data.

b) State the heights of the tallest and shortest hollyhock belonging to Mick.

c) State the heights of the tallest and shortest hollyhock belonging to Bob.

d) Compare the two sets of data and comment on whether the fertiliser is effective.

---

Measure the length from your elbow to your fingertip in mm. Record this figure.

Collect measurements from everyone in your class and draw a stem-and-leaf diagram to show the data.

## 14: Surveys

# Planning a survey

Yvette is a keen swimmer.

One morning she gets this letter.

It is from her swimming club.

> **Dear Yvette**
>
> The club committee met last night to discuss how to attract new members.
>
> As a first step we would like a survey done in your school. Could you carry it out for us?
>
> — AVONFORD SWIMMING CLUB

**Step 1**

Choose a topic that you find interesting.

---

Yvette agrees to do the survey.

She makes some notes.

> My swimming club wants me to find out whether people like swimming and if they are good at it.
>
> THINGS TO DO
> Write some questions to ask people
> Try them out on friends
> Get 60 copies made
> Get to school early + give to first 30 girls + 30 boys
> Make data collection sheet
> Write a report

**Step 2**

Write down why you are doing the survey and what you want to find out, and plan how to do the survey.

---

Yvette writes these questions.

> SWIMMING QUESTIONS
> 1 Can you swim?
> 2 Why do you like swimming?
> 3 How long do you take to do 1 length?
> 4 What stroke do you like best?

**Step 3**

Write the questions for your survey.

---

Yvette tries the questions out on some of her friends. Here are some of their answers.

*I don't know the names of the strokes.*

*I can swim 1 length of Uncle Pete's pool in 5 seconds.*

*I don't like it. I hate it.*

*I like every stroke except butterfly.*

*Anyone can swim - even my baby brother can.*

**Step 4**

Try out your questions on a few people.

# 14: Surveys

**Discussion**

**1** Look at the answers that Yvette's friends have given.

Which question is each person answering?

Do you think Yvette is getting the information she wants?

**2** Yvette decides to write some better questions so that people will give her useful answers.

Is this new set of questions better?

> SWIMMING QUESTIONS
>
> 1  Boy ☐   Girl ☐   (tick one box)
>
> 2  Is it important for everyone to be able to swim?
>    Yes ☐    No ☐    Don't know ☐
>
> 3  Can you swim at least 1 width in a swimming pool?
>    Yes ☐    No ☐
>
> 4  Which swimming stroke do you prefer?
>    Breast stroke ☐   Front crawl ☐   Back stroke ☐   Butterfly ☐
>
> 5  How many seconds do you take to swim 25 metres? _____ seconds
>
> Please return to Yvette Lee.

**Step 5**

**Write your new questions.**

**3** Why do you think Yvette asks question 1?

**4** Why does Yvette need a data collection sheet?

---

Yvette has started to design the data collection sheet for her survey.

It looks like this:

> **Data Collection Sheet**
>
> 1  Boys _____  Girls _____  Totals ☐ ☐
>
> 2  Is it important for everyone to be able to swim?
>    Yes _____ Totals ☐ ☐
>    No _____ Totals ☐ ☐
>    Don't know _____ Totals ☐ ☐

*Yvette has decided to record the boys' answers in blue and the girls' in red*

Design the rest of the data collection sheet for Yvette's survey.

## 14: Surveys

# The survey report

Yvette has now collected the answers to her questions. She has answers from 20 boys and 20 girls. Here is her data collection sheet.

**Step 6**

**Collect your data.**

```
Data Collection Sheet

1  Boys ‖‖‖‖ ‖‖‖‖ ‖‖‖‖ ‖‖‖‖    Girls ‖‖‖‖ ‖‖‖‖ ‖‖‖‖ ‖‖‖‖    Totals  20  20

2  Is it important for everyone to be able to swim?
   Yes  ‖‖‖‖ ‖‖‖‖ ‖‖‖‖ ‖‖‖‖        ‖‖‖‖ ‖‖‖‖ ‖‖‖‖ ‖‖‖‖   Totals  20  20
   No _____                      Totals   0   0
   Don't know _____                          Totals   0   0

3  Can you swim at least 1 width of a swimming pool?
   Yes  ‖‖‖‖ ‖‖‖‖ |||        ‖‖‖‖ ‖‖‖‖ ‖‖‖‖                Totals  13  15
   No   ‖‖‖‖ ||                ‖‖‖‖                         Totals   7   5

4  Which stroke do you like best?
   Breast stroke  |||            ‖‖‖‖ ||||                 Totals   3   9
   Front crawl    ‖‖‖‖ |||       ||||                       Totals   8   4
   Back stroke    ||             |                          Totals   2   1
   Butterfly                     |                          Totals   0   1

5  How many seconds do you take to swim 25 metres
   22 33 35 34 35 61 51      31 24 48 43 32 66 52 38
   55 43 25 47 32 46         38 24 41 35 36 24 27
```

Yvette talks to the swimming pool manager about her survey.

He gives her a copy of a recent magazine article.

### 90% of young people can swim

**Step 7**

**Find and use information from other sources.**

*Do you think Yvette can use this in her report?*

Now Yvette must write her report.

She starts by planning it out.

**Step 8**

**Write your report.**

```
Page 1  Say what I'm trying to find out. Stick in copy of questions.
     2  Question 1 answers. Question 2 answers. Pie chart or bar chart?
     3  Question 3 answers. Pie chart or bar chart?
     4  Question 4 answers. Pie chart or bar chart?
     5  Question 5 answers. Pie chart or bar chart?
     6  Conclusions.
```

*Look at the results from question 2 of the survey.*

*Everyone said they thought it was important to be able to swim.*

*What should Yvette say about this in her report?*

# 14: Surveys

**1** Look at the answers from question 3 of the survey.

    a) Show the girls' answers in a bar chart and a pie chart.

    b) Show the boys' answers in a bar chart and a pie chart.

    c) Which kind of chart do you think Yvette should use, and why?

**2** Look at the answers from question 4 of the survey.

Why do you think the numbers do not add up to 20 for boys and 20 for girls?

**3** Look at the times taken to swim 25 metres.

    a) What is the mean time for the boys and for the girls?

    b) What is the range of times for the boys and for the girls?

    c) What conclusions can you draw from your answers to a) and b)?

**4** a) Copy and complete this frequency table of the times taken by the girls to swim 25 metres.

    b) Make a similar table for the boys' times.

| Girls | |
|---|---|
| Seconds | Frequency |
| 20-29 | 4 |
| 30-39 | 6 |
| 40-49 | |
| 50-59 | |
| 60-69 | |

**Discussion**

What should Yvette write in her conclusions?

---

Now you are ready to do a survey of your own.

Follow the 8 steps given in this chapter.

Don't be afraid to ask other people what they think about the way you are doing it.

Here are some possible topics. Use one of these, or (even better) think up a topic of special interest to you.

- *Favourite TV programmes.*
- *Do people believe in horoscopes?*
- *Attitudes to smoking.*
- *How much time do people spend playing computer games?*

## 14: Surveys

# Social statistics

This is called the **base year**, shown by 100. All prices are compared with those in 1990

**Avonford Star** — Financial

## What is the £ in your pocket worth?

PRICES KEEP rising so wages and pensions need to increase. But do they increase enough? Our reporter tells you the truth using the Retail Price Index.

| Year | 1990 | 1995 | 2000 | 2005 |
|---|---|---|---|---|
| RPI | 100 | 127 | 165 | ? |

The RPI compares the 1995 and 2000 prices with those in 1990.

In 1995 the basket of essential groceries cost 127% of the 1990 price and in 2000 the price is 165% of the 1990 price.

This basket of groceries cost £15 in 1990.
In 1985 the basket cost
In 1995 the basket c

Martha says, 'For every £1 I paid in 1990, I paid £1.27 in 1995 and I have to pay £1.65 in 2000.'

 *Is Martha right?*

 *What did the basket cost in 1995?*

*What did the basket cost in 2000?*

Martha says 'For every £1 of pension I spent in 1990 I needed to spend £1.27 in 1995 and £1.65 in 2000. What happened in 2005?'

| Year | Single pension |
|---|---|
| 1990 | £46.90 |
| 1995 | £58.85 |
| 2000 | £67.50 |
| 2005 | ? |

 *What pension in 1995 would have been equivalent to the 1990 pension?*

*What pension in 2000 would have been equivalent to the 1990 pension?*

 *Compare Martha's actual pension with those you have just worked out.*

*Was she better off in 1990, 1995 or 2000?*

This is an example of a **time series**. The value of the same quantity, in this case the single pension, is recorded over time.

# 14: Surveys

**1** A cafe sells an 'All Day Breakfast'. In 1990 the meal cost £6.

Use the RPI values on page 406 to calculate the cost in

a) 1995
b) 2000.

**2** A sports shop sells a mountain bike. In 1990 the mountain bike cost £80.

Use the RPI values on page 406 to calculate the cost in

a) 1995
b) 2000.

**3** The diagram shows the sales of organic baby food in millions of pounds. These have been calculated at 1993 prices in order to make comparisons.

| Year | 1993 | 1994 | 1995 | 1996 | 1997 | 1998 |
|---|---|---|---|---|---|---|
| RPI | 100 | 105 | 110 | 113 | 122 | 131 |

Use the RPI values for baby food in the table above to calculate the actual sales in each year.

---

The table below shows the expected age distribution of the population in England in 2011, given in millions. The total population is expected to be 52 000 000 people.

|  | 0–14 | 15–29 | 30–44 | 45–59 | 60–74 | 75 and over | Total |
|---|---|---|---|---|---|---|---|
| Males | 4.7 | 5.1 | 5.4 | 5.4 | 3.7 | 1.6 | |
| Females | 4.3 | 4.9 | 5.2 | 5.3 | 4 | 2.4 | |
| People | | 10 | 10.6 | 10.7 | 7.7 | | |

© Crown copyright

Copy and complete the table.
Are there more males or females in England?
Find the percentage of under-15s who are

a) male
b) female.

Find the percentage difference between the numbers of males and females who are under 15. Similarly find the percentage difference between the numbers of males and females who are 75 or over.
Comment.

# Fifteen

# Using symbols

> **Before you start this chapter you should**
> ★ be familiar with the work in Chapter 13.

## Being brief

This station sign is the longest in Britain. It takes some effort to say the full name so most people call it Llanfair PG.

*There are seven 2s here*

In maths we often need to find ways to write things briefly. For example:

$$2 \times 2 \times 2 \times 2 \times 2 \times 2 \times 2 \text{ is usually written as } 2^7.$$

You say this as '2 to the power 7'. The 7 is called the **power** or **index**.

*Indices is the plural of index*

Indices are useful in algebra too. For example:

*Notice there is only one 5 here*

$$a \times a \times a \times a = a^4 \quad \text{and} \quad 5a^4 = 5 \times a \times a \times a \times a$$

How would you simplify $a^5 \times a^3$?

$$a^5 \times a^3 = a \times a \times a \times a \times a \quad \times \quad a \times a \times a$$
$$\qquad\qquad\quad 1 \quad 2 \quad 3 \quad 4 \quad 5 \qquad 1 \quad 2 \quad 3$$

*5 + 3 = 8. You have added the indices*

$$= a^8$$

 Work out $2^5$, $2^3$ and $2^8$ and so check that $2^5 \times 2^3 = 2^8$.

Here are two expressions that can be written more briefly:

$$4 \times x^3 \times 3 \qquad\qquad \text{and} \qquad\qquad 4x \times 3x^2$$

*x can be written as $x^1$*

$$= 4 \times 3 \times x^3 \qquad\qquad\qquad\qquad\qquad = 4 \times 3 \times x^1 \times x^2$$

*They turn out to be the same*

$$= 12x^3 \qquad\qquad\qquad\qquad\qquad\qquad = 12x^3$$

*Add the indices: 1 + 2 = 3*

How do you simplify $a^5 \div a^3$?

*5 − 3 = 2*

$$a^5 \div a^3 = \frac{a \times a \times \cancel{a} \times \cancel{a} \times \cancel{a}}{\cancel{a} \times \cancel{a} \times \cancel{a}} = a^2$$

*Cancelling the as leaves 1 in the denominator*

 What is $a^{106} \div a^{102}$?

*What rule did you use?*

# 15: Using symbols

**1** Write each of these using powers.
   a) $5 \times 5 \times 5$
   b) $8 \times 8 \times 8 \times 8 \times 8$
   c) $10 \times 10 \times 10$
   d) $7 \times 7 \times 7 \times 7 \times 7 \times 7 \times 7 \times 7 \times 7$
   e) $100\,000\,000$
   f) $10$
   g) $x \times x \times x \times x$
   h) $4 \times y \times y$
   i) $9 \times n \times n \times n$
   j) $3 \times a \times a \times a \times a \times a$

**2** Write each of these out in full.
   a) $2^3$
   b) $6^4$
   c) $d^3$
   d) $n^5$
   e) $5c^2$
   f) $7g^4$
   g) $10z^1$
   h) $2x \times x^5$
   i) $m^2 \times 2m^3$
   j) $4u^2 \times 2u$

**3** Write these as briefly as possible using powers.
   a) $3^2 \times 3^2$
   b) $6^7 \times 6^3$
   c) $4^3 \times 4^5$
   d) $10^6 \times 10^{11}$
   e) $a^2 \times a^3$
   f) $v^8 \times v^3$
   g) $s \times s^4$
   h) $d \times d^7$
   i) $x^2 \times x^2 \times x^3$
   j) $n^3 \times n \times n^4$

**4** Write these as briefly as possible using powers.
   a) $2x \times 3$
   b) $4x^2 \times 5$
   c) $2a^3 \times 3$
   d) $2 \times 3g^2$
   e) $2x \times 3x$
   f) $4x \times 5x$
   g) $3p \times p^2$
   h) $y^4 \times 7y$
   i) $2n \times n \times n^2$
   j) $9m \times 10m^2 \times m^3$

**5** Which is biggest, $7^2$, $2^7$ or $2 \times 7$?

Work them all out and write them in order of increasing size.

**6** Which is the biggest number? Put them in order of increasing size.

$1^{10}$, $2^9$, $3^8$, $4^7$, $5^6$, $6^5$, $7^4$, $8^3$, $9^2$, $10^1$

When you have done this, use a calculator to check your list.

**7** Work these out and write the answers using powers.
   a) $3^5 \div 3^3$
   b) $2^{10} \div 2^4$
   c) $5^{99} \div 5^{94}$
   d) $x^8 \div x^8$

**8** Find the value of $x^2 + 4$ when
   a) $x = 2$
   b) $x = 10$
   c) $x = 0$.

**9** Find the value of $2x^3$ when
   a) $x = 2$
   b) $x = 10$
   c) $x = 0$.

---

A kilometre is 1000 metres but a kilobyte is 1024 bytes.

Explain how these 2 numbers arise.

How many bytes are there in a Megabyte and a Gigabyte?

## 15: Using symbols

# Using negative numbers

By how many degrees did the temperature rise?

A change in temperature is given by

    new temperature − old temperature.

In this case, the change (in degrees Celsius) is $12 - (-8)$.

But what is the answer to $12 - (-8)$?

Look at the thermometer.

You can see by counting that the change from −8 °C to +12 °C is +20 °C.

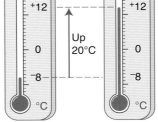

So $12 - (-8) = 20 = 12 + 8$.

You can see that $-(-8) = +8$.

Where there are two signs before one number, the signs follow these rules.

    $+ (+8) = +8 \quad + (-8) = -8 \quad - (+8) = -8 \quad - (-8) = +8$

On the Tuesday of the same week in Minneapolis, the temperature was −13 °C.

Use the rules above to work out how much the temperature rose to get to −8 °C.

Sketch a thermometer and count degrees to check your answer.

Work these out and check your answers on a thermometer scale or a number line.

(If you get a negative answer that means the temperature has dropped.)

Old temperature = 20 °C, new temperature = 6 °C. What is the change?

Old temperature = −20 °C, new temperature = −6 °C. What is the change?

Old temperature = 6 °C, new temperature = −6 °C. What is the change?

Old temperature = 20 °C, change = −6 °C. What is the new temperature?

In algebra you often need to substitute negative numbers into expressions. The same rules apply.

### Example

Work out the value of $a - b$ when $a = 25$ and $b = -8$.

### Solution

$$a + b = 25 - (-8)$$
$$= 25 + 8$$
$$= 33$$

When $a$ is 25 and $b$ is −8, $a - b = 33$.

## 15: Using symbols

Copy each expression and work down the page when you have to do more than one calculation.

**1** Simplify each of these.
   a) $-(+5)$
   b) $+(+5)$
   c) $+(-5)$
   d) $-(-5)$
   e) $-(+7)$
   f) $-(-3)$
   g) $+(-12r)$
   h) $-(+14s)$
   i) $-(-99t)$

**2** Work these out.
   a) $10 - (+5)$
   b) $12 + (+5)$
   c) $-16 + (-5)$
   d) $8 - (-5)$
   e) $2 - (+7)$
   f) $5 - (-3)$
   g) $48 + (-12r)$ when $r = 2$
   h) $10 - (+14s)$ when $s = 2$
   i) $900 - (-99t)$ when $t = 1$
   j) $9 + (-3x)$ when $x = 3$

**3** Write these as simply as possible.
   a) $9 - (-m)$
   b) $9 - (+m)$
   c) $3 + (-a)$
   d) $3 - (-2a)$
   e) $4 + (-2) \times n$
   f) $7 - (-2) \times n$

**4** Work out the value of $T - Y$ when
   a) $T = 6$ and $Y = 2$
   b) $T = -6$ and $Y = 2$
   c) $T = 6$ and $Y = -2$
   d) $T = -6$ and $Y = -2$.

**5** A train leaves London on Tuesday evening for Plymouth. The journey time should be 3 hours 35 minutes. In fact the train arrives 4 minutes early, at 0006 on Wednesday morning.

What time does the train leave London?

**6** Find the value of $3x + 4$ when
   a) $x = 1$
   b) $x = -1$
   c) $x = 2$
   d) $x = -2$.

---

Work out the value of each of these.
a) $-25 + 49$  b) $+25 - 49$  c) $+25 + 49$  d) $-25 - 49$

Use your results to help you write down simple rules about how to add and subtract any 2 numbers, positive or negative.

### 15: Using symbols

# Simplifying expressions with negative numbers

What happens when you multiply a negative number by a positive one, for example (–3) × 2?

You can see this on the number line. Multiplying by 2 doubles the distance from zero.

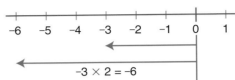

You can write this the other way round too, as 2 × (–3) = (–6).

In this, 2 means (+2), so you can see that

− × + → −

and

+ × − → −

What happens when two negative numbers are multiplied together? What is (–2) × (–2)?

This example gives you the answer. Sam goes shopping with £15. He buys two of these T shirts. Together they cost £10 so he has £5 left. You can also write it down like this.

Money left = money at start − cost of 2 T-shirts

In £:   5 = 15 − 2 × (7 − 2)

*Expand the bracket*

*2 × 7 = 14*

5 = 15 − 2 × 7 − 2 × (−2)

5 = 15 − 14 − 2 × (−2)

*15 − 14 = 1*

5 = 1 − 2 × (−2)

But 5 = 1 + 4 so this shows you that (−2) × (−2) = (+4).

When you multiply two negative numbers together the answer is positive.

This is the same for positive numbers.

− × − → +

+ × + → +

 What is the value of (−5) × (−5)?

Why is the square root of 25 written as ±5?

What is the positive square root of 25? What is the negative square root of 25?

## 15: Using symbols

**1** Use the rules on page 154 to work these out.

a) $4 \times (-2)$ b) $-4 \times 2$ c) $-4 \times (-2)$ d) $+6 \times (+4)$
e) $+8 \times (-3)$ f) $-5 \times (-4)$ g) $-3 \times (-3)$ h) $(-8)^2$
i) $-30 \div 5$ j) $-30 \div (-5)$ k) $30 \div (-5)$ l) $55 \div (-5)$
m) $-2 \times (5x)$ n) $-4 \times (-2x)$ o) $-3x \div (-1)$

**2** Simplify each of these. Remember to carry out the operations in the right order.

a) $2 \times (-3) + 15$ b) $5 - 2 \times (-3)$ c) $5 - 2 \times (+3)$
d) $3 + 12 \div (-4)$ e) $15 \div (-3) + 6$ f) $17 - 6 \div (-3)$

**3** Expand these brackets.

a) $4(2z + 7)$ b) $4(2z - 7)$
c) $-4(2z + 7)$ d) $-4(2z - 7)$
e) $10(6 + 2a)$ f) $-10(6 + 2a)$
g) $-10(6 - 2a)$ h) $7(-e - 2)$
i) $-(4 - 8y)$ j) $-(4 + 8y)$

**4** Expand the brackets in each of these, then simplify them. Check your answers.

a) $12 + 2(5 - 3)$ b) $12 - 2(5 - 3)$
c) $12 - 2(5 + 3)$ d) $14 - (7 + 4)$
e) $14 + (7 - 4)$ f) $14 - (7 - 4)$
g) $3(5 - 4) - 6$ h) $(2 + 8) + 2(1 - 3)$

**5** Expand the brackets then simplify these.

a) $21 + 2(x + 3)$ b) $21 - 2(x + 3)$ c) $6n + (4n - 10)$
d) $6n - (4n - 10)$ e) $5d + 4(1 + 2d)$ f) $19a + 3(10b - 6a)$
g) $3(5 - 3t) - t$ h) $(2 + 8c) + 2(1 - c)$ i) $23 - (5 + 4p) + 4p$

**6** Find the value of $4(p - 3q) - (7 - q)$ when

a) $p = 2$ and $q = 1$ b) $p = 2$ and $q = -1$ c) $p = -2$ and $q = 5$.

### Investigation

What happens to the sign when you keep multiplying negative numbers?

Write down the value of $(-1)^2$, $(-1)^3$, $(-1)^4$, and so on.

What is the value of $(-1)^{213}$?

Find out how to do calculations like $(-5) \times (-4)$ on your calculator.

Write some instructions to enable a friend to do them.

## 15: Using symbols

# Finishing off

**Now that you have finished this chapter you should be able to**

★ multiply and divide powers of a number
★ multiply and divide positive and negative numbers
★ use a number line to add and subtract positive and negative numbers
★ expand brackets that are multiplied by a negative number

Use the questions in the next exercise to check that you understand everything.

**Mixed exercise**

**1** Write each of these as briefly as you can.
   a) $4 \times 4 \times 4 \times 4 \times 4$
   b) $y \times y \times y$
   c) $5 \times x \times x$
   d) $7 \times a \times a \times a$
   e) $5 \times b \times b \times 3 \times b$
   f) $x^2 \times x^5$
   g) $3y^3 \div y^2$
   h) $15p^8 \div 5p^3$
   i) $4m^2 \div 4m$
   j) $d^4 \div 4d^3$

**2** Simplify each of these.
   a) $6 \times (-3)$
   b) $-4 \times (-2)$
   c) $12 \div (-4)$
   d) $-12 \div 4$
   e) $(-12) \div (-4)$
   f) $8 + 3 \times (-2)$
   g) $-24 \div 2 + 6 \times 5$
   h) $3 \times 4x - 2 \times 2x$
   i) $6 + (-2)$
   j) $-8 - (-5)$
   k) $3y - (-2y)$
   l) $-4m + (-2m)$

**3** Simplify these.
   a) $5t^2 + 4t^2$
   b) $8s^2 - 3s^2$
   c) $7r^2 - r^2$
   d) $9q^2 - 4q^2 + 2q^2$
   e) $p^2 + 2p^2$
   f) $n^2 - 4n^2 + 8n^2$

**4** Work out the value of
   a) $2 + x$ when $x = -3$
   b) $4 - m$ when $m = -6$
   c) $a - (-a)$ when $a = -1$
   d) $n + n - 4$ when $n = -2$.

## 15: Using symbols

**5** In each of these, expand the brackets and simplify the answer.

a) $2(a + 3) + a$  
b) $5b - 2(1 + b)$  
c) $3(2c - 2) - 5c$  
d) $3(d + 2) - 2(d + 3)$  
e) $12e - 6(4 + 2e)$  
f) $4(2 + f) - 3(1 - f)$  
g) $(3g + 1) - (2g + 4)$  
h) $5h - (3h + 2)$  
i) $4(a + b) - 3(a - b)$  
j) $8a + 4(b - 2a) - 5b$  
k) $3(2x + y - 4z)$  
l) $4(x + y) - 3(x + y) - (x + y)$

*Mixed exercise*

**6** Rupal went shopping with £$r$, and bought 3 presents for £$p$ each and 3 sheets of wrapping paper at £$q$ each.

Write down an expression for the number of £ she had left.

Check that this gives the right answer in the case when $r = 20$, $p = 4$ and $q = 0.5$.

**7** Amber went shopping with £$a$. She returned a skirt costing £$b$ to a shop, and bought another costing £$c$.

Write down an expression for the number of £ Amber had left.

Check that your expression gives the right answer in the case when $a = 25$, $b = 17$ and $c = 21$.

**8** Find the value of $3x^2 + 4$ when

a) $x = 1$
b) $x = -1$
c) $x = 0$
d) $x = 4$
e) $x = -4$.

### Investigation

Make ten cards with numbers and signs on like this:

Using the numbers as single digits (i.e. not as 56, 75 ...), combine some or all of the cards to make different expressions.

What are the largest and smallest numbers you can make?

Write down as many combinations as you can which give the answers
a) 0   b) 14.

# Sixteen

# Spending money

## Bills

James moves into a new flat on 27 May. He takes electricity meter readings when he moves in and again 3 months later.

 *How many units has James used?*

James's bill for this quarter is shown below.

Find the 2 readings on it.

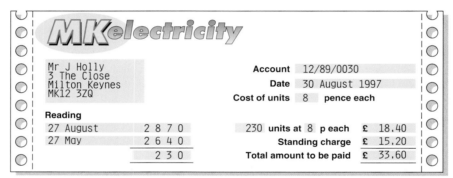

You work out the number of units used by taking 2640 away from 2870.

The cost of units is 8 pence each.

James used 230 units. You work out their total cost like this:

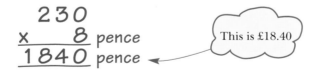

The standing charge does not depend on the number of units. It is £15.20 every quarter, however much electricity James uses.

The standing charge is added to the cost of the units to work out the total bill.

 *Do you think that James will use the same amount of electricity in the next quarter?*

# 16: Spending money

**1** Here are two of Mr Jackson's gas meter readings.

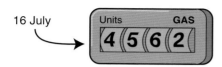

16 July — 4562
4721 — 16 October

How many units of gas has he used in this period?

**2** On 24 February Kay's water meter reading is 2698. On 24 August it is 2969.

How many units of water has she used?

**3** Find the cost of 30 units of gas at 2p per unit.

**4** Find the cost of 18 units of telephone time at 4p per unit.

**5** Find the cost of 32 units of electricity at 7p per unit.

**6** Find the cost of 12 units of water at 40p per unit.

**7** In the last quarter Ms Shah used 500 units of electricity at 9p per unit. The standing charge is £8 per quarter. Work out her total bill.

**8** In the last quarter Ryan used 250 units of telephone time each costing 5p. His quarterly rental charge was £26. Work out his total bill.

**9** On 1 March the reading on Lucy's gas meter was 4320. On 1 June it was 4720.

a) How many units had she used in this time?

b) Each unit costs 20p. How much did these units cost?

c) Lucy pays a quarterly standing charge of £16. Work out her total bill.

**10** Here are two of Stuart's water meter readings.

5 April — 5210
5873 — 5 October

a) How many units were used in this half year?

b) Each unit costs 15p. How much did these units cost?

c) Stuart pays a half-yearly standing charge of £42. Work out his total bill.

---

Find a real bill from home, for water, telephone, electricity, or gas.

Explain how each number on the bill is worked out.

# 16: Spending money

# Buy now, pay later

## Hire purchase

Harry wants to buy this satellite dish.

He does not have enough money to pay for it all at once.

He can choose to pay £30 a month for the next 12 months.

How much will this cost?

£290 OR
12 monthly payments of only £30

**Total cost = 12 × £30 = £360**

*The cash price is £290. How much extra would Harry pay?*

## Bank loan

The extra amount seems a lot to Harry. He decides to find out about a loan.

Harry already has £90 so he needs a loan of £200.

The bank will let him repay it over 6 months, 12 months or 18 months.

This table shows the monthly repayments in each case.

| Period of loan | 6 months | 12 months | 18 months |
|---|---|---|---|
| Monthly repayment | £40 | £22 | £16 |

Harry decides to repay the loan over 6 months.

What will be the cost of repaying the loan?

**Total repayment = £40 × 6 = £240**

Harry works out the total cost of the dish like this:

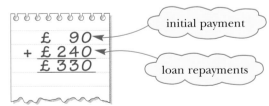

£ 90 ← initial payment
+ £240 ← loan repayments
£330

*Should Harry agree to the 12 monthly payments of £30 or take the loan?*

**Caution:** Harry must be very careful about taking on a deal like this. If he fails to make the monthly payments, he may lose both the satellite dish and the money he has already paid.

# 16: Spending money

In questions 1 to 4 work out

a) the cost of the item spread over 24 months
b) the extra amount paid.

**1**
Gold Watch £150 or £8 per month for 24 months

**2**
Gas Cooker £650 or £35 per month for 24 months

**3**
Television £300 or £16 per month for 24 months

**4**
Settee £500 £27 per month over 24 months

**5** Hannah buys a camera. She pays £100 straight away and then 12 monthly payments of £9.

How much does the camera cost her?

Use this table for questions 6, 7 and 8.
It gives the repayments per month for each £1000 of a loan.

| Time in years | 1 | 2 | 3 |
|---|---|---|---|
| Amount per month | £110 | £65 | £50 |

A £2000 loan is 2 × £1000.

To repay it over 3 years would cost 2 × £50 per month

**6** Abdul has a loan of £1000 over 2 years. How much does he pay per month?

**7** Kate has a loan of £1000 over 3 years. How much does she pay
  a) each month?
  b) over the 3 years?

**8** James has a loan of £5000 over 1 year. How much does he pay
  a) each month?
  b) over the year?

Find a catalogue which offers 'buy now, pay later' deals.

Choose an item that you would like to buy.

Compare the cash price with the total price to be paid over a period.

Can you find a better deal in the shops?

Write a short report on the different deals you can find. Which would you choose, and why?

## 16: Spending money

# Value added tax (VAT)

VAT is a tax that you pay whenever you buy something.

Only a few types of purchase are exempt from VAT.

*Find out the present rate of VAT.*

*Find out what types of item are exempt.*

### Example

Look at these two deals on the same hi-fi.

Which store offers the better deal if VAT is 17.5%?

'Ex. VAT' means 'Excluding VAT': the VAT still has to be added on. The full price is £400 + VAT

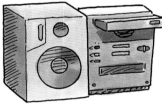  

### Solution

At Axis the VAT to be added on is 17.5% of £400.

$$17.5\% \text{ of } 400 = \frac{17.5}{100} \times 400 = \frac{35}{200} \times 400 = 70.$$

The VAT is £70.

'Incl. VAT' means that the VAT has already been added on. £450 is the full price

price ex VAT        VAT

The full price is £400 + £70 = £470.

In Jojo's the full price is £450, so Jojo's offers the better deal.

*If the rate of VAT were only 10%, which deal would be better?*

*Why do some stores display prices 'ex VAT'?*

# 16: Spending money

In questions 1 to 4 the price of each item is given excluding VAT.

VAT is to be added at 17.5%.

Work out

a) the amount of VAT   b) the price including VAT.

**5** In Store A an office chair is priced at £60 ex VAT. In Store B it is priced at £75 incl VAT at 17.5%.

Which store offers the better deal and by how much?

**6** In Store X a computer printer is priced at £215 ex VAT. In Store Y it is priced at £255 incl VAT at 17.5%.

Which store offers the better deal and by how much?

The filing cabinet in question 1 is priced at £120 ex VAT. The shop assistant works out the VAT at 17.5% as follows.

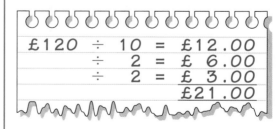

a) Why does this method work?

b) How could the assistant work out VAT if the rate were changed to 15% or 25%?

Write out your calculation for the £120 filing cabinet in each case.

# 16: Spending money

# Finishing off

**Now that you have finished this chapter you should be able to**

★ work out household bills
★ work out VAT
★ work out the costs of 'buy now, pay later' deals

Use the questions in the next exercise to check that you understand everything.

**Mixed exercise**

**1** Here are Mel's electricity meter readings at the start and end of a quarter.

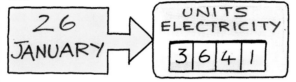 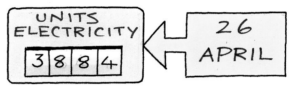

How many units of electricity has Mel used in this quarter?

**2** Find the cost of 300 units of gas at 8p per unit.

**3** Chris has used 250 units of gas at 18p per unit.
The standing charge is £6 per quarter.

Work out his total bill.

**4** Here are Laura's water meter readings for the start and end of a half-year period.

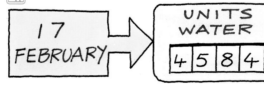 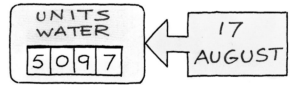

a) How many units has she used in this time?
b) Each unit costs 10p. How much do these units cost altogether?
c) Laura pays a half-yearly standing charge of £15. Work out her total bill.

**5** Adam buys this vacuum cleaner.
He chooses to make the 12 monthly payments.

a) How much does he pay in total?
b) Find the extra amount paid.

166

# 16: Spending money

**6** Jill buys a CD player priced at £99. She pays £30 straight away, then 12 monthly payments of £7.50.

How much does she pay in total?

Use this table to answer questions 7, 8 and 9.

It shows the amount to pay each month on loans of £500, £1000 and £5000.

| Repayment period | Amount of loan | | |
|---|---|---|---|
| | £500 | £1000 | £5000 |
| 2 years | £30 | £56 | £275 |
| 3 years | £22 | £41 | £200 |
| 4 years | £18 | £35 | £170 |

**7** Ms Patel has a loan of £500 to be repaid over 2 years.

How much does she pay each month?

**8** Mr Sou has a loan of £5000 to be repaid over 3 years.

How much does he pay each month?

**9** Miss Hansen has a loan of £1000 to be repaid over 4 years.

a) How much does she pay each month?

b) How much will she pay in total over 4 years?

**10** A CD midi system costs £200 excluding VAT. VAT is added at 17.5%.

Work out how much VAT is added.

**11** A video recorder costs £150 excluding VAT. VAT is added at 17.5%.

What is the total cost?

**12** A computer is offered for sale in two stores as shown. VAT is 17.5%.

Budget Bernie's £1200 ex. VAT

Quality Quentin's £1420 incl. VAT

Which store offers the better deal and by how much?

---

Using information from catalogues or magazines, work out the cost of buying the complete computer system of your choice.

Give the price both ex VAT and inc VAT.

You have a price limit of £2000.

*Mixed exercise*

# Seventeen

## Graphs

**Before you start this chapter you should**

★ be able to substitute values into a formula
★ know how to use co-ordinates

Look at the points on this straight line graph.

A is (1, 0): $x = 1, y = 0$
B is (2, 1): $x = 2, y = 1$
C is (3, 2): $x = 3, y = 2$

You can see that there is a pattern. In each case the value of $y$ is 1 less than the value of $x$.

$y = x - 1$ — This is called the **equation of the graph**

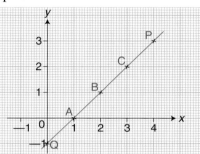

 *Do the points P and Q fit the same pattern?*

You often know the equation of a graph and want to draw it. In that case you start by making out a table of values, as in this example.

### Example

Draw the graph of $y = 2x - 4$ for values of $x$ between 0 and 4.

### Solution

| $x$ | 0 | 1 | 2 | 3 | 4 |
|---|---|---|---|---|---|
| $2x$ | 0 | 2 | 4 | 6 | 8 |
| $-4$ | $-4$ | $-4$ | $-4$ | $-4$ | $-4$ |
| $y$ | $-4$ | $-2$ | 0 | 2 | 4 |

These are the values of $x$

$2x$ is twice $x$

$-4$ stays the same whatever the value of $x$

These are the values of $y = 2x - 4$

When $x = 0$, $y = -4$

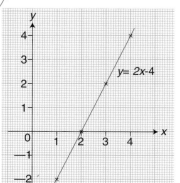

The points are plotted as crosses and then joined. In this case they lie on a straight line

In this graph the scales for $x$ and $y$ are the same. You can have different scales on the two axes

The line crosses the $y$ axis at $-4$. This is called the $y$ **intercept** or just the **intercept**

 *Work out the values of $y$ when $x = 1\frac{1}{2}$ and $3\frac{1}{4}$. Do these points also lie on the line?*

168

# 17: Graphs

**1** Look at these three sets of points.

| A | (−3, 0) | (−2, 1) | (−1, 2) | (0, 3) | (1, 4) | |
| B | (−2, −4) | (−1, −3) | (0, −2) | (1, −1) | (2, 0) | (3, 1) |
| C | (−2, −6) | (−1, −3) | (0, 0) | (1, 3) | (2, 6) | |

For each set of points,
a) choose $x$ and $y$ scales so that they fit nicely onto a piece of graph paper;
b) plot the points on a graph and join them with a straight line;
c) describe in words a formula to find $y$ for a given value of $x$;
d) write an equation for each line, of the form $y = \ldots$;
e) write down the co-ordinates of two other points on the line and check that they satisfy the equation of the line.

**2** a) Complete this table to find the co-ordinates of points on the straight line $y = 2x + 1$.

| $x$ | −2 | −1 | 0 | 1 | 2 | 3 |
|---|---|---|---|---|---|---|
| $y$ | −3 | −1 | | | | |

b) Choose suitable scales, plot the points and join them with a straight line.
c) As you read along the table, $x$ increases by 1 each time.
Describe what happens to $y$ as $x$ increases.

**3** A holiday in Spain costs £150, plus £20 per day at the Hotel Alhambra.
a) Make out a table showing the values of $H$ when $D = 0, 5, 10, 15$.
b) Choose suitable scales and draw a graph showing the values of $D$ along the $x$ axis and $H$ up the $y$ axis.
c) Use your graph to find
   (i) the cost of a 7-day holiday
   (ii) the number of days' holiday that you can get for £330.
d) Write down a formula for the total cost £$H$ of a holiday of $D$ days at the Hotel Alhambra.

---

The graph of the cost of sending a parcel against its weight looks like some steps.

Get a copy of the present postal rates for large letters and draw the graphs for 1st and 2nd class letters on the same sheet of paper.

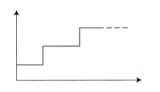

# 17: Graphs

## Gradients and intercepts

Look at Hiral's table of values for $y = 4x - 8$.

Hiral draws the graph using the same scales for his $x$ and $y$ axes.
He finds that it looks very squashed because the line is steep.
The steepness of a line is called its **gradient**.

$$\text{Gradient} = \frac{\text{increase in } y}{\text{increase in } x}$$

In this case,

$y$ increases by 32 units, from $-16$ to $+16$;

$x$ increases by 8 units, from $-2$ to $+6$.

So the gradient is $\frac{+32}{+8} = +4$.

Look at the graph and at the equation of the line,

The gradient is 4    $y = 4x - 8$    It crosses the $y$ axis at $-8$. This is the intercept

 *What are the gradient and intercept of the line $y = mx + c$?*

These diagrams show you

- a horizontal line with gradient 0
- a line which slopes down from left to right, with negative gradient.

 *What is the gradient of this line?*

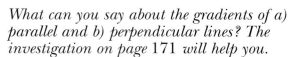

*What can you say about the gradients of a) parallel and b) perpendicular lines? The investigation on page 171 will help you.*

Hiral decides to make the graph look less squashed. He uses different scales for the $x$ and $y$ axes.

It is often helpful to use different scales like this, but you must keep the same scale all the way along the $x$ axis and all the way along the $y$ axis, for both positive and negative values.

 *Check that the gradient is still 4.*

# 17: Graphs

**1** The equation of a straight line is $y = 3x - 4$.

a) Make a table of values taking $x$ from $-2$ to $4$.

b) Choose suitable scales to draw the graph of the line.

c) Draw the graph of the line.

d) Calculate the gradient of the line.

e) What is the intercept of the line?

**2** The equation of a straight line is $y = \frac{1}{4}x + 1$.

a) Make a table of values taking $x$ from $-2$ to $4$.

b) Choose suitable scales to draw the graph of the line.

c) Draw the graph of the line.

d) Calculate the gradient of the line.

e) What is the intercept of the line?

**3** Find the gradient for each of these lines, stating whether it is positive or negative. Write down the intercept. Then write down the equation of the line.

a)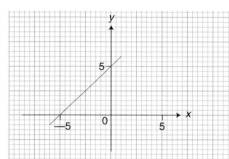

b)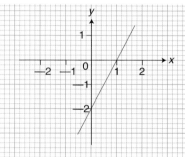

c)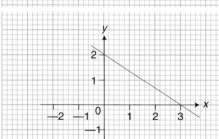

d)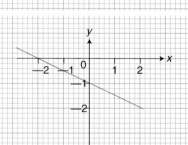

### Investigation

On graph paper draw $x$ and $y$ axes from $-5$ to $5$. Use the same scale for $x$ and $y$.

Draw the graphs of

$y = 2x, \quad y = 2x + 2, \quad y = 3x, \quad y = 3x - 2, \quad y = -\frac{1}{2}x, \quad y = -\frac{1}{3}$.

Which lines are parallel?

Which lines are perpendicular?

What can you say about the gradients of a) parallel and b) perpendicular lines?

# 17: Graphs

## Obtaining information

The graph shows how much it costs to be a member of the OK Fitness Club. There is a joining fee and then an additional monthly charge.

*Look at the graph. Can you tell what the joining fee is?*

*Can you work out the monthly charge?*

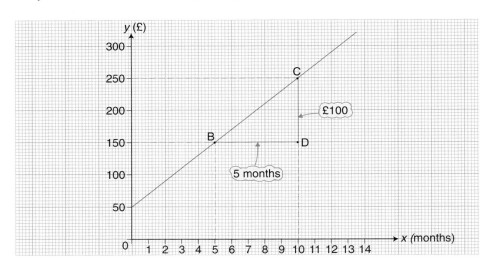

The joining fee is shown by the point where $x = 0$. This is called the **y intercept**. It is the price you pay right at the start.

Ali works out the monthly charge like this.

To find pounds per month you divide pounds by months

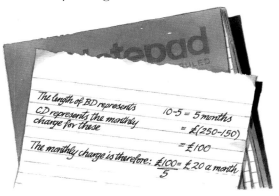

The length of BD represents
CD represents the monthly
charge for these
$10 - 5 = 5$ months
$= £(250 - 150)$
$= £100$

The monthly charge is therefore: $\frac{£100}{5} = £20$ a month

*What does the gradient of the graph represent?*

# 17: Graphs

**1** The graph shows the price of taxi fares in a city. There is a fixed charge plus a rate per mile travelled.

a) What does each small unit on the *y* axis represent?
b) What is the fixed charge?
c) Write down the co-ordinates of A, B and C.
d) Find the lengths of AC (in miles) and BC (in pounds).
e) Use your answers to d) to find the rate per mile.

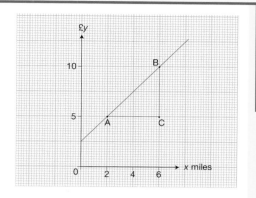

**2** The graph shows the cost of printing cards. There is a fixed cost for setting up the machine and a 'run-on cost' per 1000 cards.

a) What does each small unit on the *y* axis represent?
b) What is the fixed setting-up charge?
c) Write down the co-ordinates of A and B and so find the lengths of AC (in thousand cards) and BC (in pounds).
d) Use your answers to c) to find the run-on cost per 1000 cards.
e) Work out the cost of printing 10 000 cards.

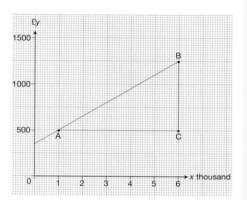

**3** The Brown family from London are planning a holiday in Dumfries. The graphs shows their estimated costs for travel to Dumfries and board for *x* days.

a) How much do they estimate for travel?
b) Find the gradient of the graph. What did they estimate for daily board?
c) Write down an equation for the total cost £*C* for a holiday of *x* days.
d) Use your equation to find the cost for 20 days.

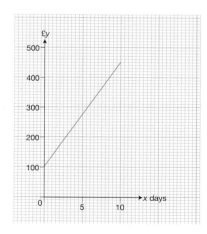

A plumber, Alf, charges a call-out fee of £40 plus £12 per hour on the job. Write this as a formula for the cost £*C* of a job lasting *h* hours.

Another plumber, Beatrix, has a call-out fee of £32 and charges £16 per hour. Write this as a formula for the cost £*C*.

On the same piece of paper draw the graphs of *C* and *h* for the two plumbers. Which plumber is more expensive?

# 17: Graphs

## Travel graphs: distance and time

Tina rides the 9 miles to work on her scooter.

One morning she sets off at 8 am as usual, but she runs out of petrol on the way.

She has to push her scooter back to a petrol station she has just passed. After filling up the tank, she continues her journey to work.

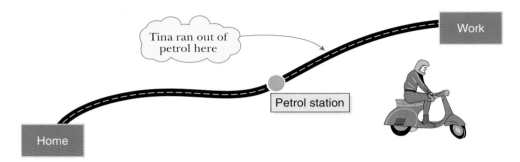

This graph shows Tina's distance from home at different times during the journey.

It is a **travel graph** or a distance–time graph.

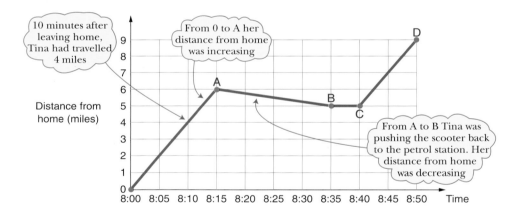

*What was happening from B to C?*

*What was happening from C to D?*

*How far had Tina travelled when she ran out of petrol at A?*

*How long did she take to travel that distance?*

From the graph, point A is 6 miles from home, and Tina reaches point A at 8.15 am.

It has taken her 15 minutes to travel 6 miles.

# 17: Graphs

Look again at the travel graph for Tina's journey to work.

**1** How long does Tina spend at the petrol station?

**2** How long does Tina's journey to work take, from beginning to end?

**3** How long do you think it would take her if she didn't run out of petrol?

Explain your answer.

This travel graph shows Jonah's journey from London to Dover.

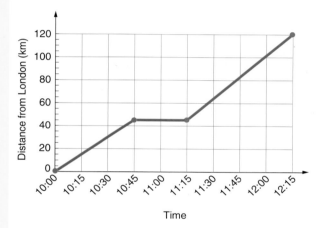

**4** Jonah travelled non-stop except for a break at a service station.

What time was it when he stopped at the service station?

**5** Jonah arrived at Dover at 12.15 pm. How far had he travelled?

**6** How long did Jonah spend

   a) driving?   b) at the service station?

**7** The graph shows Jonah's distance from London.

Sketch the graph of his distance from Dover. (Do not do any measuring.)

---

Leo plans to visit relations in Inverness. These travel graphs are for the different types of transport he could use.

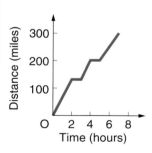 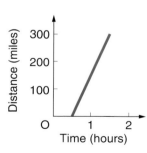 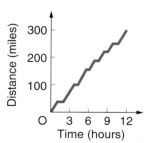

Which graph represents

a) the overnight coach, with 6 stops on the way?

b) the car journey, with 2 picnic stops?

c) the plane journey, with a half-hour wait at the airport?

## 17: Graphs

# Finding the speed from a travel graph

This graph shows Lucy's journey from home to school.

She walks from home to the bus stop, waits, then catches the bus to school.

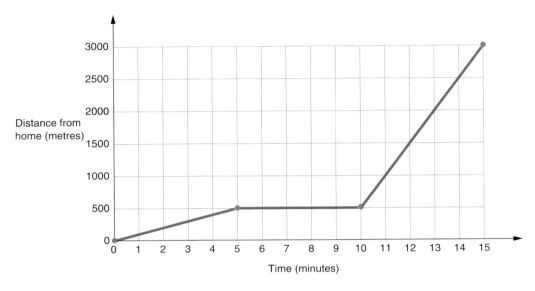

You can see that she takes 5 minutes to walk to the bus stop.

The bus stop is 500 m from home.

What is Lucy's walking speed in metres per minute?

Lucy walks 500 m in 5 minutes.

In 1 minute she walks $\frac{1}{5}$ of this.

$$\frac{1}{5} \times 500 = 100$$

**Lucy walks at 100 metres per minute.**

Another way to work this out is to use the formula

$$\text{speed} = \frac{\text{distance}}{\text{time}}$$

Lucy's speed = $\dfrac{500 \text{ m}}{5 \text{ minutes}}$

= 100 m/minute

*If you put the units in at this stage, you can see what units your answer will be in*

*How far does Lucy travel on the bus?*

*How fast does the bus go in metres per minute?*

*How fast is this in km/h?*

# 17: Graphs

Beth and her friends are on holiday.

They walk along the coast path to a nearby fishing village. They have lunch in the village and then return to their resort.

This is the travel graph for the trip.

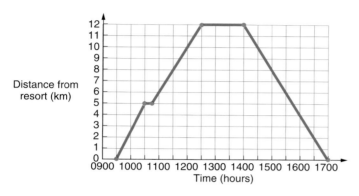

**1** At what time did Beth and her friends set off on the walk?

**2** a) How far did they walk before they had a rest?

b) How long did it take them to walk this far?

c) What was their walking speed in km/h before their rest?

**3** After their rest, they walked the rest of the way without stopping.

a) How far was this?

b) How long did it take?

c) What was their walking speed in km/h on this stretch?

**4** What was their walking speed on the return journey?

---

Look at these 3 travel graphs.

Work out the speed in each one.

Give an example of a type of transport that could do each speed.

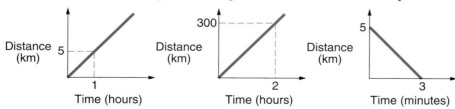

What does a horizontal line mean on a travel graph?

What does a downward-sloping line mean?

What do you think a curved line would mean?

# 17: Graphs

## The gradient of a travel graph

Catherine and her granny both live near a motorway, but at opposite ends of it. They agree to meet at a service station for lunch.

The red lines on the graph show Catherine's journey and the blue lines show Granny's.

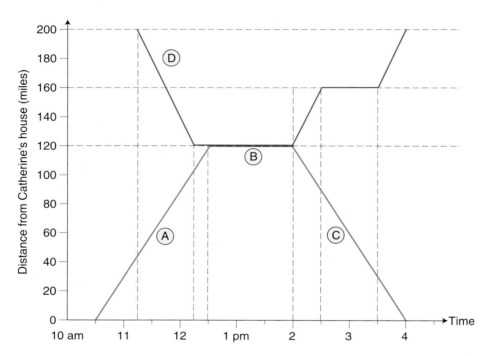

What is happening during part **B** of the graph?

Part **A** represents Catherine's journey to the service station. She travels 120 miles in 2 hours.

The gradient of the line is $\frac{120 \text{ miles}}{2 \text{ hours}}$ = 60 miles per hour, and this is her speed.

*Notice that the units can be written as miles/hour, miles hour$^{-1}$ or miles per hour (mph)*

**The gradient of a distance–time graph gives the speed.**

What is Granny's speed on the way to the service station?

What is the meaning of a negative gradient on this graph?

The lines on this graph are all straight.

What does this mean?

Is it realistic?

# 17: Graphs

**1** Use the graph on the opposite page to answer these questions.

a) When did Catherine arrive at the service station and how long did she stay?

b) How long did Granny have to wait for Catherine?

c) How long did Granny take to get home? What happened on the way?

d) Does Catherine drive more quickly, more slowly or at the same speed on the way home as she did on the way there?

**2** Karl leaves home at 12 noon for a cycle ride. The graph shows his journey.

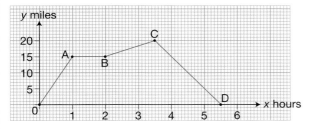

a) At what time is Karl furthest from home, and how far away is that?

b) Use the gradient to find his average speed in the part OA.

c) Describe what Karl might be doing from A to C.

d) The part CD has a negative gradient. Work out this gradient.

What does this tell you about this part of Karl's journey?

e) Karl's sister leaves home at 3.30 pm and drives at 40 mph along the same road.

Copy Karl's graph starting at 2.00 pm and add a line for his sister's journey. Where and when do they meet?

**3** A train from Oxford to London stops at Didcot and Reading. Here is its timetable.

| Oxford | Didcot | | Reading | | London |
|---|---|---|---|---|---|
| dep. | arr. | dep. | arr. | dep. | arr. |
| 1200 | 1212 | 1215 | 1240 | 1245 | 1315 |

The first leg of the journey (Oxford to Didcot) is 10 miles. The second leg is 20 miles and the third is 35 miles.

a) Draw a travel graph of the journey using 1 cm for 5 minutes and 1 cm for 5 miles.

b) Find the average speed of the train for each part of the journey.

c) A non-stop London train travelling at 100 mph passes Oxford at 1230. Draw a line on your graph for this train. Find where and when it passes the first train.

---

Speed can be measured in miles per hour, kilometres per hour or metres per second (etc.).

Work out how to convert between these three units.

Give typical speeds of ten different objects (e.g. train, bicycle, snail) in all three units.

# 17: Graphs

## Finishing off

### Now that you have finished this chapter you should be able to

- ★ construct a table of values and use it to draw a graph
- ★ find the gradient and intercept of a straight line graph
- ★ know when gradients are positive, zero and negative

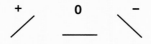

- ★ obtain information from graphs using co-ordinates of points
- ★ find a fixed value of starting value from a graph
- ★ read distance and time information form a travel graph
- ★ know that the slope of a travel graph represents speed
- ★ work out speed from a travel graph

## Mixed exercise

**1** The following equations represent straight lines.

  a) $y = x + 1$  b) $y = 2x - 3$  c) $y = \frac{1}{2}x + 2$  d) $x + y = 5$

  For each one,

  (i) construct a table of values from $x = -2$ to 4;

  (ii) choose suitable scales;

  (iii) draw the graph;

  (iv) calculate the gradient of the line;

  (v) state the intercept on the $y$ axis.

**2** Avonford Dramatic Society is going to put on a play and they need to hire some stage lights. The company Lights Galore charges £20 per day for a set of lights. Another company, Shadows, hires out an identical set of lights for £130 for up to 7 days plus £10 per day for each additional day.

  a) Draw a graph to compare the two sets of costs.

  b) Which company is cheapest if the lights are required for

   (i) 5 days?

   (ii) 8 days?

# 17: Graphs

**3** This travel graph shows Hannah's journey to work.

She runs part of the way (because she is a bit late) and walks the rest.

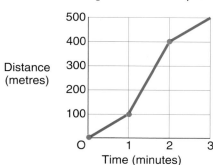

a) How far does Hannah travel to work?

b) When does she run?

c) How fast does she run? (Answer in metres per minute.)

d) How fast does she walk?

**4** Jake cycles to see his friend 13 km away.

On the way he has to cycle over Fisher's Hill.

The table shows his distance from home at 5-minute intervals.

| Time (minutes) | 5 | 10 | 15 | 20 | 25 | 30 | 35 |
|---|---|---|---|---|---|---|---|
| Distance (km) | 2 | 4 | 6 | 7 | 8 | 11 | 13 |

a) Draw the travel graph for Jake's journey.

b) When is he going up Fisher's Hill?

c) When is he going down Fisher's Hill?

d) What speed does he do on level ground?

**5** This travel graph shows a race between Anwar (red) and Greg (blue).

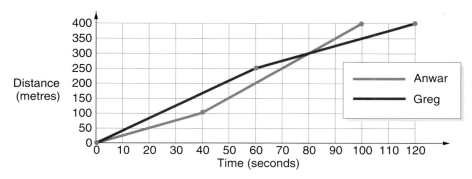

a) What distance is the race?

b) Who wins the race?

c) What does he win by (distance and time)?

# Eighteen

# Perimeter and area

## Perimeter

This is a plan of the Anderson family's living room.

Mr Anderson is going to redecorate the living room.

He wants to know how far it is round the edge of the room so that he can work out how many rolls of wallpaper to buy.

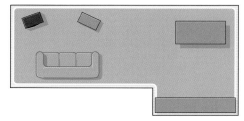

The distance around the edge of a shape is called the **perimeter**.

Mr Anderson measures the length of most of the walls and writes them on a plan of the room.

He cannot measure the wall behind the sideboard very easily.

He works out its length from the plan.

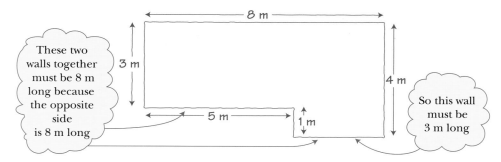

These two walls together must be 8 m long because the opposite side is 8 m long

So this wall must be 3 m long

Mr Anderson puts the 3 metre length on his diagram.

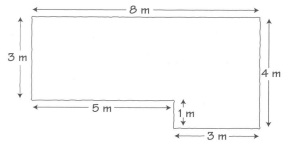

Now he can work out the perimeter of the room.

**Perimeter = 8 + 3 + 5 + 1 + 3 + 4 = 24**
**The distance round the room is 24 metres.**

# 18: Perimeter and area

**Exercise**

**1** Use a ruler to measure the perimeters of these shapes.

a)   b)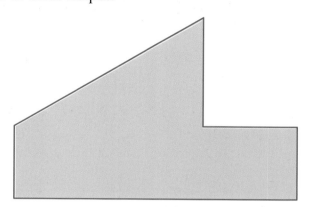

**2** Work out the perimeters of these shapes. (Do not do any measuring.)

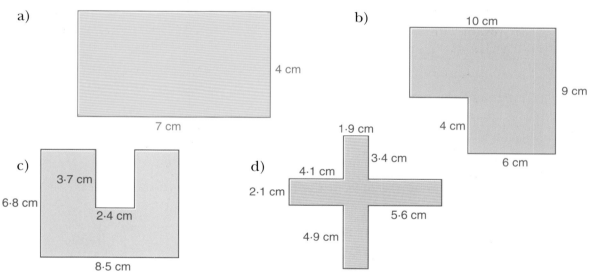

**3** A farmer wants to put a fence round this field.

Fencing costs £12.40 per metre.

How much will the fence cost?

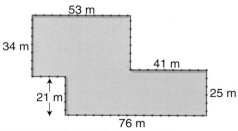

---

Find the perimeter of one of the rooms in your home.

Wallpaper is usually sold in rolls 52 cm wide.

Each roll is usually long enough to provide 3 strips of wallpaper.

Work out how many rolls of wallpaper would be needed to paper the whole room.

## 18: Perimeter and area

# Area

Sarah is laying some paving slabs to make a patio.

This is her plan for the patio.

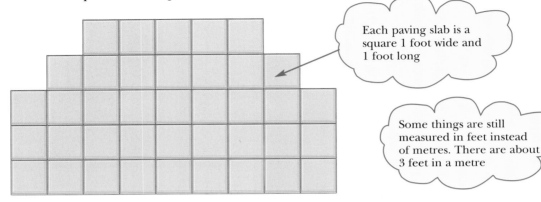

Each paving slab is a square 1 foot wide and 1 foot long

Some things are still measured in feet instead of metres. There are about 3 feet in a metre

Each paving slab has an **area** of 1 square foot.

Area is the amount of space inside a shape.

There are 39 slabs on the plan altogether.

**Area of the patio = 39 square feet.**

Sarah also wants to put some grass seed on her lawn.

The packet tells her how many handfuls to use per square metre.

She needs to know the area of the lawn to find out how much grass seed to use.

This is Sarah's plan of the lawn.

She has drawn it on a square grid.

Each square on the grid is 1 m long and 1 m wide.

Each square has an area of 1 square metre.

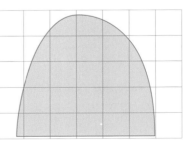

You can find an estimate of the lawn area by counting the squares that are at least half covered by the lawn.

Sarah colours in the squares that are at least half covered.

When she has finished there are 20 coloured squares.

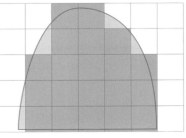

**Area of the lawn = approximately 20 square metres.**

184

# 18: Perimeter and area

**1** Find the exact areas of the shapes below by counting the squares.

Each square represents 1 square centimetre (1 cm²).

a)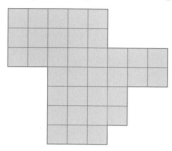

b)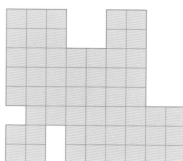

c)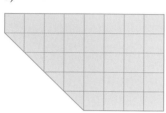

d)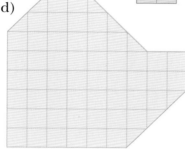

**2** Find approximate areas for these shapes.

Again each square represents 1 square centimetre.

a)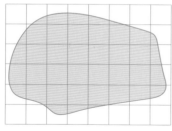

b)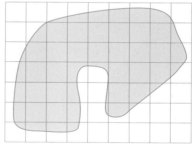

**3** Draw the rectangles below on to centimetre squared paper and count the squares to find their areas.

a) 7 cm
5 cm
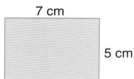

b) 6 cm
8 cm
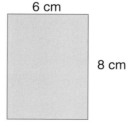

Draw round each of your hands on centimetre squared paper and estimate the area of each hand.

Which hand is bigger?

# 18: Perimeter and area

## Triangles

Jenille and Imran are taking part in a kite-making competition.

Jenille cuts out 2 right angled triangles from a rectangular piece of material for his kite.

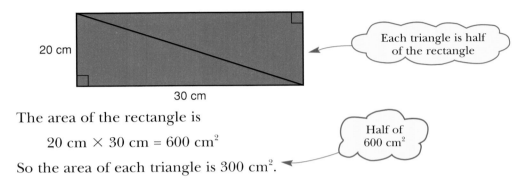

Each triangle is half of the rectangle

The area of the rectangle is

$$20 \text{ cm} \times 30 \text{ cm} = 600 \text{ cm}^2$$

Half of 600 cm²

So the area of each triangle is 300 cm².

Imran decides to make his kite from two triangles joined together as well.

Even though he does not use a right angled triangle, it is still half a rectangle.

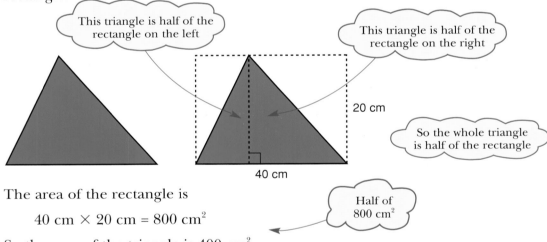

This triangle is half of the rectangle on the left

This triangle is half of the rectangle on the right

So the whole triangle is half of the rectangle

The area of the rectangle is

$$40 \text{ cm} \times 20 \text{ cm} = 800 \text{ cm}^2$$

Half of 800 cm²

So the area of the triangle is 400 cm².

To find the area of any triangle, you multiply the base by the height and halve the answer.

You can use any side of the triangle as the base, as long as the height is at right angles to it. This is the **perpendicular height**.

---

**Area of a triangle = $\frac{1}{2}$ × Base × Perpendicular Height**

$$A = \frac{1}{2} \times B \times H$$

---

*How do you think Jenille and Imran will join the triangles together to make kite shapes?*

## 18: Perimeter and area

**1** Find the area of each of these triangles by measuring the length of the base and the perpendicular height.

(Remember that you can use any side you like as the base.)

a)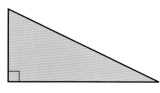

b)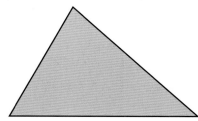

c)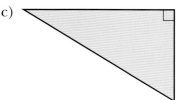

d)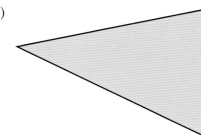

**2** Find the area of each of these triangles.

a)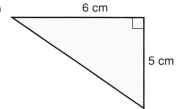

b)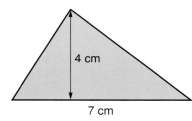

c)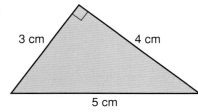

d)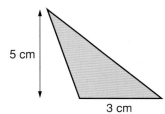

**3** This diagram shows the sail of a boat.

The area of the sail is 24 m$^2$.

The sail is 2.5 m wide at its widest point.

How high is the sail?

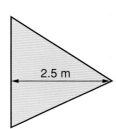

Draw as many different triangles as you can with areas of 12 cm$^2$.

## 18: Perimeter and area

# Shapes made of rectangles and triangles

 *How would you work out the area of this field?*

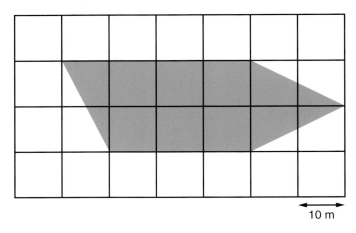

At first glance it looks like quite a complicated shape.

But it can easily be split up into simpler shapes – a rectangle and two triangles.

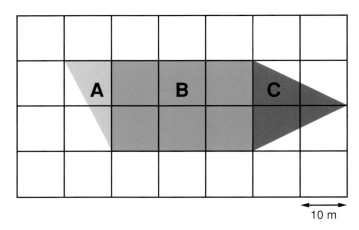

You can work out the area by finding the area of the simple shapes and adding them together.

Area of triangle A = $\frac{1}{2} \times 10 \text{ m} \times 20 \text{ m} = 100 \text{ m}^2$

Area of rectangle B = $20 \text{ m} \times 30 \text{ m} = 600 \text{ m}^2$

Area of triangle C = $\frac{1}{2} \times 20 \text{ m} \times 20 \text{ m} = 200 \text{ m}^2$

**So the area of the field = 100 m² + 600 m² + 200 m²**
**= 900 m²**

 *Do you think a real field would fit exactly in the grid like this one?*

## 18: Perimeter and area

**1** Find the area of each of these fields by splitting them up into rectangles and triangles.

a)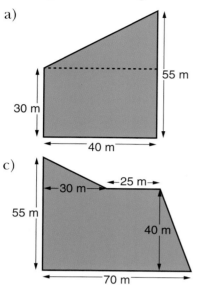

b)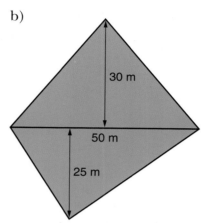

c)

**2** Find the areas of these shapes by splitting them up into rectangles and triangles and measuring the lengths you need.

a)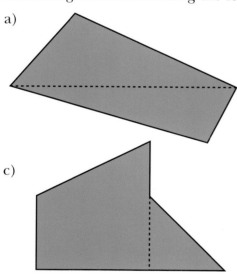

b)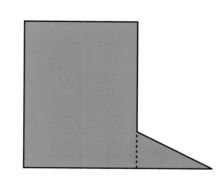

c)

Design a box in the shape of a pyramid.

Draw the net of your box and work out the area of card needed to make it.

Make the box.

## 18: Perimeter and area

# Circumference

Harry is walking around a circular lake in his local park.

He knows that the diameter is 50 m and wonders how far it is all the way round. (He does not have much time.)

The distance around the edge of a circle is called the **circumference**.

The circumference can be found by multiplying the diameter by a number called pi, which is written $\pi$.

*The value of $\pi$ is about 3.14. Your calculator may have a button for $\pi$, if not you can use 3.14*

*Also, $C = 2\pi r$ where $r$ is the radius*

**Circumference of a circle = $\pi \times$ diameter**
$C = \pi \times d$

So the circumference of the lake is
$\pi \times 50$ m = 157 m

If you know the circumference of a circle, you can work out the diameter by dividing it by $\pi$.

### Example

The circumference of the Earth is about 40 000 km.

What is its radius?

### Solution

*The radius is half the diameter*

Circumference = $\pi \times$ diameter

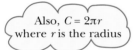

*Divide this by $\pi$*

40 000 = $\pi \times$ diameter

12 732 = diameter

radius = 12 732 ÷ 2

So the radius of the earth is about 6366 km.

# 18: Perimeter and area

**1** Find the circumference of the following circles. If you do not have a π button on your calculator, use the value 3.14 for π.

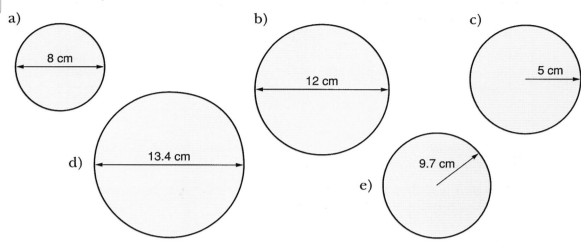

**2** Find the diameters of the circles with the following circumferences:

a) Circumference = 20 cm

b) Circumference = 42 cm

c) Circumference = 17 cm

**3** A bicycle wheel has a diameter of 70 cm.

a) How far does the bicycle travel in the time it takes for the wheel to go round once?

b) How many times does the wheel go round in the time it takes for the cyclist to travel 1 kilometre?

Measure the diameters of three different bicycle wheels.

Work out how many times the wheels go round in 1 kilometre.

Do you prefer to ride a bicycle with big wheels or small wheels? Give some reasons.

# 18: Perimeter and area

## Area of a circle

Bob works in a cake factory.

He has designed a new birthday cake.

The radius of the cake is 6 inches.

Bob wants to know the area of the cake to work out how much icing he needs.

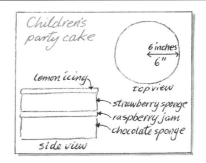

 *Can you think of any other cases where someone would want to know the area of a circle?*

The rule for working out the area of a circle uses the same number, π, that is used to work out the circumference.

**Area of circle = $\pi r^2$**
$A = \pi r^2$

*r stands for the radius*

For Bob's new cake,

Area = $\pi r^2$
     = $\pi \times 6^2$
     = $\pi \times 36$
     = 113 square inches

*You only square the radius. You work the area out by squaring the radius first, then multiplying by π*

## Finding the radius from the area

When you know the area and want to know the radius, you need to work backwards.

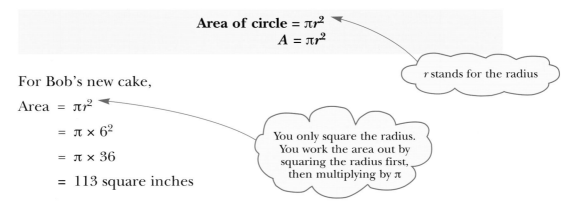

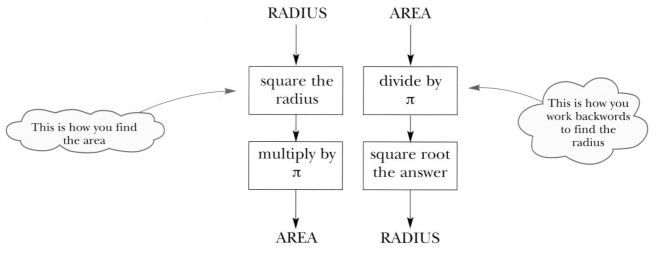

*This is how you find the area*

*This is how you work backwords to find the radius*

 *How can you work out the area if you only know the diameter?*

# 18: Perimeter and area

**1** Work out the areas of these circles:

   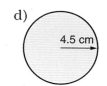 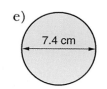

a) 8 cm   b) 12 cm   c) 10 cm   d) 4.5 cm   e) 7.4 cm

**2** Work out the radius of circles with these areas:

a) Area = 20 cm$^2$   b) Area = 38 cm$^2$   c) Area = 11 cm$^2$

**3** The diagram shows an arched window made from a rectangle with half a circle above it. Find the area of glass needed for the window.

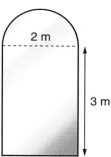

2 m

3 m

**4** This garden pond has an island in the middle of it.

The diameter of the island is 1 m.

The diameter of the pond is 5 m.

Find the area of pond that contains water.

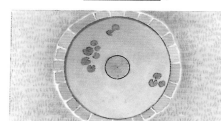

**5** Julie is making circular badges out of card.

Each badge has a diameter of 7 cm.

She is using a piece of A4 card which measures 29.7 cm by 21.0 cm.

a) How many badges will Julie be able to fit on the piece of card?

b) Find the area of one badge.

c) Work out the area of card that will be wasted.

**6** Sanjay wants a circular flower bed in his garden.

He has enough plants to fill an area of 10 m$^2$.

What is the radius of the biggest flower bed he can have?

Look at the different sizes of pizza in a supermarket or resturant.

Work out the area of each pizza.

Do you think the prices are fair?

# 18: Perimeter and area

## Finishing off

**Now that you have finished this chapter you should**

- ★ know that the perimeter of a shape is the distance round its edge
- ★ be able to find the perimeter of a shape with straight sides
- ★ know that the area of a shape is the amount of space inside it
- ★ know that area is measured in square units, like cm² or m²
- ★ be able to estimate the area of a shape drawn on a grid of squares
- ★ be able to work out the area of a rectangle
- ★ be able to find the area of a triangle from base and height measurements
- ★ be able to find the area of a shape made up of rectangles and triangles
- ★ be able to work out the circumference and area of a circle
- ★ be able to change units of length and area
- ★ know that $1 \text{ m}^2 = 100^2 \text{ cm}^2 = 10\,000 \text{ cm}^2$

**Use the questions in the next exercise to check that you understand everything.**

## Mixed exercise

**1** a) Find the area and perimeter of each of the shapes below.

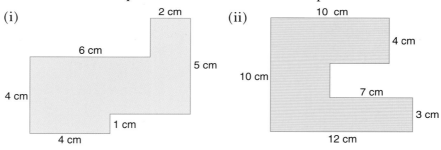

b) Change the units from centimetres to millimetres.
Find the area and perimeter of each shape.

**2** This is the floor plan of a room.

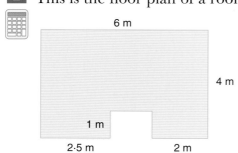

a) A decorative border is to be put round the top of the walls.
How many metres are needed?
How many centimetres are needed?

b) The room is to have a fitted carpet.
Carpet costs £16.50 per square metre.
How much does the carpet for the room cost?

c) Change the units from metres to centimetres.
Find the area of the floor in cm².

# 18: Perimeter and area

**3** The map shows a desert island.
The grid squares on the map are 1 km by 1 km.
Estimate the area of the island.

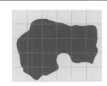

**4** The squares of a chess board are 2 cm × 2 cm.

a) What is the area of one square?

b) How many squares are there on the board?

c) What is the area of all the squares together?

d) What are the length and width of the board?

e) Check that multiplying the length and width gives the same answer as you got in c).

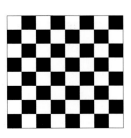

**5** A circular pond has a diameter of 4 m. It is surrounded by a path 80 cm (0.8 m) wide.

a) Find the circumference of the pond.

b) Find the area of the pond.

c) Find the area of the path.

(Hint: find the area of the larger circle first.)

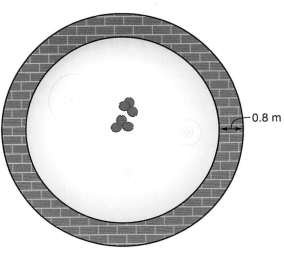

**6** Find the area of each of these shapes.

a)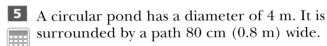
5 cm
8 cm

b)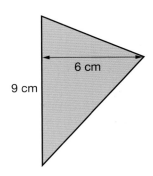
9 cm
6 cm

c)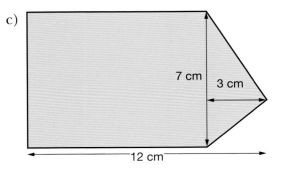
7 cm
3 cm
12 cm

*Mixed exercise*

195

# Nineteen

# Three dimensions

## Drawing solid objects

All solid objects have three dimensions: length, width and height.

A drawing has only two dimensions: length and width.

This makes solid objects difficult to draw.

There are several ways of representing solid objects on paper. Here are three different ways of drawing the solid shape called a triangular prism.

**Three-dimensional drawings**

These are drawings made to look like three-dimensional objects. They may use perspective.

**Nets**

Nets show what the shape would look like if it could be opened up and made flat.

**Views**

These show what the shape looks like from the top, the front and the side.

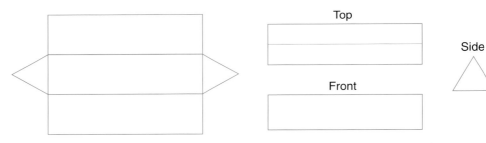

1. Which way of drawing the triangular prism do you think would be easiest to do?

   Which do you think would be hardest to do?

2. Which drawing gives the best idea of what the prism really looks like?

3. Which drawing would be best if you wanted to make a model of the prism?

4. Which drawing would be best if you wanted to show the lengths of each edge of the prism?

5. Here is a drawing of a **sphere**.

   Is it possible to draw a net of a sphere?

   What would the different views of a sphere look like?

# 19: Three dimensions

Here are drawings of some other solid shapes.

Each set of drawings shows the same 4 solid shapes.

Match up each net and each set of views with its three-dimensional drawing.

## Three-dimensional drawings

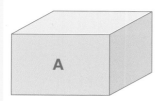

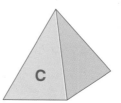

## Nets

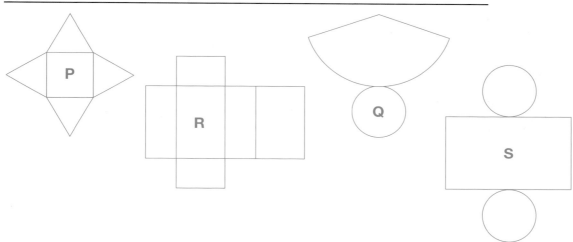

## Views

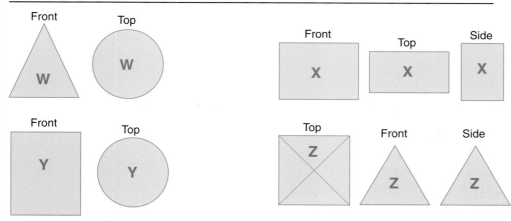

Draw the net of a model of a house. Cut it out and stick it together.

## 19: Three dimensions

# Using isometric paper

You can make drawings of three-dimensional objects using isometric paper.

It is important to make sure that you are using the paper the right way round.

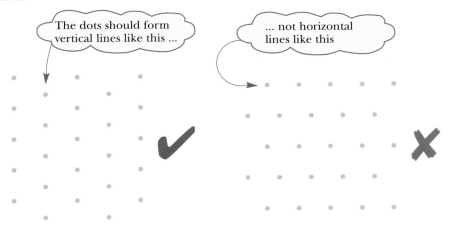

The simplest object to draw on isometric paper is a cube of side 1 cm.

Cubes and cuboids of different sizes are also easy to draw.

**Cube of side 1 cm**          **Cuboid 2 cm by 3 cm by 4 cm**

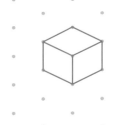
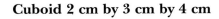
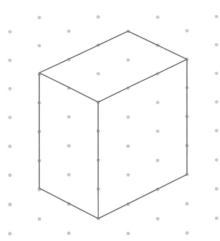

You can draw other shapes made from cubes like this.

You should only put in lines that can be seen.

This is a drawing of an arrangement of 8 cubes – 7 cubes on the bottom layer and 1 on top.

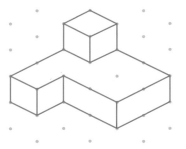

# 19: Three dimensions

**1** Draw the following shapes on isometric paper.

(Check that the paper is the right way round before you start.)

a) A cube of side 3 cm

b) A cuboid 2 cm long, 1 cm wide and 4 cm high

c) A cuboid 4 cm long, 4 cm wide and 3 cm high

**2** Draw these solid shapes on isometric paper.

For each shape, write down how many cubes have been used to make it.

a)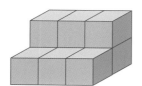

b)

c)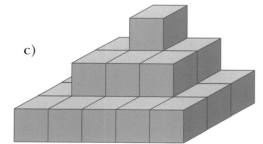

d)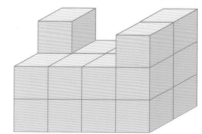

---

You can draw 'three-dimensional' letters of the alphabet using isometric paper.

The diagrams show how to draw the letter E.

　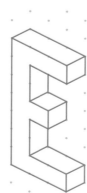

Write your name using three-dimensional lettering.

# 19: Three dimensions

## Nets

Hannah is making sweets for her friends.
She wants to put the sweets in boxes.

She has one box that is the right size.
She wants to make some more.

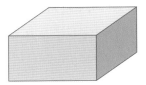

Hannah cuts down the edges of her box and flattens it out.

The flat shape that she has made is called the **net** of the box.

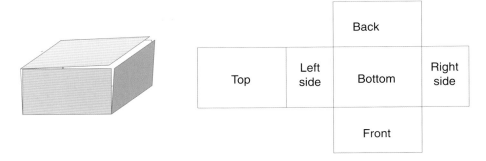

Hannah draws several of the nets on some card to make her boxes. She draws tabs on some of the sides, so that she will be able to glue the boxes together.

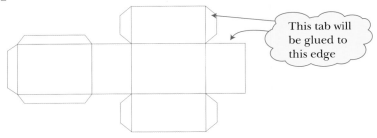

This tab will be glued to this edge

Hannah cuts out the nets, folds along the lines, and glues the tabs into place. The boxes are ready.

## Faces, edges and corners

You can see by looking at the net that Hannah's box is made of 6 rectangles. These are called **faces**.

A box with 6 rectangular faces is called a **cuboid**.

If the faces are square the box is called a **cube**.

*Find something in your classroom that is a cube or a cuboid.*

*Check that it has 6 faces.*

*Count the edges. There should be 12 edges.*

*Count the corners. A corner is called a vertex. There should be 8 vertices. (Vertices is the plural of vertex.)*

# 19: Three dimensions

**1** Which of the following diagrams are nets for a cube?

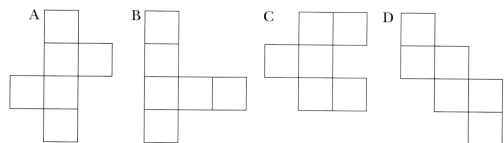

Copy your nets on to squared paper and cut them out to check your answers.

**2** The diagram shows a cuboid 4 cm long, 3 cm wide and 2 cm high.

  a) Draw a full size net for the cuboid on a piece of centimetre squared paper.
  b) Draw tabs on the net so that it can be stuck together.
  c) Cut out the net and use it to make the cuboid.

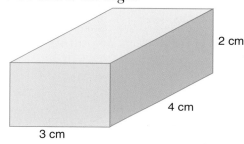

**3** The diagram shows the net of a solid shape.

You may like to make this solid shape by drawing the net yourself, cutting it out and glueing it together.

  a) What is the name of this solid shape?
  b) How many faces does it have?
  c) How many edges does it have?
  d) How many vertices (corners) does it have?

**4** The diagram shows the net of another solid shape.

Again, you may like to make the solid shape by drawing your own net.

  a) What is the name of this solid shape?
  b) How many faces does it have?
  c) How many edges does it have?
  d) How many vertices does it have?

Design some different boxes to hold sweets and draw their nets.

What are the good points and bad points of each design? (For example, would they stack well on the shelves of a shop?)

## 19: Three dimensions

# Volume

Hassan is packing boxes into a delivery lorry.

The back of the lorry is a cuboid 3 metres wide, 2 metres high and 5 metres long.

Each box is a cube 1 metre wide, 1 metre high and 1 metre long.

Each box has a **volume** of 1 cubic metre.

This is written as 1 m³.

The volume of a three-dimensional shape tells you how much space it takes up.

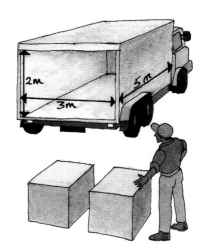

Hassan starts by putting a row of 3 boxes across the back of the lorry.

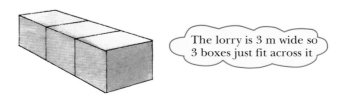

The lorry is 3 m wide so 3 boxes just fit across it

Hassan puts another row of 3 boxes on top of the row already there.

There are 2 rows of 3 boxes, making 6 boxes altogether.

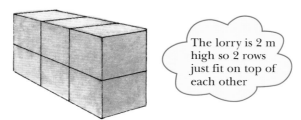

The lorry is 2 m high so 2 rows just fit on top of each other

Hassan puts 5 stacks of boxes like the one above into the lorry.

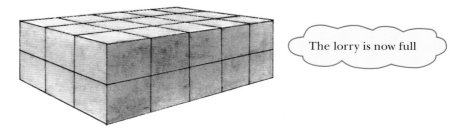

The lorry is now full

Hassan has put 30 boxes into the lorry, so the lorry has a volume of 30 cubic metres (30 m³).

The volume of any shape made up of cubes can be found by counting the number of cubes in the shape.

The volume of a cuboid is length × width × height

## 19: Three dimensions

**1** Each of these stacks is made up of boxes 1 metre long, 1 metre wide and 1 metre high.

   a) Find the volume of each stack.

   (i)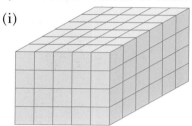

   (ii)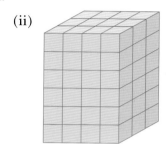

   (iii)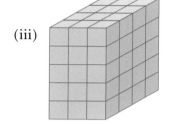

   (iv)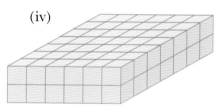

   b) Change the units from metres to centimetres and find the volume of each stack in cm$^3$.

**2** These models are made out of centimetre cubes.

   a) Find the volume of each model.

   (i)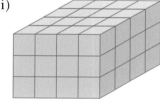

   (ii)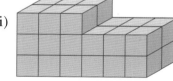

   (iii)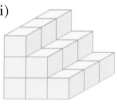

   (iv)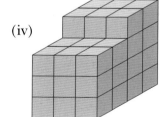

   b) Change the units of stack (i) from centimetres to millimetres. Find the volume of the model in mm$^3$.

---

Find some different ways to stack 24 stock cubes into a box. (You could use Multilink or Centicubes for the stock cubes.)

Which box would use the least cardboard?

## 19: Three dimensions

# Surface area of a prism and dimensions of formulae

Fran brings home some wooden offcuts from her technology class and covers them with plastic. They will be a toy for her little sister. One shape is an L-shaped prism.

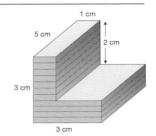

 *What part of the L-shaped piece does this rectangle cover?*
*What is the complete surface area of the L-shaped piece?*

Another piece is this prism. The cross section is a right-angled triangle.

 *What is the area of the triangle?*
*What is the perimeter of the triangle?*

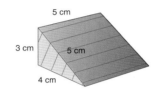

Fran writes this.

 *Explain Fran's work.*

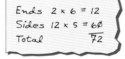

The general formula for the surface area of a prism is

> **Surface area = 2 × area of cross section + perimeter of cross section × length**

You have met a large number of formulae for perimeters, areas and volumes. You can tell whether a particular formula is for length, area or volume by considering the number of dimensions in the formula.

The number of dimensions in a formula is the number of measurements of length that are multiplied together in the formula. (Numbers, like 2 or π, do not count as dimensions.)

Formulae for lengths have one dimension, for those area have two and for volume have three.

**Examples**

$lwh$ ← This formula has three dimensions, $l$, $w$ and $h$. It is a formula for volume. What solid shape has this volume?

$\pi r^2$ ← This formula has two dimensions as $r^2$ means $r \times r$. It is a formula for area. What solid shape has this surface area?

$2(l + h)$ ← This formula has one dimension. The $l$ and the $h$ have been added together so they count as one dimension. It is a formula for length. What shape has this perimeter?

# 19: Three dimensions

**1** Find the surface area of each of these bricks. Each brick is 8 cm long.

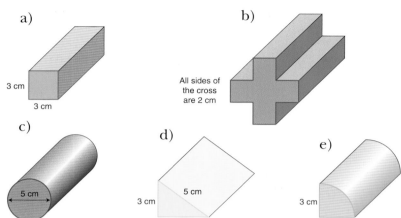

**2** The cross section of a swimming pool is shown in this diagram.

The pool is 10 m wide. It needs re-tiling. Find the area to be tiled.

(Hint: remember that there are no tiles on the top surface.)

**3** The diagram shows the cross section of a tunnel. The top is a semi-circle.

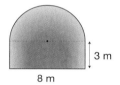

The tunnel is 50 m long. Find the surface area of the sides and roof of the tunnel.

**4** Draw a table with four columns, headed 'Length', 'Area', 'Volume' and 'None of these'. Check the dimensions of each formula below, then write it in the appropriate column.

a) $\frac{4}{3}\pi p^3$  b) $\pi p(p+r)$  c) $\frac{1}{3}p^2 q$

d) $p+q+r$  e) $pq^2 r$  f) $\pi pq^2$

g) $\pi p^2$  h) $\frac{1}{2}pq$  i) $\frac{1}{3}\pi p^2 q$

j) $2\pi q(p+q)$  k) $2\pi p$  l) $pq + pr^2$

m) $pqr$  n) $4\pi p^2$  o) $2(pq+qr+pr)$

Here is the cross section of a shape 5 cm long.

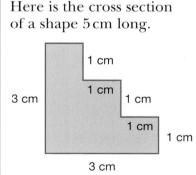

Draw this shape on isometric paper.

Calculate its surface area and draw its net.

## 19: Three dimensions

# Finishing off

**Now that you have finished this chapter you should**

★ recognise different ways of drawing solid shapes

★ be able to draw three-dimensional objects on isometric paper

★ be able to draw nets for a cube, cuboid, prism and pyramid

★ know what is meant by *faces*, *edges* and *vertices* of a solid shape

★ know that volume is measured in cubic units such as cm³ and m³

★ know that $1 \text{ m}^3 = 100^3 \text{ cm}^3 = 1000000 \text{ cm}^3$

★ be able to find the volume and surface area of a prism

Use the questions in the next exercise to check that you understand everything.

## Mixed exercise

**1** On centimetre squared paper, draw a full size net for this cuboid.

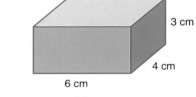

**2** Use isometric paper to draw each of these solid shapes.

Each one is made up of cubes.

a) 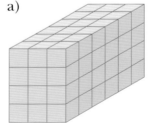   b)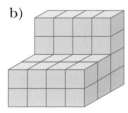

**3** Each of the shapes in question 2 is a stack of centimetre cubes.

Find the volume of each one.

**4** This sweet box is in the shape of a prism.
a) Draw a full size net for the prism on centimetre squared paper.
b) How many faces does the prism have?
c) How many edges does it have?
d) How many vertices (corners) does it have?

# 19: Three dimensions

**5** A company that packs tennis balls is trying out this new pyramid-shaped box.

a) How many tennis balls do you think will fit in the box?

b) Draw a net for the pyramid. Use 1 cm to represent 4 cm.

c) How many faces does the pyramid have?

d) How many edges does it have?

e) How many vertices does it have?

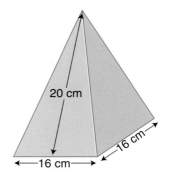

**6** Find the volume and surface area of each of these prisms.

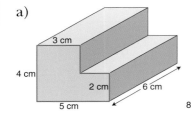
a)

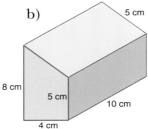
b)

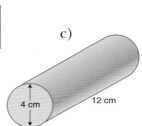
c)

**7** In this question, $h$ and $w$ represent lengths.

Does the formula $\frac{1}{3}\pi hw^2$ respresent a perimeter, an area, a volume or none of these? Give a reason for your answer.

## Investigations

**1** Make a table like this.

Fill in the numbers of faces, edges and vertices for any solid shapes with straight edges you can find. (You could use your answers to questions 3 and 4 on page 199. If you have Multilink or Polydron, you could use this to make some more shapes of your own.)

Try to find a rule linking the numbers of faces, edges and vertices for any shape.

| Shape | Faces | Edges | Vertices |
|---|---|---|---|
| Cuboid | | | |

Mixed exercise

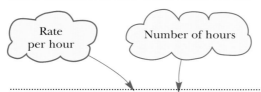

# Earning money

## Wages

Julie applies for this job and gets it.
How much will she earn each week?

**Supermarket assistant**

Suit school leaver
38 hour week
£4 per hour

*Rate per hour*  *Number of hours*

**Weekly wage = £4 × 38 = £152**
**Julie will earn £152 per week.**

Another supermarket pays its assistants £8000 *a year.*

*Is Julie's job better paid?*

A member of staff is ill, so Julie works 5 hours' overtime. She is paid at 'time and a half' for overtime. How much will she earn?

For 5 hours' overtime work she is paid for $\;5 \times 1\frac{1}{2} = 7\frac{1}{2}\;$ hours

**Overtime pay £4 × $7\frac{1}{2}$ = £30**

*'Time and a half' is $1\frac{1}{2}$ times*

You may find it easier to do it this way:

For one hour, normal pay is £4 but 'time and a half' pay is

£4 × $1\frac{1}{2}$ = £6

**Overtime pay £6 × 5 = £30**

*'Time and a half' is $1\frac{1}{2}$ times*

Here is a quicker way to work out overtime pay.

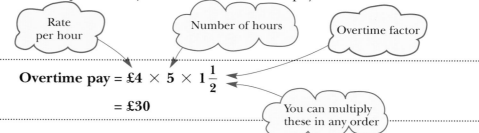

**Overtime pay = £4 × 5 × $1\frac{1}{2}$**
= £30

On the Sunday before Christmas Julie works 6 *hours' overtime which is paid at 'double time'. Use the quick method to work out how much she is paid.*

## 20: Earning money

**1** Work out how much each person earns per week.

a) Debbie: £5 an hour for a 36-hour week.

b) Joseph: £4.50 an hour for a 20-hour week.

c) Maria: 5 days a week, £40 a day.

d) Jamie: £30 per half day, for Monday morning, Tuesday afternoon and all day Friday.

**2** Rosie works 3 days a week for 4 hours each day.

a) How many hours does she work in a week?

b) She is paid £5 an hour. What is her weekly wage?

**3** This timesheet shows Kezi's hours of work for last week.

a) How many hours did she work on Monday?

b) How many hours did she work in the week?

c) Her pay is £4 an hour. How much did she earn for the week?

|           | Start | Finish | Hours |
|-----------|-------|--------|-------|
| Monday    | 0800  | 1200   |       |
| Tuesday   | 0800  | 1200   |       |
| Wednesday | 0800  | 1200   |       |
| Thursday  |       |        |       |
| Friday    | 0800  | 1200   |       |
| Saturday  | 0700  | 1200   |       |
|           |       | Total  |       |

**4** Victoria is paid £4 an hour and works 6 hours at 'double time'.

How much does she earn?

**5** Ian is paid £4.50 an hour and works 5 hours at 'time and a half'.

How much does he earn?

**6** Gary earns £6.50 an hour for a 36-hour week. He does 6 hours' overtime at 'double time'.

How much does he earn in total?

**7** Tara earns £5 an hour for a 37-hour week. She does 4 hours' overtime at 'time and a half'.

How much does she earn in total?

---

Find out how much a paperboy/girl earns per day.

How many papers are there on each round?

Is the rate of pay different on a Sunday?

How much does the customer pay each week for delivery?

## 20: Earning money

# Salaries

Gemma has an annual salary of £12 600. This means that Gemma earns £12 600 per year.

How much is this each month?

Gemma's pay = $\frac{£12\,600}{12}$ = £1050

*There are 12 months in a year*

## *Salary scales*

Mitesh starts work as a trainee environmental health officer. He is on point 1 of the scale shown below.

| Point | Salary (£) |
|-------|------------|
| 1 | 8400 |
| 2 | 8800 |
| 3 | 9200 |
| 4 | 9600 |
| 5 | 10 000 |

*Starting salary*

*Salary after 1 year*

*Salary after 2 years*

*How much will he earn after 4 years?*

## *Commission*

Ann gets this job.

Part of her salary is based on the value of the goods she sells. This is called **commission**.

Ann sells £30 000 worth of goods one year.

What is her total salary for the year?

Ann's commission is 25% of £30 000.

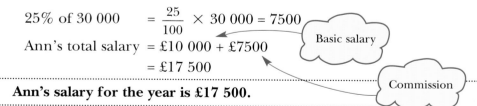

Salesperson

£10 000
plus 25% commission

$$25\% \text{ of } 30\,000 = \frac{25}{100} \times 30\,000 = 7500$$

Ann's total salary = £10 000 + £7500
= £17 500

*Basic salary*

*Commission*

**Ann's salary for the year is £17 500.**

*How much does Ann earn in total in a year when she sells £60 000 of goods?*

*What are the advantages of being on commission?*

*What are the disadvantages of being on commission?*

## 20: Earning money

**1** Helen's salary is £9000 a year. How much is this per month?

**2** Mitchell earns £820 a month. What is his annual salary?

**3** Claudette and Sean do this job on a job-share basis.

This means that they each work half a week and get half the salary.

How much does Claudette earn each month?

**Science Technician**
**Salary £9600**
**Suitable for job share**

**4** Sacha's annual salary is £14 040.

His friend Tamara earns £255 a week.

Who earns more in a year and by how much?

**5** Danielle is a trainee housing manager. She starts work on point 1 (£8400) of the salary scale opposite.

She gains one salary point each time she completes a year with her firm, and each time she passes a stage in her training.

Work out Danielle's salary

a) after 9 months

b) after 18 months, having passed Stage 1 of her training

c) after 27 months, having passed Stage 2 of her training.

**6** Wesley's basic salary is £600 a month and he gets 25% commission on his sales.

His sales figures for May to August are shown below.

| Month | May | June | July | August |
|---|---|---|---|---|
| Sales (£) | 1000 | 1100 | 960 | 660 |

Calculate Wesley's total pay for each month.

**7** Last year Ginette and Mark both earned a basic salary of £8000.

They both received 30% commission on their sales.

a) Ginette made sales worth £20 000. How much did she earn in total?

b) Mark made sales worth £17 500. How much did he earn in total?

Some job advertisers say *attractive salary plus benefits*.

Find out 4 examples of benefits.

## 20: Earning money

# Simple interest

Martyn has saved £600 and wants to invest it for 2 years in a building society. These are the interest rates offered.

| Amount | Interest rate |
|---|---|
| £100–£499 | 3.5% p.a. |
| £500–£2499 | 4% p.a. |
| £2500 and over | 4.25% p.a. |

p.a. means 'per annum' (each year)

*Why do you think the interest rate is higher on larger amounts?*

What interest will Martyn get on his £600?

£600 is between £500 and £2499 so the interest rate will be 4% p.a.

Each year Martyn's interest is

$$4\% \text{ of } £600 = \frac{4}{100} \times £600 = £24$$

(Interest rate) × (Martyn's savings)

This £24 interest is sent to Martyn at the end of the first year.

The £600 in the account will earn another £24 during the second year.

The simple interest Martyn earns over 2 years will be

$$2 \times £24 = £48$$

This is called **simple interest**. If the building society added the interest to the account instead, Martyn would get **compound interest**

You can use this formula to calculate simple interest:

$$I = \frac{P \times R \times T}{100} = \frac{PRT}{100}$$

In algebra you can leave out the '×' sign

where  $I$ is the interest,
$P$ is the money invested the 'principal',
$R$ is the present interest rate p.a.,
$T$ is the time in years.

In Martyn's case $P = 600$, $R = 4$, $T = 2$:

$$I = \frac{600 \times 4 \times 2}{100} = 48$$

Martyn's simple interest over 2 years is £48.

*How much simple interest would Martyn earn over 5 years?*

## 20: Earning money

**1** Work out the simple interest on each of these.

a) £500 at 6% p.a. for 1 year

b) £4000 at 5% p.a. for 3 years

c) £200 at 3% p.a. for 2 years

d) £5000 at 7% p.a. for 4 years

e) £1500 at 4% p.a. for 2 years

f) £4500 at 7.5% p.a. for 5 years

**2** Work out the simple interest on each of these amounts using the interest rates in the table.

| Amount | Interest rate |
|---|---|
| up to £100 | 1% p.a. |
| £100–£999 | 4% p.a. |
| £1000–£4999 | 4.5% p.a. |
| £5000 and over | 5% p.a. |

a) Edward invests £700 for 3 years.

b) Frances invests £8500 for 2 years.

c) Jack invests £3000 for 4 years.

d) Zoe invests £80 for 1 year.

---

Building societies and banks compete with each other. They offer you different rates of interest on your savings. This rate depends on the amount you have to invest and whether you want to be able to take out your money without giving advance notice. (This is called *instant access*.)

Go to at least one building society or bank.

Find out the rates they offer on instant access accounts.

Compare your information with others in your group.

What is the highest rate of interest you can get on £200?

What is the highest rate you can get on £500, £1000 and £5000?

For each amount, work out the highest amount of simple interest it could earn in 2 years.

## 20: Earning money

# Tax

 Tax rates often change. The rates given here will probably not be quite the same as the real rates on the day you are reading this page.

Jo's salary is £8000. This is her **gross salary**.

She does not get all this money. She has to pay **tax**.

To see how this is calculated, look at this diagram. (Each square is £1000.)

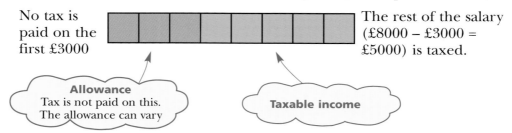

No tax is paid on the first £3000

The rest of the salary (£8000 − £3000 = £5000) is taxed.

**Allowance** Tax is not paid on this. The allowance can vary

**Taxable income**

Jo pays tax at 20 pence for every pound of her taxable income.

She pays 20% of £5000
$= \dfrac{20}{100} \times £5000$
$= £1000$

20p for every pound or 20p in the pound is the same as 20%

**Jo pays £1000 tax.**

She keeps the rest of her salary.

The money she keeps is called the **net salary**.

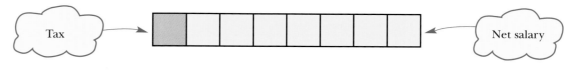

Tax

Net salary

**Jo's net salary is £7000.**

Jo's sister gets £13 000 salary.

What is her net salary?

She works it out like this:

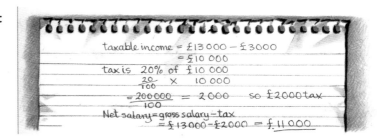

taxable income = £13 000 − £3000
= £10 000
tax is 20% of £10 000
$\dfrac{20}{100} \times 10\,000$
$= \dfrac{200\,000}{100} = 2000$ so £2000 tax
Net salary = gross salary − tax
= £13 000 − £2000 = £11 000

**So her net salary is £11 000.**

# 20: Earning money

**1** Work out the taxable income for a salary of:

   a) £9000 with an allowance of £3000
   b) £12 000 with an allowance of £4000
   c) £18 000 with an allowance of £3000.

**2** Work out the tax paid on a taxable income of:

   a) £16 000 at 20p in the pound
   b) £15 000 at 25p in the pound
   c) £12 000 at 22p in the pound.

**3** Stewart is a chef.

   His salary is £16 000, and his allowance is £3000.

   a) What is his taxable income?

   He pays tax at 20p in the pound.

   b) How much tax does he pay?

   c) What is his net salary?

**4** Poppy sees these two jobs.

   The allowance for both of them is £4000.

   Tax is paid at 25p in the pound for both of them.

   a) Work out the net salary for each job.

   b) How much more is the net salary of the second job?

**5** Julie's allowance is £3000.

   She pays tax at 25p in the pound.

   Julie says 'I only want the job if the net salary is more than £15 000.'

   Does Julie want the job?

Find out the allowance for a single person and the rates of tax.

Look at job adverts in a newspaper or job centre. Choose three and work out the net salary for each of them.

# 20: Earning money

## Finishing off

**Now that you have finished this chapter you should be able to**

★ work out wages
★ work out monthly earnings and salaries
★ use salary scales
★ work out commission
★ work out simple interest

Use the questions in the next exercise to check that you understand everything.

## Mixed exercise

**1** What is the weekly wage of

  a) Hassan who works a 5-day week and is paid £70 a day?

  b) Abigail who earns £5.50 an hour and works a 37-hour week?

**2** Emma's normal rate of pay is £5 an hour. She works 4 hours' overtime at 'double time'.

How much does she earn?

**3** Steven works 36 hours a week and is paid £8.50 an hour. He also does 2 hours' overtime which is paid at 'time and a half'. Find his total wage for the week.

**4** Here is Kevin's time sheet for last week. Copy it and complete the last column.

|  | Start | Finish | Hours |
|---|---|---|---|
| Monday | 0700 | 1300 |  |
| Tuesday | 0700 | 1300 |  |
| Wednesday | 0700 | 1300 |  |
| Thursday | 0700 | 1300 |  |
| Friday | 0700 | 1300 |  |
| Saturday | 0800 | 1300 |  |
| Sunday | 0900 | 1300 |  |
|  |  | Total |  |

From Monday to Saturday Kevin is paid £6 an hour. On Sunday he is paid 'double time'.

How much did Kevin earn last week?

**5** a) Caroline's salary is £12 864 a year. How much does she earn in a month?

  b) Matthew earns £350 a month. How much does he earn in a year?

# 20: Earning money

**Mixed exercise**

**6** Meena gets a job as a building inspector and starts on point 2 of this salary scale.

Her salary goes up one point after each year with the company.

a) What is her starting salary?

b) What is her salary after $1\frac{1}{2}$ years?

c) What is her salary after $3\frac{1}{2}$ years?

| Point | Salary (£) |
|---|---|
| 1 | 14 200 |
| 2 | 14 650 |
| 3 | 15 100 |
| 4 | 15 550 |
| 5 | 16 000 |

**7** Mick earns £800 a month plus commission. His commission is 20% of the value of the goods he sells.

These are Mick's sales figures for the first 3 months of the year.

| Month | January | February | March |
|---|---|---|---|
| Sales | £1000 | £1500 | £1200 |

Work out Mick's total pay for
a) January   b) February   c) March
d) the whole 3-month period.

**8** Katie invests £300 at 4% p.a. for 2 years. How much simple interest will she get?

**9** Alice and Jack both want to open an account at a building society.

They are offered these rates.

| Amount | Interest rate |
|---|---|
| up to £999 | 3% p.a. |
| £1000–£2499 | 4% p.a. |
| £2500 and over | 5% p.a. |

Alice has £900 and Jack has £1600. They both want to invest their money for 3 years.

a) How much simple interest will Alice get?

b) How much simple interest will Jack get?

c) How much extra money would they get by investing their £2500 in a single account?

**10** Terry's salary is £20 000.

His tax-free allowance is £3000 and he pays tax at 25p in the pound.

a) What is Terry's taxable income?

b) How much tax does he pay?

c) What is his net salary?

d) Terry gets a pay rise of £5000.

How much extra tax does he pay?

# Twenty one

# Estimation

## Approximations

Abigail is driving to a job interview.
She finds herself in a traffic jam.
A radio bulletin says that the traffic jam is 8 miles long.

*Does this mean that it is exactly 8 miles long?*

Very often, numbers that you read or hear are **approximations**.

The radio report means that the traffic jam is approximately 8 miles long.
It means they think that 8 is the nearest whole number.
The traffic jam might be 7.7 miles long, or 8.4 miles.

This number line shows the range of numbers that can be described as 'approximately 8' or 'about 8'.

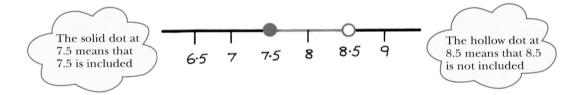

*What is the nearest whole number to 7.5?*

Usually, a number ending in .5 is rounded up to the next whole number. So 7.5 is rounded up to 8, and 8.5 is rounded up to 9.

*A traffic jam is 5.2 miles long. What would the radio bulletin call this, to the nearest whole number?*

Abigail drove about 600 miles in her car last month.
This figure is rounded to the nearest 100.

Look at this number line. It shows the range of numbers that can be described as 'about 600'.

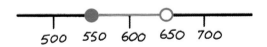

*What is the smallest number of miles that can be described as 'about 900'?*

*What is the greatest number of miles that can be described as 'about 900'?*

# 21: Estimation

**1.** If a number has only one figure and place-keeping noughts, then the number is said to be correct to 1 significant figure.

Correct to 1 significant figure, 3694 is 4000 and 0.04295 is 0.04.

0.04 is given to 2 decimal places but 1 significant figure

Write the following numbers correct to 1 significant figure.

a) 2144  b) 923  c) 41
d) 0.0049  e) 0.00016  f) 0.0672

**2.** a) Write the following numbers, stating how many significant figures are in each answer.

(i) 56.8 seconds to the nearest second
(ii) £64.23 to the nearest pound
(iii) 241 miles to the nearest 10 miles
(iv) 77 km/h to the nearest 10 km/h
(v) 575 kilocalories to the nearest 100 kilocalories
(vi) 2368 m² to the nearest 100 m²
(vii) £26 788 to the nearest £1000
(viii) 0.753 to the nearest whole number

b) Write each of the answers in a) correct to 1 significant figure.

**3.** Estimate

a) the number of hours you have been alive,
b) the length of time you have spent asleep,
c) the number of times your heart has beaten today,
d) the number of meals you have eaten,
e) the weight of food you have eaten.

### Investigation

Measure your height in centimetres.
Write it to the nearest centimetre.
Write it to the nearest 10 cm.
Write it to the nearest metre.

---

Collect 8 newspaper articles that include some approximation statements.

Examples: They played to a crowd of 30 000.

The bullet missed by a fraction of an inch.

For each statement, give the range of numbers that you think it might mean.

Is the meaning always clear?

---

Abigail is in an 8-mile-long traffic jam on the motorway. There are 3 lanes on the motorway.

a) Estimate how many vehicles are involved.
b) Estimate how many people are involved.

## 21: Estimation

# Decimal places

Connor is doing a survey to find out how well known leading twentieth-century figures are. He asks 35 people to identify these two.

Of the people he asks, 19 recognise ① to be Mother Teresa.

Connor works this out as a percentage.

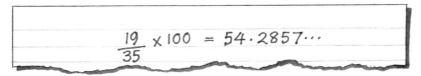

He has to decide how accurately to state this in his report.

 *How many decimal places, if any, do you think Connor should use?*

Look at this number line.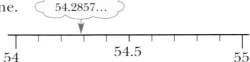

54.2857 . . . is nearer to 54 than to 55.
It is 54 correct to the nearest whole number.

Now look at this number line.

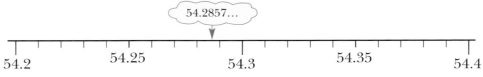

54.2857 . . . is nearer to 54.3 than to 54.2.
It is 54.3 correct to 1 decimal place.

 *Copy the last number line. Colour the range of numbers which would be written as 54.3 correct to 1 decimal place.*

 *Out of 35 people, 15 recognise ② to be Nelson Mandela.*

*What is this as a percentage*

*a) correct to the nearest whole number?*
*b) correct to 1 decimal place?*

# 21: Estimation

**1** The value of π is 3.141 592 . . . Write this value correct to

   a) 1 decimal place   b) 2 decimal places   c) 3 decimal places.

**2** Use your calculator to find the square root of 20.

   Write this value correct to

   a) 1 decimal place   b) 2 decimal places   c) 3 decimal places.

**3** How many decimal places do you give if your answer is correct to the nearest

   a) tenth?  b) hundredth?

**4** For each of these, write down your estimate of the reading to the nearest hundredth, and then write the reading correct to the nearest tenth.

   a)     b)

**5** Write these fractions as decimals correct to 3 decimal places.

   a) $\frac{1}{6}$     b) $\frac{1}{12}$     c) $\frac{2}{3}$     d) $\frac{3}{7}$

**6** Calculate 12.6% of 43.8, giving your answer correct to 2 decimal places.

**7** a) Measure the length and width of this rectangle in centimetres, giving your answers correct to 1 decimal place.

   b) Use these values to calculate the area of the rectangle, giving your answer correct to 1 decimal place.

**8** Work out the mean of these numbers giving your answer correct to 2 decimal places.

   4, 4, 5, 7, 8, 12, 15

**9** A circle has radius 7.3 m. Using π = 3.14 calculate, correct to 1 decimal place,

   a) its circumference  b) its area.

Make the following measurements and decide how many decimal places it is sensible to use.

a) The height of a friend in metres.

b) The length and width of a sheet of paper in centimetres.

c) The length and width of a car in metres.

## 21: Estimation

# Significant figures

 Look at these two newspaper headlines. Which do you think is better?

Sam has actually won £9 124 167, but the Avonford Star describes this as £9 million, which is £9 000 000. They decided their readers were not interested in the exact amount and so they rounded it to 1 **significant figure**.

£9 million   or £9 000 000   has 1 significant figure: it is the 9.

£9.1 million   or £9 100 000   has 2 significant figures: 9 and 1.

£9.12 million   or £9 120 000   has 3 significant figures: 9, 1 and 2.

**Example**

Write   a) 294 217 to 3 significant figures (3 s.f.),

b) 0.004 297 to 2 significant figures.

**Solution**

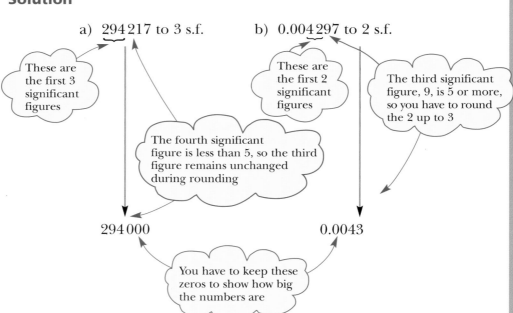

 In the example above, 0.004 297 is rounded to 2 significant figures.

What do you get if you round it to 2 decimal places?

Which way of rounding do you think is better?

Write 0.004 297 correct to 3 significant figures.

# 21: Estimation

**1** Write each of these to the number of significant figures (s.f.) shown.

   a) 77 328 to 2 s.f.
   b) 42.195 to 3 s.f.
   c) 5372 to 2 s.f.
   d) 758 423 to 3 s.f.
   e) 3780 to 1 s.f.
   f) 61.977 to 3 s.f.
   g) 6.7394 to 2 s.f.
   h) 53 660 to 1 s.f.
   i) 33.830 7 to 3 s.f.
   j) 0.005 38 to 1 s.f.

**2** For each of these, write down your estimate of the reading to 4 significant figures, then write the reading correct to 3 significant figures.

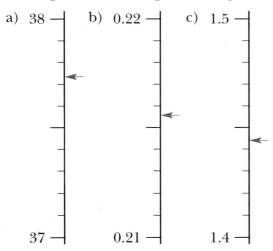

a) 38 / 37   b) 0.22 / 0.21   c) 1.5 / 1.4

**3** Repeat question 2 giving your answers correct to 2 significant figures.

**4** Work out 15% of 175 000 correct to 2 significant figures.

**5** A football pitch is 114 m long and 73 m wide. Work out, correct to 2 significant figures

   a) the perimeter   b) the area.

**6** The profits of a business for the four quarters of a year are

£55 000    £79 000
£127 000   £98 000

   a) Work out the average quarterly profit correct to 2 significant figures.

   b) Work out the total profit for the year correct to 2 significant figures.

**7** Tickets for a play are £4.25 each. This table shows the number of tickets sold for each of the three performances.

| Day | Thursday | Friday | Saturday |
|---|---|---|---|
| **Tickets sold** | 257 | 319 | 348 |
| **Income** | | | |

   a) Copy and complete the table, writing each day's ticket income correct to 3 significant figures.

   b) Write down the total income correct to 2 significant figures.

**8** A newspaper reported that the number of mobile phone text messages sent in the UK during March 2002 was 1530 million.

How many was this per second, on average? Give your answer to a suitable degree of accuracy.

---

Measure the length and width of a page of this book in centimetres. Use your answers to work out the area of the page.

What is a sensible number of significant figures for your answer?

## 21: Estimation

# Estimating costs

Vicky buys these clothes.

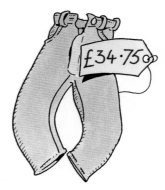

Vicky does not want to be overcharged so she makes a rough estimate of the total cost.

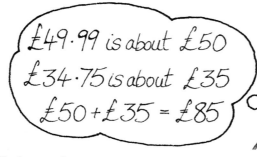

Work out the exact cost.

Is Vicky's estimate close to it?

Dean and his two brothers go to a Pizza Bar. They each order a pizza costing £5.95, and Dean orders an ice cream costing £2.10.

The waitress gives Dean a bill for £23.95.

Dean is surprised by the amount so he makes a rough estimate.

Do you think the bill was correct?

Work out the exact cost of 3 pizzas and an ice cream.

# 21: Estimation

**1** Use this menu to estimate the cost of each of these meals.

(Write down the cost of each item to the nearest pound, then add them up.)

a) Andrew has steak pie and chips.

b) Clare has chicken, chips and a large juice.

c) Marcus has plaice, chips and ice cream.

d) Emily has cod, chips, a cream cake and a small juice.

**MENU**

- Chicken £2.90
- Cod £2.85
- Plaice £2.95
- Steak Pie £1.85
- Chips £1.10
- Ice cream £1.90
- Cream Cake £2.85
- Juice (small) £1.10
- Juice (large) £2.10

**2** Estimate the total cost of these purchases.

a)

b)

c)

d)

**3** Mel's weekly rent is £49.

Estimate how much rent she pays in a year.

You start your first job next month in an office.

Estimate the cost of buying a new set of clothes.

## 21: Estimation

# Using your calculator

You should always check that answers you get on your calculator are sensible.

Mr Harris wants to reseed his lawn. He asks his 4 children to work out the area of the lawn.

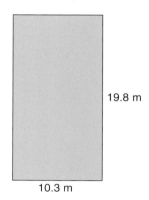

They each try to work out 19.8 × 10.3 on their calculators. Here are their answers.

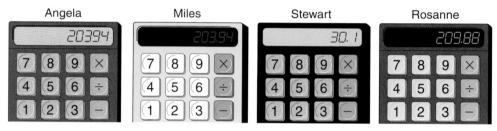

*Which of these answers seem sensible?*

One way to decide is to say 19.8 is about 20 and 10.3 is about 10, so the answer should be roughly 200.

Miles and Rosanne both have sensible answers. It is not easy to judge which of them is correct without doing an accurate calculation, but Mr Harris might decide that 200 is accurate enough for his purpose.

*Do the calculation on your calculator and find out which is correct.*

*What mistake did Angela make?*

*What mistake did Stewart make?*

*Can you tell what mistake Rosanne made?*

Rosanne actually keyed in 10.6 instead of 10.3 by mistake.

As it only made a small difference to her answer, a rough check did not show up the fault. However, by doing his rough check Mr Harris did avoid getting the answer completely wrong.

*Lawn seed costs £8.99 for a packet that will cover about 100 m$^2$.*

*What is the cost of reseeding Mr Harris's lawn?*

# 21: Estimation

*Do the questions in this exercise quickly, using your calculator. Then check that none of your answers is silly by doing rough calculations by hand.*

**1** Estimate the area of these shapes.

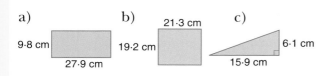

**2** Geoff employs 11 people who earn between £180 and £220 each per week. Estimate

a) Geoff's total wage bill for a week

b) Geoff's total wage bill for a year.

**3** Marcus is touring the world. He changes £80 into the local currency in each country he visits.

Estimate how much he gets in

a) Malaysia
b) Thailand
c) Fiji
d) Australia

**4** One Saturday the attendances at five football grounds in one county are

7927    26 371    9925    12 304    13 849

a) Estimate the average attendance.

b) In a season each team plays 21 home games. Estimate the total season's attendance at each ground.

c) Suggest a reason why the figures in b) may be poor predictions.

**5** For each of the calculations, do the following.

i) Estimate the answer.

ii) Write down the display when you do it on your calculator.

iii) Round your calculator answer to 2 significant figures.

a) $2.91 \times 4.63$

b) $\dfrac{26.93}{8.12}$

c) $79.6 - 31.4 \times 2.3$

d) $\dfrac{3.91}{7.81 - 2.61}$

e) $\sqrt{19}$

f) $\dfrac{6.72 + 0.451}{0.312}$

g) $\sqrt{2.55 \times 43.4}$

h) $\sqrt{23.6 \times 0.41 \times 0.112}$

## 21: Estimation

# Finishing off

**Now that you have finished this chapter you should be able to**

★ approximate to the nearest whole number, nearest 10, nearest 100 and so on

★ estimate prices to the nearest pound

★ estimate total costs

★ make rough calculations

★ round to a given number of decimal places

★ round to a given number of significant figures

Use the questions in the next exercise to check that you understand everything.

## Mixed exercise

**1** a) Write £17.85 to the nearest pound.

b) Write 83 minutes to the nearest 10 minutes.

c) Write 3826 to the nearest 100.

d) Write 63 883 to the nearest 1000.

**2** Tony buys one CD that costs £7.75 and another that costs £11.15.

Find the total cost to the nearest pound.

**3** Paul is given this bill after he has eaten in a cafe.

a) Write down the price of each item to the nearest pound.

b) What is your rough estimate of the bill?

c) Should Paul pay the £6.85?

| | |
|---|---|
| Chicken pie | £3.85 |
| Chips | £1.05 |
| Orange | £0.95 |
| Total | £6.85 |

**4** A school play is held in the hall.

There are 12 rows of chairs with 18 chairs in each row.

a) Estimate the number of chairs in the hall.

Tickets cost £3 each and the hall is nearly full.

b) Estimate the total receipts.

**5** Ebrahim drives 241 miles.

His car does 31 miles per gallon.

a) Estimate the number of gallons used.

The journey takes 4 hours.

b) Estimate his average speed in miles per hour.

## 21: Estimation

**6** Forest Park is open to visitors from May until September. This table shows the number of visitors last year.

| Month | May | June | July | August | September |
|---|---|---|---|---|---|
| Number of visitors | 7780 | 8200 | 12 130 | 13 840 | 6319 |

a) Write each figure to the nearest thousand.

b) Use these figures to estimate the total number of visitors.

c) The entry fee was £2.95. Estimate the total receipts.

**7** For each of these rectangles write down each length to the nearest 10 cm.

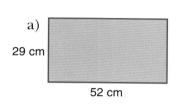
a) 29 cm, 52 cm

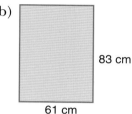
b) 83 cm, 61 cm

i) Use your rounded numbers to estimate the perimeter of the rectangles.

ii) Work out the exact perimeter and compare this result with your estimate.

**8** Use your calculator to find the square root of 28.

Write this value correct to

a) 1 decimal place    b) 2 decimal places    c) 3 decimal places.

**9** Estimate each of these readings to 2 decimal places.

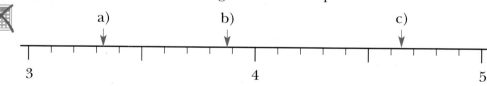

**10** a) Use your calculator to find

$$\frac{(3.11 \times 2.69)}{(9.21 - 7.75)}$$ to 1 decimal place.

b) Jayne says

'3.11 × 2.69 = 8.4 to 1 decimal place.
9.21 − 7.75 = 1.4 to 1 decimal place.
8.4 ÷ 1.4 = 6.0 to 1 decimal place.
So the answer is 6.0.'

What mistake has Jayne made?

**11** Write each of these to the number of significant figures (s.f.) shown.

a) 8397 to 1 s.f.

b) 764 729 to 3 s.f.

c) 14.7528 to 3 s.f.

d) 0.06527 to 2 s.f.

Estimate the number of words in a typical paperback novel.

*Mixed exercise*

# Twenty two
# Equations

## Solving equations (1)

Look at this picture.

The bag contains an unknown number of marbles.

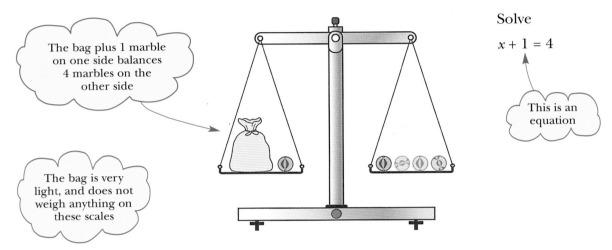

The bag plus 1 marble on one side balances 4 marbles on the other side

The bag is very light, and does not weigh anything on these scales

Solve

$x + 1 = 4$

This is an equation

What happens when you take a marble from both sides?

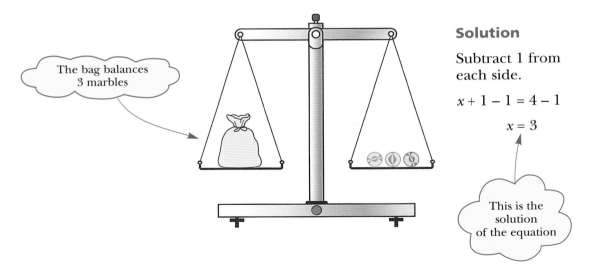

The bag balances 3 marbles

**Solution**

Subtract 1 from each side.

$x + 1 - 1 = 4 - 1$

$x = 3$

This is the solution of the equation

Now check your answers.

There are $3 + 1 = 4$ marbles on the left.

There are 4 marbles on the right.

It balances, so $x = 3$ is correct.

## 22: Equations

**1** Work out how many marbles are in each bag.

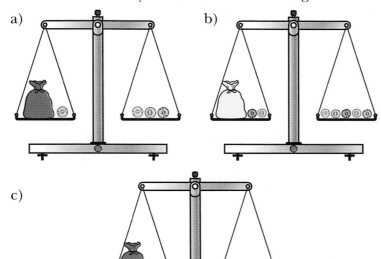

a)

b)

c)

**2** Solve these equations by subtracting the same number from both sides. Check your answers.

a) $x + 1 = 3$

b) $x + 2 = 5$

c) $x + 1 = 11$

d) $x + 5 = 6$

**3** Solve these equations by adding the same number to both sides. Check your answers.

a) $x - 1 = 2$

b) $x - 2 = 6$

c) $x - 5 = 1$

d) $x - 3 = 10$

Find out how the scoring works in snooker, if you don't already know.

A snooker player scores 12 points in a 5-ball break. List all the different ways in which this can happen.

## 22: Equations

# More equations

To solve an equation, you need to get just the unknown on one side.

Always do the same thing to both sides when you solve an equation.

How do you solve $2x = 10$?

This is how Nat does it:

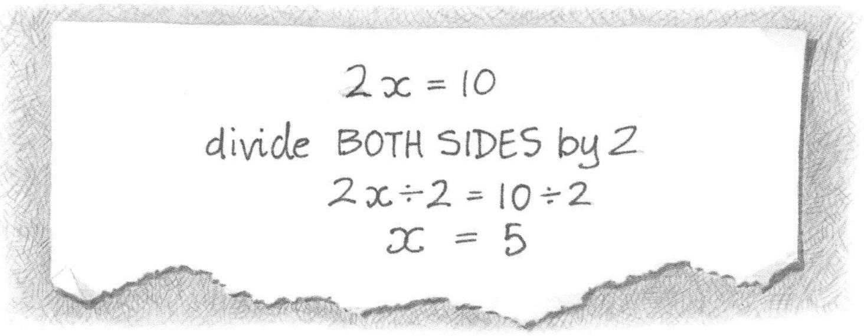

$2x = 10$
divide BOTH SIDES by 2
$2x \div 2 = 10 \div 2$
$x = 5$

Sometimes you have to start by tidying up the unknown terms.

The equation $5x - 3x = 10$ is the same as $2x = 10$. Nat has solved this equation above.

Sometimes it takes more than one step to get just the unknown on one side.

At each step, you need to do the same thing to both sides.

How do you solve $2x + 1 = 9$?

Nat does it like this:

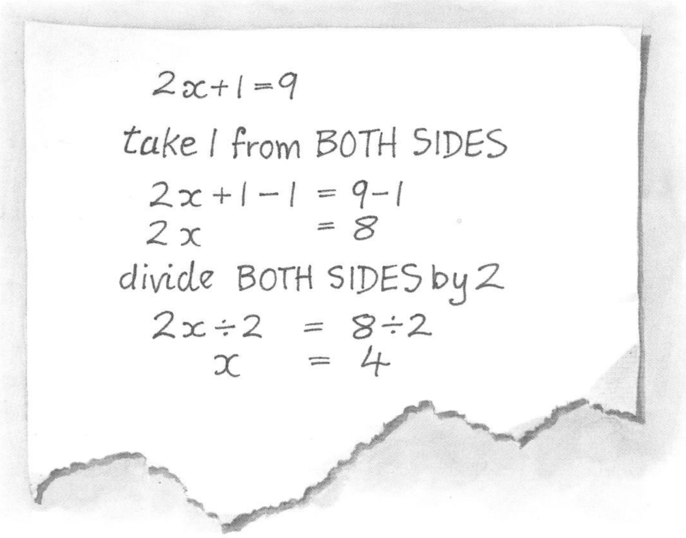

$2x + 1 = 9$
take 1 from BOTH SIDES
$2x + 1 - 1 = 9 - 1$
$2x = 8$
divide BOTH SIDES by 2
$2x \div 2 = 8 \div 2$
$x = 4$

Is Nat correct? Check by putting $x = 4$ in the left side.

You get $2 \times 4 + 1 = 8 + 1 = 9$.

It is the same as the right side, so Nat is correct.

## 22: Equations

**1** Solve these equations by dividing both sides by the same number.

a) $2x = 6$

b) $3x = 9$

c) $2x = 20$

d) $10x = 20$

**2** Solve these equations by doing the same thing to both sides.

a) $x + 6 = 10$

b) $x - 3 = 2$

c) $5x = 15$

d) $7x = 14$

**3** Solve these equations. Check your answers.

a) $2x + 1 = 9$  b) $3x - 5 = 10$

c) $5x - 2 = 8$  d) $6x + 3 = 15$

e) $4x - 1 = 11$  f) $3x + 2 = 20$

g) $5x + 4 = 39$  h) $2x - 1 = 4$

**4** This equilateral triangle has a perimeter of 12 cm.

This can be written as $3x = 12$.

Find the length of each side by solving the equation.

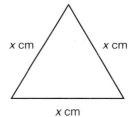

**5** This square has a perimeter of 20 cm.

a) Write this fact as an equation.

b) Find the length of each side by solving the equation.

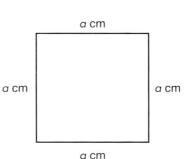

Working with a partner write some statements with missing numbers and challenge your partner to find the missing numbers.

## 22: Equations

# Using equations

Mrs Singh uses this car hire company.
She pays them £140.

*How many days' car hire does she get?*

*How did you get your answer?*

There are several ways to work out the answer to this question.
One of them is to use algebra to set up an **equation** and then **solve** it.

Start by letting $n$ stand for the number of days. Then the cost is given by

$$20 + 15n = 140$$ ← This is the equation

Subtract 20 from both sides $\quad 20 + 15n - 20 = 140 - 20$

(Tidy up) $\quad\quad\quad\quad\quad\quad\quad\quad\quad 15n = 120$

Divide both sides by 15 $\quad\quad\quad 15n \div 15 = 120 \div 15$

Remember to do the same to both sides

$\quad\quad\quad\quad\quad\quad\quad\quad\quad\quad n = 8$

Now we have solved the equation

**So Mrs Singh has the car for 8 days.**

**Check** by substituting $n = 8$ in the left-hand side of the original equation:
$\quad 20 + 15 \times 8 = 140$ ✓ It is right.

**Example** Solve the equation $\quad 5x - 6 = 44$

Add 6 to both sides $\quad\quad\quad\quad\quad 5x = 44 + 6$

Addition is the opposite of subtraction

(Tidy up) $\quad\quad\quad\quad\quad\quad\quad\quad\quad 5x = 50$

Divide both sides by 5 $\quad\quad\quad\quad\quad x = 50 \div 5$

Division is the opposite of multiplication

$\quad\quad\quad\quad\quad\quad\quad\quad\quad\quad\quad x = 10$

*Some of the lines of this solution have been written more concisely than those in the example above. How?*

**Check** by substituting $x = 10$:

$\quad 5 \times 10 - 6 = 44$ ✓ It matches the right-hand side, so it works.

**The solution is $x = 10$.**

## 22: Equations

**1** Solve each of these equations, writing out the steps.

Check your answer by substituting it in the left-hand side of the original equation.

a) $n + 13 = 67$   b) $v - 4 = 11$
c) $3c - 5 = 19$   d) $2f + 15 = 99$
e) $3x + 7 = 37$   f) $2y - 4 = 16$

**2** Solve each of these equations. Check your answers.

a) $2x + 4 = 10$
b) $3y - 5 = 13$
c) $7z + 100 = 135$
d) $5x + 12 = 37$

**3** Solve each of these equations.

Check your answer by substituting it in the left-hand side of the original equation.

a) $4y - 16 = 24$       b) $5x + 12 = 24$       c) $5x + 7 = 17$
d) $3c - 4 = 8.3$        e) $2y - 8 = 12$        f) $10n + 4 = 13$
g) $x + 1 = 2$           h) $n - 42 = 36$        i) $12 + m = 23$
j) $13 + u = 16$         k) $2t + 5 = 25$        l) $5x - 6 = 4$
m) $3x - 2.6 = 4.9$      n) $12 + 2x = 14.6$     o) $7y + 4 = 4$

**4** Make equations for the angles $x$, $y$ and $z$ shown in these diagrams.

Solve your equations to find the angles.

a)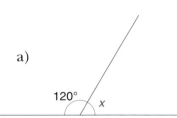

b)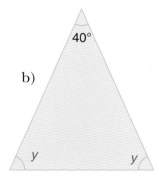

c)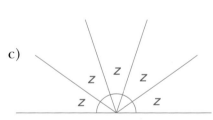

A company pays its employees a mileage rate when they use their own cars for work. This is 40p per mile for the first 100 miles and 30p per mile after 100 miles. Write this as two formulae for the payment £$P$ for a journey of $m$ miles. One formula is for $m \leq 100$ and the other for $m > 100$.

Shomeet is paid £67 for one journey. Write down an equation for $m$ and solve it.

# 22: Equations

## Solving equations (2)

*Can you think of a number which, if you multiply it by 6 then subtract 10, gives you the original number again?*

You can write an equation to help you do this.

Call the number $x$. Then the equation is $6x - 10 = x$

> Multiplying $x$ by 6 and subtracting 10 gives $x$

You can solve this using the same method as before.

| | |
|---|---|
| Start with | $6x - 10 = x$ |
| Add 10 to both sides | $6x = x + 10$ |
| Subtract $x$ from both sides | $6x - x = 10$ |
| (Tidy up) | $5x = 10$ |
| Divide both sides by 5 | $x = 10 \div 5$ |
| (Tidy up) | $x = 2$ |

> Add and subtract to get all the $x$ terms on the left and the numbers on the right

> Multiply or divide once you have separated the $x$ terms from the numbers

**Check** by substituting $x = 2$ in both sides of the original equation.

Left-hand side $= 6 \times 2 - 10 = 2$

Right-hand side $= 2$     ✓     Both sides are equal so $x = 2$ is correct.

> This is called **back-substitution**

**The solution is $x = 2$.**

Sometimes you need to solve equations with brackets in. To do this, you just expand the brackets and continue as before.

### Example

Solve the equation     $2(5 + x) = -2$

### Solution

$2(5 + x) = -2$

| | |
|---|---|
| (Expand the brackets) | $10 + 2x = -2$ |
| Subtract 10 from both sides | $2x = -2 - 10$ |
| (Tidy up) | $2x = -12$ |
| Divide both sides by 2 | $x = (-12) \div (+2)$ |
| | $x = -6$ |

> In expanding the brackets we have just written the left-hand side differently: the right-hand side isn't affected

> Dividing a negative by a positive gives a negative

Check by back-substitution that $x = -6$ satisfies the original equation.

What is the solution of the equation $-5x = -40$?

## 22: Equations

**1** Solve these equations. Write them out carefully and say what you have done at each step. Check your answers by back-substituting.

a) $h + 5 = 13$
b) $n + 3n = 12$
c) $5 + 2x = 22 + x$
d) $5y = 3 - y$
e) $k + 24 = 35 - 10k$
f) $2x - 7 = 19 - 2x$
g) $4x - 7 = 14 + 2x$
h) $8y + 1.2 = 7.2 + 3y$
i) $8x + 20 = 6x + 14$

**2** Solve these equations. Check each answer by back-substituting.

a) $4 - 2x = 2$
b) $4 - 2a = 8$
c) $28 - 3x = 11x$
d) $22 - 6y = 13 + 3y$
e) $d + 20 = 4d$
f) $8 - 2x = 0$
g) $101 - 10k = 1 + 10k$
h) $1 - 10k = 101 + 10k$

**3** Solve these equations. Check each answer by back-substituting.

a) $3(a - 2) = 2a$
b) $2(t - 3) = 8$
c) $4y + 2 = 3(y + 2)$
d) $11(c + 1) = 17 - c$
e) $5(x + 2) = 10 - x$
f) $17 - d = 9(1 - d)$

**4** Equations do not always involve simple numbers, but so long as you know the method for solving them, all you need is a calculator to help. Try these, using a calculator when you need to.

a) $2.4c = 9$
b) $7y + 2 = 11$
c) $1.5 = 2.3 - 3x$
d) $9 + 99p = 333$
e) $4.5x = 13 - 2.5x$
f) $3.8y - 21 = 1.7y$
g) $6.2b + 3 = 5.5 + 4b$
h) $62 - 70a = 44 - 10a$

**5** The formula for converting a temperature $F\,°$Fahrenheit into $C\,°$Celsius is given by

$$C = \frac{5}{9}(F - 32)$$

a) Find the value of $C$ when $F = 68$.
b) Find the value of $F$ when $C = 30$.
c) Copy and complete these steps to make $F$ the subject of the formula.

| Formula | $\frac{5}{9}(F - 32) = C$ |
|---|---|
| × 9 | $5(F - 32) = 9C$ |
| ÷ 5 | $F - 32 =$ |
| + 32 | $F =$ |

### Investigation

One of these equations has no solution.

For the other equation, every value of $x$ is a solution.

(i) $3(x + 6) - 2x = x + 18$

(ii) $5(2 + x) + 3x = 4(2x + 3)$

Which equation is which and how can you tell?

*When an equation is true for every value of $x$, it is called an **identity***

# 22: Equations

## Finishing off

**Now that you have finished this chapter you should be able to**

★ understand the terms *unknown* and *equation*

★ solve equations such as $2x + 1 = 5$

★ solve equations using brackets and fractions

Use the questions in the next exercise to check that you understand everything.

### Mixed exercise

**1** Fill in the missing numbers.

a) $\square \times 5 = 35$    b) $7 \times \square = 63$    c) $\square \times 4 = 36$

d) $\square \times 6 = 66$    e) $2 \times \square = 26$

**2** Find the unknowns in these sums.

a)
```
   4 7 □
 + 1 □ 6
 ─────────
   □ 2 9
```

b)
```
   2 9 □
 - 1 □ 6
 ─────────
   □ 1 7
```

**3** Solve these equations by subtracting the same number from both sides.

a) $x + 4 = 11$    b) $x + 7 = 40$    c) $x + 12 = 24$

d) $x + 6 = 18$    e) $x + 9 = 12$    f) $x + 7 = 21$

**4** Solve these equations by adding the same number to both sides.

a) $x - 3 = 7$    b) $x - 1 = 2$    c) $x - 8 = 30$

d) $x - 13 = 11$    e) $x - 6 = 4$    f) $x - 11 = 2$

**5** Solve these equations by dividing both sides by the same number.

a) $2x = 12$    b) $5x = 25$    c) $9x = 18$

d) $10x = 4$    e) $5x = 55$    f) $3x = 24$

## 22: Equations

**6** Solve these equations by doing the same thing to both sides.

a) $x + 15 = 35$   b) $x - 12 = 17$   c) $7x = 707$
d) $x + 26 = 2$    e) $x - 7 = 9$     f) $3x = 126$

**7** Solve these equations.

a) $3(x + 2) = 9$   b) $4(1 + x) = 20$   c) $6(x - 7) = 12$
d) $5(x + 5) = 30$  e) $4(x - 2) = 4$    f) $7(x + 10) = 70$

**8** Solve these equations.

a) $5x + 3x = 32$    b) $4x = 6 + 2x$        c) $3x + 7 = x + 9$
d) $8(x - 1) = 4x$   e) $19 - x = 6(4 - x)$  f) $x + 2x = 30 - 3x$

**9** Find the values of $x$.

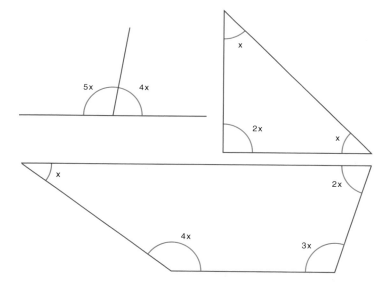

*Mixed exercise*

Look in a cookery book for a recipe for roasting a chicken. You will find that the cooking time depends on the weight of the bird. Write out the instructions in words and as a formula.

Give some examples of how you use the formula.

A chicken takes 2 hours to cook: how big is it?

# Twenty three

# Probability

## Calculating probabilities

Katijah is playing snakes and ladders.

She is on square 21 and throws a single die.

How likely is she to land on square 25, which has a ladder?

To land on square 25, Katijah must throw a 4.

When she throws the die there are six possible outcomes: 1, 2, 3, 4, 5 or 6. They are all equally likely.

*6 because there are six equally likely outcomes*

In everyday English you say

*1 because there is just one way of getting a 4*

**'There is a 1 in 6 chance of getting a 4.'**

In mathematics you say

*P is short for probability*

**'The probability of getting a 4 is $\frac{1}{6}$.'**

You can write this as $P(4) = \frac{1}{6}$

Here are two other examples of calculating probabilities.

Tossing a fair coin:

*1 head*  $P(\text{head}) = \frac{1}{2}$  *2 sides*

Choosing a card at random from a pack:

$P(\text{ace}) = \frac{4}{52} = \frac{1}{13}$

*4 aces*

*52 cards*

 *The probability of winning a game is $\frac{3}{5}$. What is the probability of losing it?*

Probability is a number on a scale between 0 (impossible) and 1 (certain).

It can be written as a fraction, a decimal or a percentage, so a probability of $\frac{1}{2}$ can also be written as 0.5 or 50%.

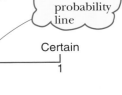

*This is a probability line*

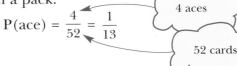

Impossible — 0 — Unlikely — Evens or 50—50 — $\frac{1}{2}$ — Likely — Certain — 1

*Katijah throws a 4.*

# 23: Probability

**1** You throw a die. What is the probability that you get

a) a 6?

b) an even number?

c) an odd number?

**2** A fish is caught at random from this tank. What is the probability that it is

a) red?

b) blue?

c) green?

**3** What is the probability that the next light that fails on this Christmas tree will be

a) yellow?

b) pink?

c) blue?

d) red?

**4** The diagram shows a game of draughts.

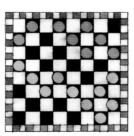

At random red picks one of the pieces that can move.

What is the probability that it can be taken, when he moves it?

**5** 2 out of the 32 teeth in Arthur's mouth are bad; the rest are good.

He goes to an incompetent dentist who pulls out a tooth at random.

What is the probability that the tooth pulled out is

a) bad?

b) good?

**6** Amir has three flavours of ice cream to choose from and two different types of toppings.

| Ice cream | Toppings |
|---|---|
| Chocolate chip (C) | Nuts (N) |
| Toffee fudge (T) | Chocolate flakes (F) |
| Raspberry (R) | |

a) List all the possible ways he can choose one ice cream and one topping.

One way has been done for you.

You may need more lines.

| Ice cream | Toppings |
|---|---|
| C | N |
| | |
| | |

b) Amir chooses one of these at random.

What is the probability he chooses raspberry?

### 23: Probability

# Working with probabilities

At a fête Tim chooses a number on the Wheel of Fortune.

If his number ends up at the bottom he wins a prize.

He chooses number 7.

What is the probability that Tim wins a prize?

> **There are 10 equally likely numbers. Tim chooses 1 of them.**
> **The probability that he wins is $\frac{1}{10}$.**

You can also work out the probability that Tim does not win.

There are 9 non-winning numbers, so the probability that he does not win is $\frac{9}{10}$.

You will notice that

$$P(\text{Tim wins}) + P(\text{Tim does not win}) = 1$$
$$\frac{1}{10} + \frac{9}{10} = 1$$

If you add the probabilities of all the possible outcomes the answer always comes to 1.

Tim plays 20 times. How many prizes does he expect to win?

Tim expects to win 1 in every 10 times.

He plays 20 times so he expects to win

$$\frac{1}{10} \times 20 = 2 \text{ times}$$

That does not mean he always wins exactly twice every 20 times.

Sometimes he wins more than that (and we say he is lucky), sometimes fewer.

*Fairground games are not always fair.*

*What could be done to make the Wheel of Fortune unfair?*

# 23: Probability

**1** Bella is playing Scrabble.

These 20 tiles are in the bag and she chooses one without looking.

What is the probability that she chooses:

a) the letter A?

b) a tile which scores 1 point?

c) a tile which scores more than 1 point?

d) a tile which scores 5 points?

e) a tile which is not an N?

**2** In a raffle 400 tickets have been sold. Alex has bought 5 of them.

The tickets are put in a hat and one is selected for first prize.

What is the probability that

a) Alex wins first prize?

b) Alex does not win first prize?

In fact Ron wins first prize. His ticket is not put back.

Another ticket is now drawn for second prize.

c) What is the probability that Alex wins second prize?

**3** In a gambling game you buy a scratch ticket for 50p.

Out of every 40 tickets, one is marked WINNER – COLLECT £15, the others LOSER – TRY AGAIN.

a) What is the probability that a ticket is a winner?

b) What is the probability that a ticket is a loser?

Hamish buys 2 tickets a week for 60 weeks.

c) How many tickets does Hamish buy, and how much do they cost?

d) How many of Hamish's tickets can he expect to be winners?

e) How much money can he expect to win?

f) What is his profit or loss? Explain your answer.

g) Do you think that if Hamish goes on playing he will end up making a profit?

---

Make a circular spinner like the Wheel of Fortune on the opposite page but design it to be unfair.
Test it and see if it really is unfair.

## 23: Probability

# Estimating probabilities

In many situations you cannot calculate probability exactly, but you can sometimes estimate it using past data.

*What is the probability that you will be ill next Tuesday?*

When James was asked this question he looked at his diary for last year. He found 4 days when he was ill.

| 19 and 20 January | 1 March | 10 August |
| Heavy cold | Cough | Sunstroke |

There is no exact answer to this question but James's data does allow him to make an estimate:

*This is called the **relative frequency***

$$\text{Probability of being ill one day} = \frac{\text{number of days ill}}{\text{total number of days}} = \frac{4}{365} = 0.01096$$

Because this is only an estimate it is better to give the answer rounded: about 0.01 (or 1%).

*You can only estimate probability well if all the possible outcomes are equally likely.*

- *Are you more likely to be ill on some days than others?*
- *Can you do anything to change your probability of being ill?*

  *Decide which of the following will decrease the probability of your being ill, which will increase it and which will have no effect.*

| Have plenty of sleep | Eat fruit and vegetables |
| Eat junk food | Smoke |
| Take exercise | Drink heavily |
| Miss breakfast | Laugh and have fun |

Although estimating probability may not be exact, it is very important for some people.

*Why do the following people need to estimate probabilities?*
- *Insurance company managers*
- *Bookmakers (bookies)*

# 23: Probability

**1** June drives across a major junction on her way to work every day.

She notes that, out of 20 days, the traffic lights are green on 6 days and red on the rest.

Use relative frequency to estimate the probability that next time she will find the lights
a) green   b) red.

**2** In a survey of 7000 children, 1004 were found to be left handed.

A baby is about to be born.

Estimate the probability that the baby is
a) left handed   b) not left handed.

Give your answers as fractions with the numbers rounded sensibly.

**3** Roy plants a packet of 40 seeds. They all grow into plants.

10 of these plants have pink flowers, the rest white.

Next year Roy plants 5 similar packets.
a) Estimate the probability that a seed chosen at random from these will grow into a pink-flowering plant.
b) About how many of the seeds would Roy expect to grow into plants with pink flowers?
c) Estimate the probability that a seed chosen at random will grow into a white-flowering plant.
d) About how many of the seeds would Roy expect to grow into plants with white flowers?

**4** Last year there was no rain in Avonford on 30 out of 90 days in January, February and March.
a) Estimate the probability that it will rain on 6 March next year.
b) Estimate the probability that it will not rain on 6 March next year.
c) Do you trust your answers?

**5** Ela keeps a record of how many letters she receives each day (except Sunday) for 10 weeks and then draws this line chart.

Use Ela's data to estimate the probability that on Thursday next week she will receive
a) 1 letter   b) 3 letters
c) 3, 4 or 5 letters   d) 12 letters.
Do you believe your answer to part d)?

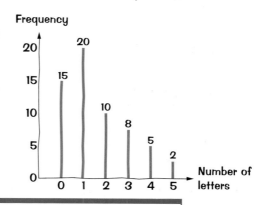

Solitaire is a well-known card game that you can play on many computers.
Play enough games for it to come out completely several times and so estimate the probability that this happens.

# 23: Probability

## Finishing off

**Now that you have finished this chapter you should be able to**

★ when possible calculate the probability of an outcome using theory

★ calculate the probability of an outcome not happening

★ use probability to estimate the number of times an outcome will occur

★ use past data to estimate probability

Use the questions in the next exercise to check that you understand everything.

### Mixed exercise

**1** Draw a probability line and mark on it points A, B, C, D, E to show the probabilities that

A  A cat will catch a mouse somewhere today
B  You will live to the age of 145 years
C  When you toss a coin it will come down heads
D  A football match, selected at random, will end in a draw
E  You will see the sun on Friday next week.

**2** A playing card is chosen at random from an ordinary pack of 52.

What is the probability that it is
a) a diamond?
b) a red card?
c) a queen?
d) the queen of diamonds?

**3** There are 28 dominoes in a standard set. This one is 2-4.

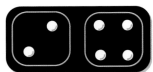

Either list or draw the full set of dominoes.

You choose a domino at random from the complete set. What is the probability that

a) it has a 6 on it?
b) it is a double?
c) the two numbers on it are different?

**4** Two children play a game with dice. Tan throws a red die and then Lim throws a green die.

The one with the higher score is the winner.
If the two dice have the same score, the game is a draw.

In one game Tan throws a 3. What is the probability that

a) Lim wins?
b) Tan wins?
c) The game is drawn?

Add your three answers together.
d) What do you notice?
In the next game Tan throws a 6.
e) What is the probability of Lim winning?

# 23: Probability

**5** Kevin is the supervisor on a production line making mugs.
Yesterday he noticed that 15 out of 500 mugs were chipped and so were rejected.
a) Estimate the probability that any mug made today will chipped.
b) Estimate the probability that any mug made today will not be chipped.
c) The company makes 50 000 mugs one day. How many of these would Kevin expect to be chipped?

**6** The police do a survey of 100 motorists going under a particular bridge over a motorway, using a speed camera.

| Speed (mph) | under 60 | 60–69 | 70–79 | 80–89 | 90+ |
|---|---|---|---|---|---|
| Number of cars | 8 | 28 | 34 | 26 | 4 |

They use these data to estimate the probability that a car chosen at random will be doing a certain speed.

Find the probability that a randomly selected car is travelling at
a) under 60 mph
b) under 80 mph
c) 80 mph or over.

They decide to prosecute all those doing 90 mph or over.
During one afternoon 12 000 cars go by.
d) Estimate how many of them will be prosecuted.

**7** There are 5 coloured balls in a bag. They are yellow, light blue, dark blue, green and red. A ball is taken from the bag at random. Find the probability that
a) it is red
b) it is not red
c) it is a shade of blue
d) it is not blue.

**8** Pat has a fair five-sided spinner numbered 1 to 5.
She spins it twice and adds the numbers to get her score.
a) Copy and complete this table to show her possible scores.

|  |  | Number on second spin | | | | |
|---|---|---|---|---|---|---|
|  |  | 1 | 2 | 3 | 4 | 5 |
| Number on first spin | 1 | | 3 | | | |
| | 2 | | | | | |
| | 3 | | | | | |
| | 4 | | | | 8 | |
| | 5 | | | | | |

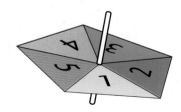

b) Find the probability that her score is
 (i) 5
 (ii) 8 or more
 (iii) 1.

# Twenty four

# Using indices

### About this chapter

★ This chapter builds on and revises your number work.

★ Make sure that you really understand it.

## Number patterns

Colin runs a small business.

He sells cans of home-made lemonade in packs of 8.

*How many cans are there in 2 packs?*

*How many cans in 3 packs?* $1 \times 8$    $2 \times 8$

The **multiples** of 8 are 8, 16, 24, ...

Colin packs 8 cans like this:    $3 \times 8$

4 and 2 are **factors** of 8.

2 cans along here

4 cans along here

*How else can he pack 8 cans using one layer?*
*What are the other factors of 8?*

*How can 7 cans be packed in a rectangular box?*

The only factors of 7 are 1 and 7. We say 7 is a **prime** number.

*How can 9 cans be packed in a rectangular box?*

9 is a **square number** because $3 \times 3 = 9$.

The **square root** of 9 is 3.

$3 \times 3$ can be written as $3^2$ (3 squared). $\sqrt{\phantom{x}}$ means square root. So $\sqrt{9} = 3$

$2 \times 2 \times 2$ can be written as $2^3$ (2 cubed)

Now Colin packs 8 cans like this

8 is a **cube number** because $2 \times 2 \times 2 = 8$.

*What is the cube root of 8?*
*What is the next cube number after 8?*
*What is its cube root?*

*How can you use your calculator to find $\sqrt{9801}$ and $25^3$?*

All calculators have a square root key $\boxed{\sqrt{\phantom{x}}}$.
Most use the symbols $\boxed{\wedge}$ to indicate power.

## 24: Using indices

**1** List all the factors of

a) 12   b) 16   c) 15   d) 28
e) 25   f) 30   g) 48   h) 60.

**2** List the common factors of

a) 16 and 24   b) 24 and 36   c) 16 and 36.

**3** Find the value of

a) $7^2$   b) $\sqrt{64}$   c) $20^2$   d) $5^3$
e) $\sqrt{100}$   f) $30^2$   g) $3^3$   h) $\sqrt{144}$.

**4** The first three primes are 2, 3 and 5. Write down the next five primes.

**5** A pack contains 6 bags of crisps. How many bags of crisps are there in

a) 2 packs?   b) 3 packs?   c) 5 packs?

**6** 20 cartons can be packed like this.

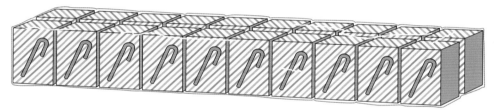

a) List all the other ways of packing 20 cartons in a rectangular box.

b) Now list all the factors of 20.

**7** Two square cakes are made in tins which are 30 cm by 30 cm.

The chocolate cake is cut into 5 cm by 5 cm pieces.

a) How many pieces is the chocolate cake divided into?

b) Now the lemon cake is cut into 3 cm by 3 cm pieces.

How many pieces does the lemon cake make?

 Check your answers to question 3 using the power and square root keys on your calculator.

# 24: Using indices

## Prime factorisation

 *Which of these are prime numbers?*

    15, 17, 19, 21

If a number is not itself prime then it can be written as the product of primes.

    e.g.  $20 = 2 \times 2 \times 5$    *Each of these is a prime*

    and  $84 = 2 \times 2 \times 3 \times 7$    *Each of these is a prime*

This is called **prime factorisation** (or **prime factor decomposition**).

Ian and Lin are both finding the prime factorisation of 20.

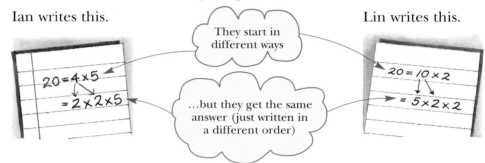

Ian writes this.    *They start in different ways*    Lin writes this.

$20 = 4 \times 5$      $20 = 10 \times 2$
$= 2 \times 2 \times 5$      $= 5 \times 2 \times 2$

*...but they get the same answer (just written in a different order)*

Remember that you must go on factorising until all the numbers are primes. Sometimes it may take several lines of working.

 *What is the prime factorisation of 360?*

## Highest common factor (HCF)

You can find the HCF of 12 and 20 like this:

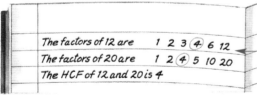

The factors of 12 are    1   2   3   ④   6   12
The factors of 20 are    1   2   ④   5   10   20
The HCF of 12 and 20 is 4

*The common factors are 1, 2, 4*

*4 is the highest factor in the list*

The HCF of 12 and 20 is 4.

## Lowest common multiple (LCM)

You can find the LCM of 9 and 6 like this:

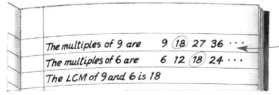

The multiples of 9 are    9   ⑱   27   36 ...
The multiples of 6 are    6   12   ⑱   24 ...
The LCM of 9 and 6 is 18

*The common multiples are 18, 36 ...*

*18 is the lowest multiple in the list*

# 24: Using indices

**1** Find the prime factorisation of each of these numbers.
  a) 14    b) 15    c) 28    d) 36
  e) 30    f) 27    g) 90    h) 126
  i) 150   j) 210   k) 539   l) 1540

**2** Find the HCF of each of these.
  a) 6, 4      b) 6, 15      c) 18, 12     d) 12, 4
  e) 10, 25    f) 3, 8       g) 18, 45     h) 14, 10
  i) 21, 49    j) 22, 33     k) 63, 36     l) 56, 126
  m) 12, 24, 54    n) 25, 35, 75    o) 56, 24, 32    p) 60, 80, 100

**3** Find the LCM of each of these.
  a) 10, 4     b) 5, 6      c) 4, 8      d) 12, 9
  e) 6, 10     f) 3, 7      g) 27, 18    h) 16, 8
  i) 14, 35    j) 8, 20     k) 20, 30    l) 45, 10
  m) 2, 3, 4    n) 5, 15, 2    o) 9, 12, 8    p) 6, 10, 15

**4** a) Write down the LCM of 5 and 7.
  b) Write $\frac{2}{5}$ as a fraction with denominator 35.
  c) Write $\frac{3}{7}$ as a fraction with denominator 35.
  d) Which is larger, $\frac{2}{5}$ or $\frac{3}{7}$?

**5** Look at these gear wheels.

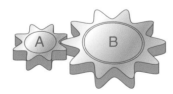

  a) A completes 10 turns. How many turns does B complete?
  b) What is the least number of turns that A can complete so that B also completes an exact number of turns?
  c) B completes 30 turns. On how many occasions will both A and B have been back in their starting position at the same time?

Roger has 4 hens.

Ingrid lays every second day. Ferdie lays every third day. Chookle lays every fourth day. Mumtie lays every fifth day. They all lay an egg on 1 January.

What is the date when they next all lay an egg?

On how many days in the year do they all lay an egg?

# 24: Using indices

## Index notation

$10 \times 10$ can be written $10^2$ (10 squared). *The 2 is an* **index** *(or power)*

$10 \times 10 \times 10$ can be written $10^3$ (10 cubed). *The 3 is an index (or power)*

What is the value of $10^2$?

What is the value of $10^3$?

What do you think $10^4$ means?

What is the value of $10^4$?

Index notation is used so we don't have to write out

$$10 \times 10 \times 10 \times 10 \times 10 \times 10$$

How would you write this in index notation?

What is its value?

## Standard form

Copy and complete this table.

| 4 | 4 | 4 |
|---|---|---|
| $4 \times 10$ | $4 \times 10$ | 40 |
| $4 \times 10 \times 10$ | $4 \times 10^2$ | 400 |
| $4 \times 10 \times 10 \times 10$ | | |
| $4 \times 10 \times 10 \times 10 \times 10$ | | |
| $4 \times 10 \times 10 \times 10 \times 10 \times 10$ | $4 \times 10^5$ | 400 000 |

*These are in* **standard form**. *The leading number (4) is between 1 and 10*

Here, the number is written in standard form and then worked out.

$$6 \times 10^3 = 6 \times 10 \times 10 \times 10 = 6000$$

*The leading number is between 1 and 10*

Here the number has been changed into standard form.

$$300\,000 = 3 \times 10 \times 10 \times 10 \times 10 \times 10 = 3 \times 10^5$$

*The leading number is between 1 and 10*

## 24: Using indices

**1** Work out the value of
   a) $4^3$    b) $2^5$    c) $3^4$    d) $2^6$
   e) $6^3$    f) $6^4$    g) $1.5^2$    h) $2.5^3$.

**2** In this question, the number is given in standard form. Work out the value of the number.
   a) $5 \times 10^2$    b) $7 \times 10^4$    c) $8 \times 10^3$    d) $2 \times 10^5$
   e) $6.5 \times 10^2$    f) $5.8 \times 10^3$    g) $2.4 \times 10^6$    h) $8.75 \times 10^4$.

**3** In this question, write the number in standard form.
   a) 7000    b) 300    c) 90 000    d) 6000
   e) 8600    f) 57 000    g) 750    h) 290 000.

**4** Using index notation, ten can be written as $10^1$.
   Write the following in index notation.
   a) A hundred    b) A thousand    c) A million

**5** Most calculators have either an $\boxed{x^y}$ key or a $\boxed{\wedge}$ key.
   Do question 1 again using this key.

### Investigation

Look at this list of powers of 6.

$6^1 = 6$

$6^2 = 36$

$6^3 = 216$

$6^4 = 1296$

$6^5 = 7776$

$6^6 = 46\,656$

What do you notice about the last digit of these numbers?
Do any other numbers apart from 6 make a pattern like this?
Make a list like the one above for all the numbers less than 10 and see if you can spot any more patterns.

---

Look in a book containing lots of numbers, such as *The Guinness Book of Records*.

Find 4 very large numbers, and write them in standard form.

# 24: Using indices

## Rules of indices

Look at this table.

| | | |
|---|---|---|
| ÷2 ↘ | $16 = 2 \times 2 \times 2 \times 2$ | $2^4$ | ↘ ÷2 |
| ÷2 ↘ | $8 = 2 \times 2 \times 2$ | $2^3$ | ↘ ÷2 |
| ÷2 ↘ | $4 = 2 \times 2$ | $2^2$ | ↘ ÷2 |
| ÷2 ↘ | $2 = 2$ | $2^1$ | ↘ ÷2 |
| ÷2 ↘ | $1 = 1$ | $2^0$ | ↘ ÷2 |
| ÷2 ↘ | $\frac{1}{2} = \frac{1}{2}$ | $2^{-1}$ | ↘ ÷2 |
| ÷2 ↘ | $\frac{1}{4} = \frac{1}{2 \times 2} = \frac{1}{2^2}$ | $2^{-2}$ | ↘ ÷2 |
| ÷2 ↘ | $\frac{1}{8} = \frac{1}{2 \times 2 \times 2} = \frac{1}{2^3}$ | $2^{-3}$ | ↘ ÷2 |

You can see the meaning of 2 to the power zero and to a negative power. Notice that

$$2^0 = 1 \quad \text{and} \quad 2^{-3} = \frac{1}{2^3}$$

*What are the values of $10^0$ and $10^{-2}$?*

When you write 16 as $2^4$ then $2^4$ is called **index form**.

When you multiply two numbers given in index form you add the powers.

$$2^4 \times 2^3 = 2^{4+3} = 2^7$$

$2^4 \times 2^3 = (2\times2\times2\times2) \times (2\times2\times2) = 2^7$

*What is $2^2 \times 2^3 \times 2^4$?*

When you divide one number by another you subtract the powers.

$$2^5 \div 2^3 = 2^{5-3} = 2^2$$

$\frac{2^5}{2^3} = \frac{2 \times 2 \times 2 \times 2 \times 2}{2 \times 2 \times 2} = 2 \times 2 = 2^2$

*What is $2^4 \div 2$? (Hint: 2 is the same as $2^1$).*

*What is $(2^3)^4$ in index form?*

You can write the rules on this page as general laws.

$$a^0 = 1 \qquad\qquad a^{-n} = \frac{1}{a^n}$$

$$a^m \times a^n = a^{m+n} \qquad\qquad a^m \div a^n = a^{m-n}$$

## 24: Using indices

**1** Write each value as a fraction. For example $2^{-2} = \frac{1}{4}$

a) $4^{-2}$   b) $10^{-3}$   c) $5^{-2}$   d) $8^{-1}$   e) $3^{-3}$   f) $6^{-2}$

g) $10^{-2}$   h) $3^{-4}$   i) $4^{-1}$   j) $2^{-4}$   k) $6^{-3}$   l) $10^{-4}$

**2** Work out the value of

a) $7^2$   b) $3^{-2}$   c) $10^{-1}$   d) $4^3$   e) $9^{-2}$   f) $6^0$

g) $2^5$   h) $5^{-3}$   i) $10^6$   j) $9^1$   k) $5^{-1}$   l) $7^0$

m) $6^3$   n) $8^{-2}$   o) $4^1$   p) $5^4$.

**3** Work these out giving your answer in index form.

For example $4^3 \times 4^2 = 4^5$

a) $5^2 \times 5^4$   b) $2^6 \times 2^3 \times 2^2$   c) $6 \times 6^3$   d) $10^6 \div 10^3$

e) $2^8 \div 2$   f) $(10^2)^3$   g) $(3^4)^2$   h) $4^5 \times 4^{-3}$

i) $10^3 \div 10^{-1}$   j) $(5^{-1})^2$   k) $3^4 \times 3^{-2} \times 3$   l) $2^{-1} \div 2^{-2}$

m) $\dfrac{4^9}{4^3 \times 4^4}$   n) $\dfrac{3^4 \times 3^2}{3^8}$   o) $\dfrac{(2^3)^2 \times 2}{2^7}$   p) $\dfrac{10^4 \times 10^6 \times 10}{10^5 \times 10^3}$

**Investigation**

**1** Use your calculator to work out

a) $7 \times 10$

b) $7 \times 10 \times 10$

c) $7 \times 10 \times 10 \times 10$

. . . and so on until you get an answer in standard form.

How many digits can your calculator display?

**2** Use your calculator to work out

a) $7 \div 10$

b) $7 \div 10 \div 10$

c) $7 \div 10 \div 10 \div 10$

. . . until the form of answer changes.

For which calculation does the form of your answer change?

What does the display show?

What do you think this display means?

## 24: Using indices

# Calculators

Without using a calculator write down the value of $4 \times 10^6$.

Do the same for $4 \times 10^{11}$.

Now work out $4 \times 10^{11}$ on your calculator, by pressing

$$4 \times 10 \times 10 \times 10 \times 10 \times 10 \times 10 \times 10 \times 10 \times 10 \times 10 \times 10$$

What do you get?

Instead of showing 400 000 000 000 your calculator probably shows something like one of these:

> Your calculator uses this to mean $4 \times 10^{11}$
> It does not mean $4^{11}$

Your calculator can only show a limited number of digits (probably 8 or 10).

It does not have enough space to display 400 000 000 000.

William is a tax inspector.

He wants to know the total number of hours of work people in Britain do in a year. He works it out like this:

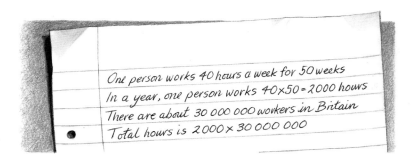

One person works 40 hours a week for 50 weeks
In a year, one person works 40×50 = 2000 hours
There are about 30 000 000 workers in Britain
Total hours is 2000 × 30 000 000

He does this on a calculator and gets

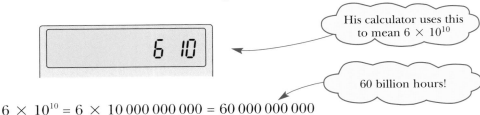

> His calculator uses this to mean $6 \times 10^{10}$

> 60 billion hours!

$6 \times 10^{10} = 6 \times 10\,000\,000\,000 = 60\,000\,000\,000$

*How can you make your calculator write 6000 as* ⎡ 6 3 ⎤ *?*

*What happens when you do* ⎡ 6 3 ⎤ × ⎡ 5 2 ⎤ *?*

Most calculators have an ⎡EXP⎤ key.

$6 \times 10^3 \times 5 \times 10^2 = 3 \times 10^6$, so you should get ⎡ 3 6 ⎤

## 24: Using indices

**1** Work out the value of the numbers displayed.

a)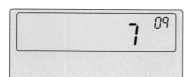

b)

c)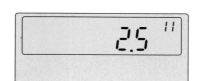

d)

**2** Using standard form, work out the value of

a) $3\,000\,000 \times 200\,000$

b) $80\,000\,000 \times 50\,000$

c) $20\,000\,000 \times 400\,000$

d) $5000 \times 6\,000\,000$

e) $25\,000 \times 600\,000$

f) $5\,500\,000 \times 400\,000$.

**3** The population of Russia was estimated as 146 000 000.

a) Write 146 000 000 in standard form.

The electricty consumption in a year for Russia was $7.02 \times 10^{11}$ kilowatt hours.

b) Calculate the average consumption per person. Give your answer to a suitable degree of accuracy.

### Investigation

Work out the powers of 3 starting with $3^2$.

Stop when your calculator goes into standard form.

Work out the sum of the digits in each answer.

Write your results in a table like this.

| Power of 3 | Sum of the digits |
|---|---|
| $3^2 = 3 \times 3 = 9$ | $9 = 9$ |
| $3^3 = 3 \times 3 \times 3 = 27$ | $2 + 7 = 9$ |

*To find the sum of the digits add them together. For 6561, $6 + 5 + 6 + 1 = 18$*

a) Look at the answers in the first column.

What do you notice about the last digits?

b) Look at the answers in the second column.

What do you notice?

---

The Earth is believed to be 4500 million years old. How many seconds is this?

# 24: Using indices

# Brackets

What is $1 + 2 \times 3$?

Did you add or multiply first?

John does it like this:

Lisa does it like this:

Why is 7 *right* and 9 *wrong*?

Using brackets avoids confusion.

You always work out brackets first.

$(1 + 2) \times 3$  $\qquad\qquad\qquad$  $1 + (2 \times 3)$

This means work out $1 + 2$ first and then times by 3, the way John does

This means work out $2 \times 3$ first and then add it to 1, the way Lisa does

You should always work out operations in this order:

work out brackets → work out the powers → divide and multiply → add and subtract

This is how a scientific calculator works out $1 + 2 \times 3$:

no brackets → no powers → multiply $2 \times 3 = 6$ → add $1 + 6 = 7$

Using brackets, key in $1 + (2 \times 3)$ and check that you get 7.

Using brackets, key in $(1 + 2) \times 3$ and check that you get 9.

What is $(20 - 8) \div 2$?

What is $20 - (8 \div 2)$?

What is $20 - 8 \div 2$?

# 24: Using indices

**1** Work out the value of

a) $4 + (5 \times 3)$  b) $(4 + 5) \times 3$  c) $(11 - 2) \times 4$

d) $11 - (2 \times 4)$  e) $2 \times (7 - 3)$  f) $(2 \times 7) - 3$.

**2** Work out the value of

a) $5 + 3 \times 4$  b) $8 \times 3 - 1$  c) $12 \div 2 + 4$  d) $16 - 2 \times 3$.

**3** Work out the value of

a) $(3 + 2) \times (4 - 1)$  b) $3 + (2 \times 4) - 1$

c) $(5 \times 2) + (3 \times 4)$  d) $(20 \div 4) - 2$.

**4** Complete the following by using $+$, $-$, $\times$ or $\div$ in each box.

You can use brackets as well to show which operation is done first.

For example: $3 \square 4 \square 2 = 14$.

One solution is $(3 + 4) \times 2 = 14$.

Another is $(3 \times 4) + 2 = 14$.  *These brackets are not essential here*

a) $4 \square 5 \square 3 = 6$  b) $7 \square 5 \square 3 = 4$  c) $2 \square 6 \square 3 = 15$

d) $5 \square 2 \square 2 = 14$  e) $3 \square 5 \square 1 = 12$  f) $3 \square 2 \square 2 = 8$

g) $1 \square 6 \square 2 = 4$  h) $6 \square 6 \square 4 = 9$  i) $8 \square 2 \square 3 = 7$

j) $10 \square 8 \square 2 = 9$

## Investigation

Look at these calculations.

$4 + 4 + \frac{4}{4} = 9$     $4 - \frac{(4+4)}{4} = 2$

($\frac{4}{4} = 1$)   ($\frac{4+4}{4} = \frac{8}{4} = 2$)

$(4 \times 4) + (4 \times 4) = 32$

($4 \times 4 = 16$)

Each calculation uses exactly four 4s.

Using exactly four 4s invent 10 calculations of your own which give different answers.

**PRICES**
Fish . . . . . . . . . . . £1·60
Burger . . . . . . . . . £1·50
Veggie Burger . . . £1·50
Chips . . . . . . . . . . ·90

Buy a meal for your family or a group of friends. Write down the total cost using the signs $\times$, $+$, and ().

Write down the cost per person, if it is shared equally. You will need to use the $\div$ sign now as well.

# 24: Using indices

## Finishing off

**Now that you have finished this chapter you should be able to**

★ recognise primes and work out prime factorisations

★ recognise factors and find the HCF of two or more numbers

★ recognise multiples and find the LCM of two or more numbers

★ write very large and very small numbers using powers of 10

★ write a large number in standard form

★ do calculations with numbers in standard form

Use the questions in the next exercise to check that you understand everything.

## Mixed exercise

**1** Write down all the factors of

a) 24  b) 21  c) 13  d) 36.

**2** Which of 24, 21, 13 and 36 are a) prime? b) square?

*Hint: use your results from the last question*

**3** Work out

a) $8^2$  b) $\sqrt{81}$  c) $6^3$  d) $3^5$.

**4** Work out these numbers:

a) The speed of light is $3 \times 10^8$ metres per second.

b) The speed of sound is $3.3 \times 10^2$ metres per second.

**5** Write these numbers in standard form.

a) The distance round the Equator is about forty thousand km.

b) The Sun is ninety three million miles from Earth.

**6** Work out these calculations giving your answer in standard form.

a) 200 000 × 4 000 000   b) 350 000 × 70 000

**7** Work out the value on this calculator display.

```
  7  12
```

Find out the world record for 100 m, 1 mile and a marathon. In each case what is the average speed of the runner in metres per second?

Time a friend over one of these distances (or get a friend to time you). What is the average speed?

**8** Work out

a) (8 + 3) × 2   b) 8 + (3 × 2)

c) 8 + 3 × 2   d) (6 × 3) + (5 × 2).

## 24: Using indices

**Mixed exercise**

**9** Kelly goes to a disco every fourth Saturday, and ten-pin bowling every third Sunday. She does both during the weekend of 1 and 2 March.

a) How many weeks pass before the two outings again occur in the same weekend?
b) What will be the dates?
c) How many more times before the year end will Kelly have the two outings in the same weekend?

**10** Work out the prime factorisation of
a) 18   b) 48   c) 100   d) 120.

**11** Write down the highest common factor of
a) 15, 20       b) 16, 36
c) 80, 30       d) 24, 36, 60.

**12** Write down the lowest common multiple of
a) 4, 6         b) 24, 8
c) 20, 50       d) 6, 10, 18.

**13** Work out
a) $8^2$        b) $10^3$       c) $6^{-1}$
d) $4^0$        e) $2^7$        f) $\sqrt{36}$
g) $12^2$       h) $\sqrt[3]{8}$    i) $3^{-2}$
j) $6^4$        k) $5^0$        l) $4^{-3}$
m) $\sqrt{225}$ n) $7^{-2}$     o) $\sqrt[3]{64}$
p) $2^{10}$.

**14** Work these out, giving your answer in index form.

For example $3^2 \times 3^5 = 3^7$

a) $2^5 \times 2^3$     b) $7^4 \div 7$
c) $(5^3)^2$            d) $4^2 \times 4^6$
e) $6^5 \div 6^2$       f) $(4^2)^2$
g) $3^5 \times 3 \times 3^{-2}$   h) $\sqrt{3^2}$

**15** The numbers in this question are in standard form.
Write them out in full.

a) The radius of the Earth is $6.4 \times 10^6$ metres
b) A capillary tube has radius $2 \times 10^{-4}$ metres
c) The wavelength of mercury green light is $5.4 \times 10^{-7}$ metres
d) The density of mercury is $1.36 \times 10^4$ kg/m$^3$

**16** Write these numbers in standard form.

a) A train has a mass of **200 000** kg.
b) The thickness of a piece of cardboard is **0.0015** metres.
c) Thorium-230 has a half life of **83 000** years.
d) The linear expansivity of aluminium is **0.000 026** per degree Kelvin.

**17** Work out the value of these calculations, and give your answer in standard form.

a) $(5 \times 10^4) + (8 \times 10^5)$
b) $(3.1 \times 10^{-2}) - (7 \times 10^{-3})$
c) $(4 \times 10^8) \times (9.7 \times 10^{13})$
d) $(3.6 \times 10^{12}) \div (9 \times 10^4)$
e) $(3.2 \times 10^{14}) \times (7.5 \times 10^{-9})$
f) $(4.9 \times 10^{11}) \div (2.8 \times 10^{-5})$
g) $(4.5 \times 10^7)^2$
h) $\sqrt{1.6 \times 10^{13}}$

# Twenty five
# Measuring and drawing

**Before you start this chapter you should be able to**

- ★ use a scale on a map or scale drawing
- ★ use the eight compass directions
- ★ use a protractor or angle measurer to measure angles in degrees.

## Triangles

Gary is an architect.

He needs to draw this triangle accurately.

This is how he does it.
He draws a line exactly 8 cm long.

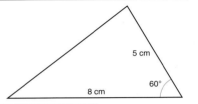

He puts his protractor so that the centre is on the right-hand side of his line and the zero line is exactly on top of the line he has just drawn.

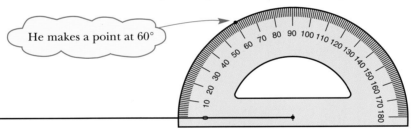

He makes a point at 60°

He takes the protractor away and joins up the point and the end of the line.

He extends the line until it is 5 cm long.

Then he joins up the two lines.

He measures this line – it is exactly 7 cm

Now do this yourself.
Check that you get 7 cm as well.

# 25: Measuring and drawing

**1** In this question, you will draw a triangle when you know the length of one side and the sizes of two angles.

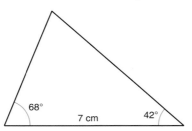

a) Draw a line exactly 7 cm long.

b) Mark off an angle of 42° at the right-hand end of the line.

   Join this end of the line to your mark.

c) Mark off an angle of 68° at the left-hand end of the 7 cm line. Join this end of the line to your mark.

d) Extend both lines if necessary until they meet.

e) Measure your lines.

**2** This diagram shows three of the fields at Springdale Farm.

The diagram is not drawn to scale but some of the lengths and angles are marked on.

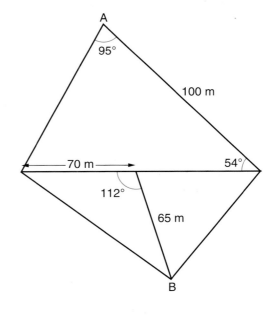

a) Make an accurate drawing of the three fields, using a scale of 1 cm to 20 m. (Hint: start by drawing the line which is 100 m long.)

b) The farmer is thinking of making the three fields into one big field. He wants to know how far it would be across the whole field from corner A to corner B. Measure the distance on your scale drawing, and work out what the distance would be in real life.

---

Cut out the triangle you drew in question 1.

Stack all the triangles in the class.

Are they all the same? — These are called **congruent**

# 25: Measuring and drawing

## More triangles

Roger draws a line exactly 8 cm long.

He uses his protractor to mark an angle of 40° and draws a second line.

 Which side is this?

How long does Roger draw it?

He measures 6 cm with his compasses.

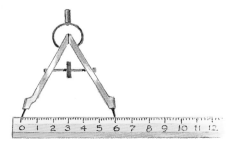

 Why does he do this?

He puts the compass point at C and draws a large arc.

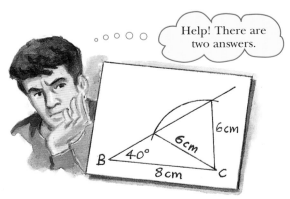

 How many triangles has Roger drawn?

Which triangle will make the better tent?

Are any of these triangles congruent?

## 25: Measuring and drawing

**1** In this question you will draw a triangle with sides 6 cm, 7 cm and 8 cm.

   a) Draw a line exactly 8 cm long.

   b) Take a length of 6 cm in your compasses.

   c) Put the compass point in the left-hand end of the line and draw a large arc passing through the line.

   d) Take a length of 7 cm in your compasses.

   e) Put the compass point in the right-hand end of the line and draw a large arc passing through the line to cut your other arc in two places.

   f) Join each end of your line to the points where the arcs cross.

   g) Cut out your two triangles and compare all the triangles in the class. Are they congruent?

**2** In this question you will draw a triangle with one side 8 cm, another side 10 cm and an angle of 90° opposite to the 10 cm side.

   a) Draw a line exactly 8 cm long.

   b) Use your protractor to mark an angle of 90° at the left-hand end of the line and draw a second line at right angles to the first.

   c) Measure 8 cm with your compasses.

   d) Put the compass point in the right-hand end of the first line and draw a large arc cutting the second line in two places.

   e) Join each end of your first line to the points where the arcs cross the second line.

   f) Measure these two lengths.

   g) Cut out your two triangles and compare all the triangles in the class. Are they congruent?

**3** Draw a line of length 6 cm and construct an equilateral triangle.

**4** Sketch an isosceles triangle of base 6 cm and height 4 cm.

Now draw it accurately and measure the other sides.

---

Give 5 examples of where you see triangles in everyday life.

In each case, roughly how long are the sides of the triangle?

## 25: Measuring and drawing

# Using bearings

An aeroplane is flying from London to Manchester.

The pilot needs to know exactly what direction to take.

She can do this by measuring the angle on the map between the direction she needs to fly in and a line going north from London.

The angle is measured clockwise from the North line.

Use an angle measurer to measure this angle.

You should find that it is 330 degrees.

This angle is called the **bearing** of Manchester from London.

Remember that bearings are always measured clockwise from North.

Bearings are always written with three figures.

A bearing of 62° is written as 062°.

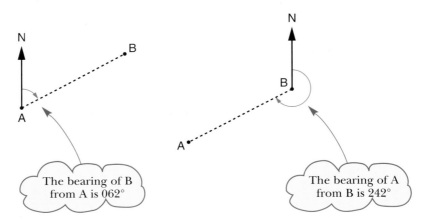

The bearing of B from A is 062°

The bearing of A from B is 242°

# 25: Measuring and drawing

**1** A, B, C and D are 4 ships.

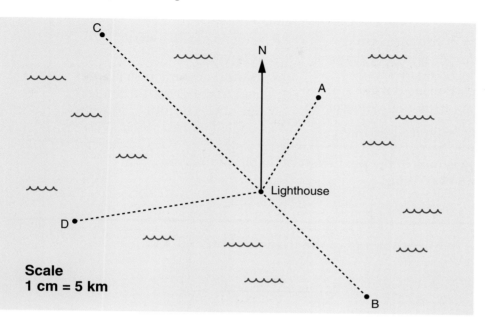

Scale
1 cm = 5 km

a) Measure the bearing of each of the ships from the lighthouse.

b) Find the distance of each ship from the lighthouse.

**2** This diagram shows three towns, A, B and C.

Measure the bearing of

a) A from B    b) B from A

c) A from C    d) C from A

e) B from C    f) C from B

g) What do you notice about each pair of bearings?

**3** Andrew and Ranjit are on a hike in the hills.

They find that a nearby hill, Mill Crag, is on a bearing of 158°.

They walk for 4 km north-east and then find that Mill Crag is on a bearing of 205°. Make a scale drawing of their journey, using a scale of 1:50 000.

Use your drawing to find out how far they are from Mill Crag at each of the points where they take a bearing.

Plan a walk through open country.

# 25: Measuring and drawing

# Finishing off

**Now that you have finished this chapter you should be able to**

★ draw, full size or to scale, accurate drawings of shapes such as rectangles and triangles, using a ruler and protractor or angle measurer

★ use three-figure bearings

★ understand the word congruent

★ know that some instructions lead to congruent triangles but others do not

Use the questions in the next exercise to check that you understand everything.

## Mixed exercise

**1** Make accurate full size drawings of the triangles ABC and XYZ shown below.

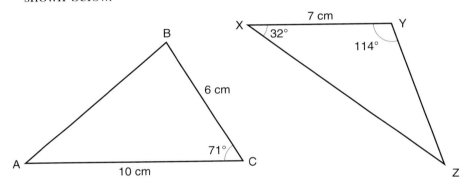

a) Measure the length of side AB on your first drawing.

b) Measure the lengths of sides XZ and YZ on your second drawing.

**2** a) Make a scale drawing of the field PQRST shown here, using a scale of 1:2500.

(Hint: start by drawing the rectangle OPQR, then add the triangle OPT, then the triangle RST.)

b) Measure the lengths of sides ST and TP on your drawing.

c) What are the lengths of sides ST and TP of the real field?

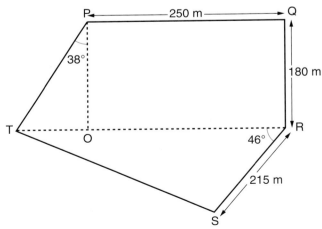

# 25: Measuring and drawing

**3** The map shows some ferry routes across the English Channel.

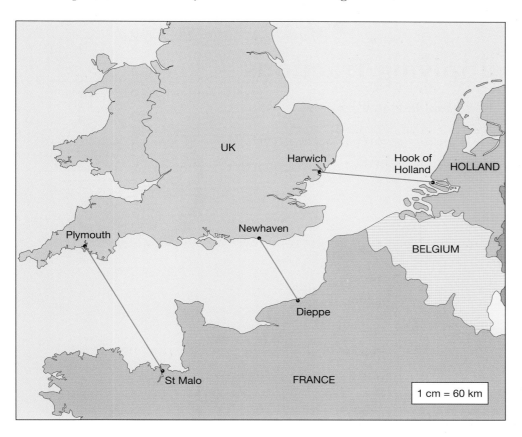

a) What bearing must a ferry sail on to travel from

  (i) Plymouth to St Malo?

  (ii) Dieppe to Newhaven?

  (iii) Harwich to Hook of Holland?

b) How far is each of the journeys above in real life?

---

Measure the length and width of your living room.

Make a scale drawing of the room on squared paper. Put in the positions of the doors and windows, and any fixed furniture.

Make drawings on squared paper, using the same scale, of the furniture in the room. Cut them out and use them to design a new arrangement for your living room.

When you are happy with your design, stick it down.

This is the way designers work in real life.

# Twenty six

# Using fractions

## Multiplying fractions

 Dave, Becky and Ravi share a pizza.

*Dave has a quarter of it. How much is left?*

Becky and Ravi share the other three quarters.

Each has half of it.

Each gets $\frac{1}{2}$ of $\frac{3}{4}$:

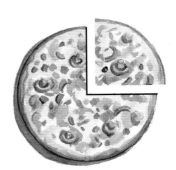

$$\frac{1}{2} \times \frac{3}{4} = \frac{3}{8}$$

> Multiply out:
> top $1 \times 3 = 3$
> bottom $2 \times 4 = 8$

 *Dave only eats half of his piece. What is $\frac{1}{2} \times \frac{1}{4}$?*

The pizza costs £8.00, but Ravi has a voucher for 30% discount.

Dave and Becky each work out how much this is.

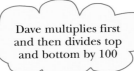

> Dave multiplies first and then divides top and bottom by 100

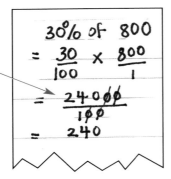

> Becky divides top and bottom by 100 before multiplying. This is called **cancelling**

 *How much does Ravi pay for the pizza?*

The next two examples show you how to multiply mixed numbers.

> Change $4\frac{1}{2}$ into an improper fraction

$$\frac{2}{3} \times 4\frac{1}{2}$$

$$2\frac{1}{3} \times 3\frac{3}{4}$$

> Change the mixed numbers into improper fractions

> Cancel

$$= \frac{\cancel{2}^1}{\cancel{3}_1} \times \frac{\cancel{9}^3}{\cancel{2}_1}$$

$$= \frac{7}{\cancel{3}_1} \times \frac{\cancel{15}^5}{4}$$

> Cancel

> Multiply out and change back into a mixed number

$$= \frac{3}{1} = 3$$

$$= \frac{35}{4}$$

$$= 8\frac{3}{4}$$

> Multiply out and change back to a mixed number

## 26: Using fractions

**1** Work out

a) $\frac{1}{2} \times \frac{1}{3}$  b) $\frac{1}{2} \times \frac{3}{8}$  c) $\frac{1}{4} \times \frac{3}{5}$  d) $\frac{3}{4} \times \frac{5}{6}$

e) $\frac{3}{8} \times \frac{2}{3}$  f) $\frac{6}{7} \times \frac{7}{10}$  g) $\frac{3}{8} \times \frac{5}{8}$  h) $\frac{3}{4} \times \frac{10}{1}$.

**2** Work out

a) $\frac{1}{2}$ of 7  b) $\frac{3}{4}$ of 6  c) $\frac{1}{3}$ of 8  d) $\frac{2}{5}$ of 4

e) $\frac{5}{8}$ of 20  f) $\frac{2}{3}$ of 14  g) $\frac{3}{8}$ of 10  h) $\frac{5}{6}$ of 9.

**3** Work out

a) $\frac{1}{2} \times 6\frac{1}{2}$  b) $\frac{3}{4} \times 4\frac{1}{2}$  c) $\frac{1}{3} \times 2\frac{5}{8}$  d) $2\frac{1}{2} \times \frac{7}{10}$

e) $2\frac{4}{5} \times \frac{5}{8}$  f) $\frac{3}{4} \times 5\frac{1}{3}$  g) $1\frac{1}{2} \times 2\frac{1}{2}$  h) $2\frac{1}{4} \times 3\frac{1}{2}$

i) $3\frac{2}{3} \times 1\frac{1}{2}$  j) $5\frac{1}{3} \times 3\frac{3}{4}$  k) $1\frac{3}{8} \times 3\frac{1}{2}$  l) $6\frac{2}{5} \times 1\frac{7}{8}$.

**4** Amanda lives $2\frac{3}{4}$ miles from work. She works 5 days a week. How many miles does she cover, travelling to and from work, in a week?

**5** (i) 14 cm    (ii) 28 cm

a) Using $\pi = \frac{22}{7}$ find the circumference of each circle.

b) The radius of the second circle is twice the radius of the first. What has happened to the circumference?

c) Using $\pi = \frac{22}{7}$ find the area of each circle.

d) The radius of the second circle is twice the radius of the first. What has happened to the area?

**6** Paula buys 60 lbs of boiled sweets. She makes up twenty $\frac{1}{4}$ lb bags, fifteen $\frac{1}{2}$ lb bags and fifteen $\frac{3}{4}$ lb bags. How much has she left over?

Find out how to enter a fraction into your calculator. Now use your calculator to check your answers to questions 1 to 3.

# 26: Using fractions

## Dividing fractions

Becky and Ravi share $\frac{3}{4}$ of a pizza.

$$\frac{1}{2} \text{ of } \frac{3}{4} = \frac{1}{2} \times \frac{3}{4} = \frac{3}{8}$$

They each have $\frac{3}{8}$ of the pizza.

Another way of working this out is to say $\frac{3}{4}$ of a pizza is divided between 2 people.

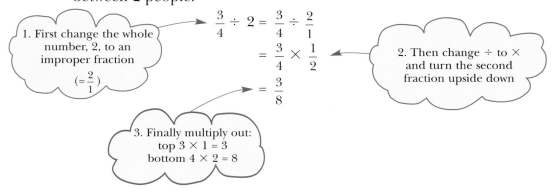

1. First change the whole number, 2, to an improper fraction ($=\frac{2}{1}$)

$$\frac{3}{4} \div 2 = \frac{3}{4} \div \frac{2}{1}$$
$$= \frac{3}{4} \times \frac{1}{2}$$
$$= \frac{3}{8}$$

2. Then change ÷ to × and turn the second fraction upside down

3. Finally multiply out:
top 3 × 1 = 3
bottom 4 × 2 = 8

Dave, Becky and Ravi share $4\frac{1}{2}$ chocolate bars equally.

How much does each person get?

Ravi works it out like this:

Dave does it like this:

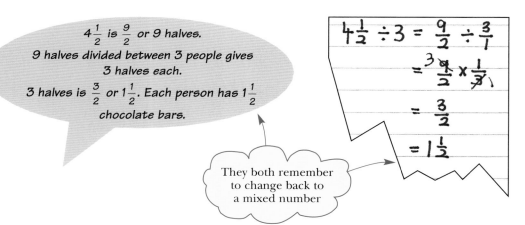

$4\frac{1}{2}$ is $\frac{9}{2}$ or 9 halves.
9 halves divided between 3 people gives 3 halves each.
3 halves is $\frac{3}{2}$ or $1\frac{1}{2}$. Each person has $1\frac{1}{2}$ chocolate bars.

They both remember to change back to a mixed number

Here are two more examples of division.

$$5 \div \frac{3}{4} = \frac{5}{1} \div \frac{3}{4}$$
$$= \frac{5}{1} \times \frac{4}{3}$$
$$= \frac{20}{3} = 6\frac{2}{3}$$

$$4\frac{1}{2} \div 1\frac{1}{4} = \frac{9}{2} \div \frac{5}{4}$$
$$= \frac{9}{1\cancel{2}} \times \frac{\cancel{4}^{2}}{5}$$
$$= \frac{18}{5} = 3\frac{3}{5}$$

# 26: Using fractions

**1** Work out

a) $5 \div 4$  b) $\dfrac{1}{5} \div 2$  c) $1\dfrac{3}{5} \div 4$  d) $\dfrac{1}{3} \div 3$

e) $2\dfrac{1}{2} \div 5$  f) $2\dfrac{1}{4} \div 3$  g) $\dfrac{5}{8} \div 2$  h) $1\dfrac{1}{2} \div 6$

**2** Work out

a) $4 \div \dfrac{1}{3}$  b) $3 \div \dfrac{1}{2}$  c) $12 \div \dfrac{3}{4}$  d) $12 \div \dfrac{2}{5}$

e) $3\dfrac{1}{2} \div 4$  f) $2\dfrac{1}{4} \div 1\dfrac{1}{4}$  g) $3\dfrac{3}{4} \div \dfrac{3}{8}$  h) $2\dfrac{3}{16} \div 1\dfrac{1}{4}$

i) $2\dfrac{5}{8} \div 3\dfrac{1}{2}$  j) $8\dfrac{3}{4} \div 1\dfrac{1}{4}$  k) $6\dfrac{7}{8} \div 2\dfrac{3}{4}$  l) $12 \div 3\dfrac{1}{3}$

**3** A grocer buys pieces of cheese weighing 5 kg.

a) How many $\dfrac{1}{2}$ kg pieces can he get from this?

b) How many $\dfrac{1}{4}$ kg pieces can he get from it?

**4** A box is $12\dfrac{1}{2}$ inches long, 5 inches wide and $1\dfrac{1}{4}$ inches high.

Toy bricks are cubes with edges $1\dfrac{1}{4}$ inches long.

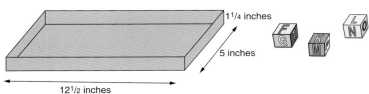

How many toy bricks can fit into the box?

**5** Jermaine's car does 35 miles per gallon and he has 6 gallons of petrol in the tank.

Jermaine's house  Fiona's house

How many times can he go to Fiona's house and back?

**6** A bookshelf is $29\dfrac{1}{4}$ inches long.

How many books can fit on the shelf if each book is

a) $\dfrac{3}{4}$ inch thick?  b) $1\dfrac{1}{8}$ inches thick?  c) $\dfrac{5}{8}$ inch thick?

## 26: Using fractions

# Fractions to decimals

You know that $\frac{1}{4} = 0.25$, $\frac{1}{2} = 0.5$ and $\frac{3}{4} = 0.75$.

What about $\frac{1}{8}$?

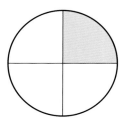 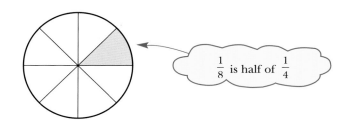

$\frac{1}{8}$ is half of $\frac{1}{4}$

What decimal is half of 0.25?

$0.25 \div 2 = 0.125$

**So $\frac{1}{8} = 0.125$**

Also, $\frac{1}{8}$ is $1 \div 8 = 0.125$

So again, $\frac{1}{8} = 0.125$.

We can do this to change any fraction into a decimal.

 What is $\frac{1}{5}$ as a decimal?

Something interesting happens when we work out $\frac{1}{3}$ as a decimal.

$\frac{1}{3} = 1 \div 3 = 0.333\ldots$

the 3s go on forever!

'0.3 recurring'

This is a **recurring decimal**.

Instead of writing out all the 3s (that would take forever!) we write $0.\dot{3}$.

The same thing happens for $\frac{1}{6}$.

$\frac{1}{6} = 1 \div 6 = 0.166\ldots$

**So $\frac{1}{6} = 0.1\dot{6}$**

'0.16 recurring'

# 26: Using fractions

**1** Change the fraction to a decimal:

a) $\frac{3}{5}$  b) $\frac{3}{8}$  c) $\frac{9}{20}$  d) $\frac{5}{4}$

e) $\frac{8}{25}$  f) $\frac{1}{16}$  g) $\frac{7}{2}$  h) $\frac{13}{5}$

**2** Change the fraction to a recurring decimal:

a) $\frac{4}{3}$  b) $\frac{1}{9}$  c) $\frac{7}{6}$  d) $\frac{4}{9}$

## Investigation

What is $\frac{1}{11}$ as a decimal?
Work out $1 \div 11$.

So $\frac{1}{11} = 0.0909...$

a) In the same way, find the decimal form of
   (i) $\frac{2}{11}$  (ii) $\frac{6}{11}$.

These results have been put into this table.

| Fraction | Decimal |
| --- | --- |
| $\frac{1}{11}$ | 0.0909... |
| $\frac{2}{11}$ | 0.1818... |
| $\frac{3}{11}$ | |
| $\frac{4}{11}$ | |
| $\frac{5}{11}$ | |
| $\frac{6}{11}$ | 0.5454... |
| $\frac{7}{11}$ | |
| $\frac{8}{11}$ | |
| $\frac{9}{11}$ | |
| $\frac{10}{11}$ | |

b) What do you think the missing decimals are?

**3** George has to work out

$\frac{1}{3} + \frac{1}{3} + \frac{1}{3}$.

George wrote down:

0.3 + 0.3 + 0.3 = 0.9  ✗

You've rounded your answer too soon

What is the right answer?

Find out which temperatures are whole numbers of degrees in both Celsius and Fahrenheit.

What temperature is the same number of degrees on both scales?

## 26: Using fractions

# Fractions to percentages

You know that $\frac{1}{4} = 25\%$, $\frac{1}{2} = 50\%$ and $\frac{3}{4} = 75\%$.

What about $\frac{3}{5}$?

One way to change $\frac{3}{5}$ into a percentage, is to work out an equivalent fraction with 100 on the bottom.

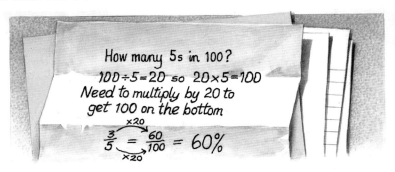

Another way to do it is like this:

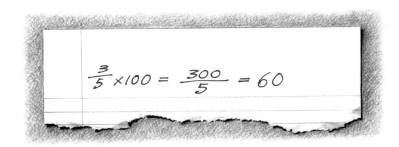

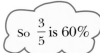

So $\frac{3}{5}$ is 60%

What is $\frac{9}{20}$ as a percentage?

Work it out both ways to check you get the same answer.

What is $\frac{1}{8}$ as a percentage?

It is easier to use the second method for this.

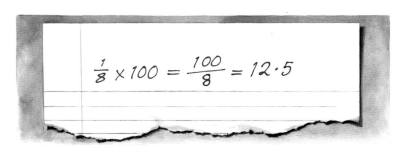

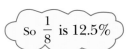

So $\frac{1}{8}$ is 12.5%

Why is it easier to do it this way for $\frac{1}{8}$?

## 26: Using fractions

**1** Change the fraction to a percentage:
a) $\dfrac{1}{5}$
b) $\dfrac{7}{20}$
c) $\dfrac{9}{25}$
d) $\dfrac{3}{8}$
e) $\dfrac{39}{50}$
f) $\dfrac{33}{40}$
g) $\dfrac{68}{200}$
h) $\dfrac{75}{120}$

**2** In a survey of 40 oversixties (20 women and 20 men) 12 say they smoke.

a) What percentage of the oversixties smoke?

8 of the oversixties who smoke are women, 4 are men.

b) What percentage of the women smoke?

c) What percentage of the men smoke?

**3** For each person, work out what percentage of their weekly wage is spent on housing.

a) Karen earns £80 a week and spends £40 a week on student accommodation.

b) Marie earns £180 a week and spends £60 a week on rent.

c) Jo earns £350 a week and spends £80 a week on her mortgage.

**4** This table shows the profits of a CD shop over the four quarters of last year.

| Quarter    | Spring | Summer | Autumn | Winter |
|------------|--------|--------|--------|--------|
| Profit (£) | 21 000 | 34 000 | 38 000 | 27 000 |

a) What was the total profit last year?

b) What percentage of the total profit was made in summer?

c) What percentage of the total profit was made in autumn?

**5** Mike bought an old car for £250. He repaired it and sold it, making a profit of £50.

What was the profit as a percentage of the cost?

---

Look at the postmarks on the first class letters you receive during one week.

How many days does each letter take to arrive?

What percentage of the letters arrive the day after posting?

## 26: Using fractions

# Making comparisons

Ranjit does a survey of people's opinions of their local bus and train services. Here are his results.

| Service | Number Satisfied | Number questioned |
|---------|------------------|-------------------|
| Bus     | 79               | 111               |
| Train   | 37               | 59                |

 Which service is satisfying more of its customers?

Ranjit works out what proportion of the customers he asked are satisfied with each service.

$$\text{Bus: } \frac{79}{111} \qquad \text{Train: } \frac{37}{59}$$

These figures are still not easy to compare, so he writes them as percentages.

$$\text{Bus: } \frac{79}{111} = 0.711\ldots = 71\% \text{ (to nearest 1\%)}$$

$$\text{Train: } \frac{37}{59} = 0.627\ldots = 63\% \text{ (to nearest 1\%)}$$

The bus service satisfies more of its customers.

Sometimes the proportions you need to compare might be very close together. For example, which is larger, $\frac{95}{212}$ or $\frac{47}{105}$?

$$\frac{95}{212} = 0.4481132\ldots \qquad \frac{47}{105} = 0.447619\ldots$$

> The fractions give the **exact** value

You can see from the third decimal place that $\frac{95}{212}$ is slightly larger.

Writing these numbers as whole number percentages would not show the difference – they both round to 45%.

 The only way to write $\frac{95}{212}$ or $\frac{47}{105}$ exactly is to use fractions. Why is this?

 You can write some fractions, like $\frac{1}{2}$ and $\frac{3}{10}$ exactly as decimals or percentages. What makes these special? Do the investigation on page 277 to learn more about recurring decimals.

# 26: Using fractions

**1** Write down a decimal equal to each of these fractions.

a) $\frac{4}{5}$  b) $\frac{11}{25}$  c) $\frac{3}{8}$  d) $\frac{2}{3}$  e) $\frac{1}{6}$  f) $\frac{7}{12}$

**2** Arrange these numbers in order of size, starting with the smallest.

0.83    $\frac{17}{20}$    $\frac{21}{25}$    0.09    $\frac{5}{6}$

**3** Ella does a survey to find out people's opinions on two brands of personal stereo. These are her results.

| Brand | Number questioned | Number reporting faults |
|---|---|---|
| A | 129 | 13 |
| B | 186 | 21 |

Which brand is the more reliable?

**4** Harriet manages her company's training centres. She draws up this table to compare the success of each centre.

| Centre | No. of passes | No. of recruits |
|---|---|---|
| Northhill | 35 | 44 |
| Heartland | 39 | 71 |
| Southdown | 41 | 52 |

a) Work out the percentage of recruits at each centre who passed.
b) How would you interpret these results?

**5** This chart shows a manufacturer's daily output of bicycle frames from 3 production lines.

| Line | Total output | Rejects |
|---|---|---|
| A | 800 | 36 |
| B | 840 | 41 |
| C | 625 | 29 |

a) Work out the percentage of rejects for each production line.
b) Which production line is the most efficient?
c) Suggest a possible reason why the output of machine C was lower than A or B.

---

You know that $\frac{1}{3} = 0.33\ldots$ (going on for ever).

You say this as '0.3 recurring' and write it as $0.\dot{3}$.

Some fractions give a pattern of recurring digits.

For example, $\frac{2}{11} = 0.1818\ldots$, written $0.\dot{1}\dot{8}$, a pattern of length 2.

Investigate the patterns for $\frac{1}{7}, \frac{2}{7}, \frac{3}{7}, \frac{4}{7}, \frac{5}{7}$ and $\frac{6}{7}$.

What is the fraction with the longest pattern you can find?

# 26: Using fractions

## Finishing off

**Now that you have finished this chapter you should be able to**

★ multiply and divide fractions
★ change fractions to decimals and percentages
★ make comparisons involving fractions, decimals and percentages

Use the questions in the next exercise to check that you understand everything.

### Mixed exercise

**1** Work these out.

a) $2\frac{7}{8} + 1\frac{3}{4}$   b) $5\frac{1}{16} - 4\frac{1}{4}$   c) $5\frac{2}{3} - 1\frac{1}{6}$   d) $2\frac{4}{5} + 6\frac{7}{10}$

e) $\frac{1}{4} \times 2\frac{2}{3}$   f) $6\frac{3}{4} \times 1\frac{1}{2}$   g) $4\frac{5}{8} \times 1\frac{1}{4}$   h) $1\frac{1}{3} \times 3\frac{3}{4}$

i) $1\frac{1}{2} \div 2$   j) $2\frac{3}{4} \div \frac{1}{4}$   k) $1\frac{1}{8} \div 4\frac{1}{2}$   l) $6\frac{1}{4} \div 1\frac{2}{3}$

**2** Work these out.

a) $5\frac{1}{2} - 3\frac{3}{4} - 1\frac{1}{8}$   b) $(\frac{2}{3} - \frac{1}{6}) \div 5$   c) $6 \div (\frac{1}{2} + \frac{1}{4})$

d) $2\frac{1}{2} \times 1\frac{3}{5} \times 1\frac{1}{4}$   e) $2\frac{1}{4} + 3\frac{2}{3} + 4\frac{1}{2}$   f) $(4\frac{1}{2} - 1\frac{1}{6}) \div \frac{3}{4}$

**3** Isabel is testing children's mental arithmetic skills using a 20 question test.

a) The pass mark is 16. What percentage is this?

Isabel tests children in 3 schools and gets these results.

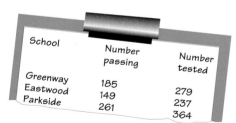

| School | Number passing | Number tested |
|---|---|---|
| Greenway | 185 | 279 |
| Eastwood | 149 | 237 |
| Parkside | 261 | 364 |

b) Which school has the highest pass rate?

c) Which school has the lowest pass rate?

**4** What fraction is midway between $\frac{1}{2}$ and $\frac{3}{4}$?

**5** Place these numbers in order of size, starting with the smallest.

$\frac{27}{8}$   $\sqrt{11}$   $\frac{10}{3}$   3.3   3.04

# 26: Using fractions

**6** This bar chart shows how the sales of a large company are spread across 3 countries.
  a) Write down the sales (in £m) made in each country.
  b) Write down the fraction of the total sales that occur in each country. Write each fraction in its simplest form.

**7** There are 180 applicants for these crew positions. One quarter of them are interviewed. Of those interviewed, two thirds are offered places.
  a) How many of the applicants are offered places?
  b) Write, in its simplest form, the fraction of the applicants who are offered places.

**8** Express each fraction as a percentage.
  a) $\dfrac{14}{25}$   b) $\dfrac{7}{8}$   c) $\dfrac{5}{6}$   d) $\dfrac{371}{500}$

**9** Nita and Mark run a business. They have drawn up this table of their profits last year.
  a) What was the total profit for last year?
  b) What percentage of the total profit was made in the first quarter?
  c) What percentage of the total profit was made in the second half of the year?
  d) The expected profit for this year is £200 000. Write down the expected increase in profit as a sum of money and as a percentage of last year's profit.

| Quarter | 1 | 2 | 3 | 4 |
|---|---|---|---|---|
| Profit (£ thousands) | 32 | 44 | 50 | 34 |

**10** This table shows the numbers of votes cast in 4 parish council elections. Which parish had
  a) the highest percentage turnout?
  b) the lowest percentage turnout?
  c) a turnout of approximately 2 voters in 5?

| Parish | Votes cast | No. of voters |
|---|---|---|
| Waterbeach | 105 | 267 |
| Oakington | 154 | 309 |
| Witchford | 187 | 393 |
| Northwood | 137 | 291 |

**11** Ethel leaves £4800 to be divided between her grandchildren.
Ben gets a quarter of this.
Jo gets 40% of the remainder.
Guy and Tanya share the rest equally.
  a) How much does each grandchild get?
  b) What percentage of Ethel's money does Jo get?

---

A snail is making a journey of 2 km. It has travelled 40 cm.
Write this as a) a fraction b) a percentage of the whole journey.
What percentage of the whole journey is 1 mm?

# Twenty seven

# Angles and shapes

## Parallel lines

A trellis like this is often used in gardens.

It is made of two sets of parallel lines.

(Parallel lines are lines that go in the same direction and never meet.)

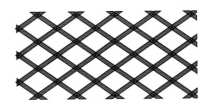

Here is a larger diagram of part of the trellis with some of the angles marked with letters.

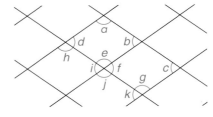

*Measure all the angles marked with letters.*

*Which ones are the same as each other?*

*Find some rules for working out which angles are the same.*

## The rules

You may have found the rules below.

**Where two lines cross, the angles opposite each other are equal. These are opposite angles.**

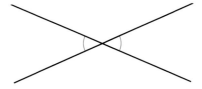

**Where a line crosses two parallel lines, these two angles are equal. These are corresponding angles. Look for the shape like a letter F.**

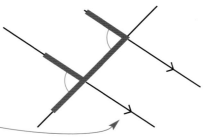

Parallel lines are shown with arrows

**Where a line crosses two parallel lines, these two angles are equal. These are alternate angles. Look for the shape of a letter Z.**

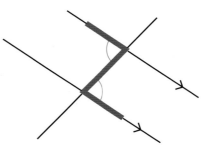

# 27: Angles and shapes

**1** a) Which angles in the diagram are the same as angle *x*?

b) Which angles are the same as angle *y*?

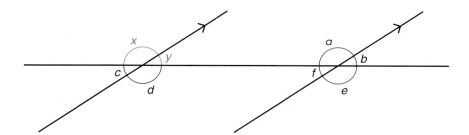

**2** Work out the angles marked with letters in these diagrams.

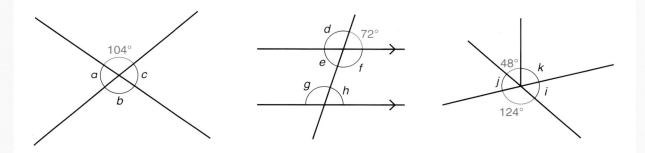

**3** The diagram shows a pair of parallel lines and a triangle.

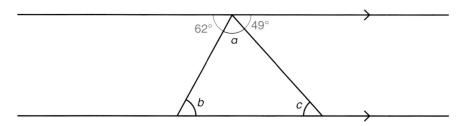

a) Work out the size of angles *a*, *b* and *c*.

b) Angles *a*, *b* and *c* are the three angles in a triangle.
   Add up these three angles.

List six places where you can see parallel lines.

## 27: Angles and shapes

# Angles and triangles

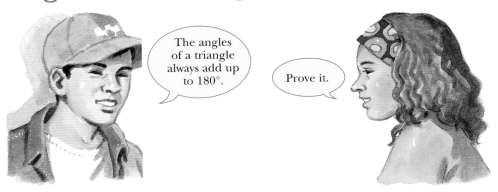

Look at this triangle. One side has been extended.

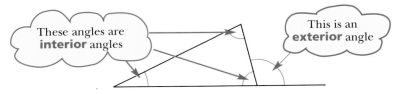

 What do the words interior and exterior mean?

Notice the arrows on two of the lines.

 What do they tell you?

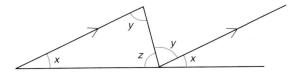

 Why are the angles marked x the same size?

Why are the angles marked y the same size?

The exterior angle of the triangle is $x + y$.

**The exterior angle of the triangle is equal to the sum of the interior opposite angles.**

Look at the straight line which forms the exterior angle.

 What do you know about $x + y + z$?

**The angles of a triangle always add up to 180°.**

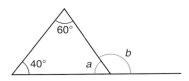

 What are the sizes of a and b in this diagram?

# 27: Angles and shapes

**1** Find the angles marked with letters in these diagrams.

a)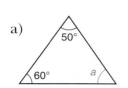

b)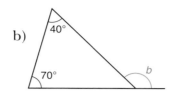

c)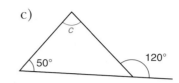

d)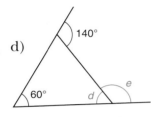

e)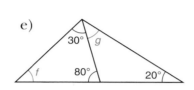

f)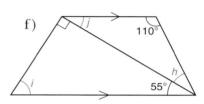

g)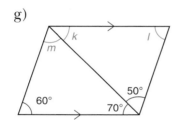

h)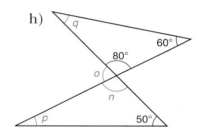

**2** a)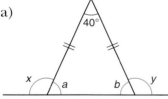

b)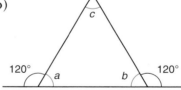

(i) Find the angles marked with letters in the diagrams.

(ii) In each case state the special name which is given to the triangle.

---

Draw any triangle. Colour the three angles in different colours.

Cut off the corners of your triangle. Fit the coloured angles together to make a straight line.

Does it matter how you fit them together? Why is this?

## 27: Angles and shapes

# Quadrilaterals

A **quadrilateral** is any four-sided shape.

Any quadrilateral can be split up into two triangles by drawing in a diagonal, like this:

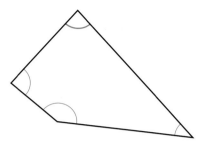 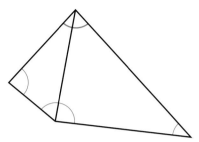

The angles in the quadrilateral must add up to the same as the angles in the two triangles.

So the angles in any quadrilateral must add up to $2 \times 180° = 360°$.

**Angles in a quadrilateral add up to 360°.**

 What are the names of special types of quadrilateral?

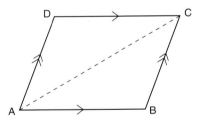

This quadrilateral is a **parallelogram**.

Sides AB and AC are parallel. So are sides AD and BC.

 Which sides are equal?

Which angles are equal?

AC is a **diagonal** of the parallelogram.

 What does AC do to the parallelogram?

# 27: Angles and shapes

**1** Find the angles marked with letters in these triangles and quadrilaterals.

They are not drawn accurately so you will need the rules on the opposite page.

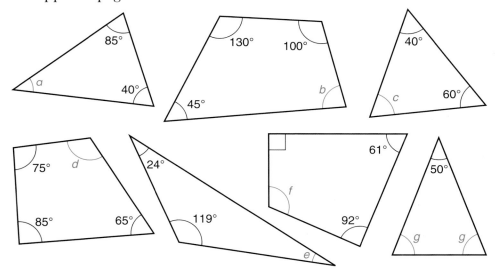

**2** Find the angles marked with letters in the diagrams below.

You will need to use the rules about angles that you have learnt so far in this chapter.

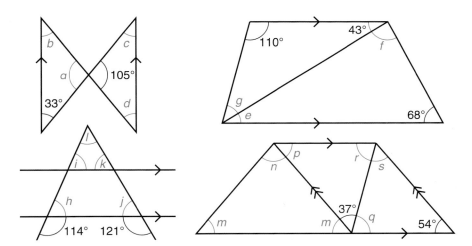

Road signs are made in a number of shapes.

What do the different shapes tell you?

Give some examples.

# 27: Angles and shapes

# Interior angles of polygons

A **polygon** is a shape with several sides.

The angles inside a polygon are called the **interior angles**.

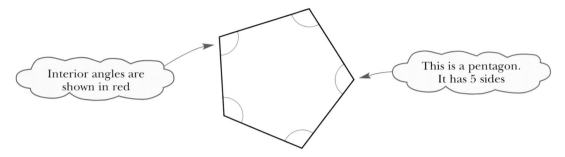

Interior angles are shown in red

This is a pentagon. It has 5 sides

A quick way of working out the sum of the interior angles of any polygon is by splitting it into triangles. The pentagon can be split into 3 triangles.

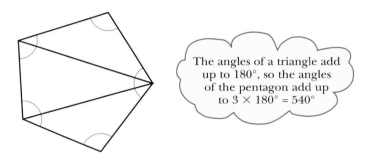

The angles of a triangle add up to 180°, so the angles of the pentagon add up to 3 × 180° = 540°

*Try splitting some other polygons into triangles. What is the rule for the number of triangles you can make?*

Another way of finding the sum of the interior angles is to use the formula

**Angle sum of a polygon = (Number of sides − 2) × 180°**

*Why do you think this formula works?*

A polygon is **regular** if all its sides are the same length and all its angles are equal. You can find the size of each interior angle by dividing the angle sum by the number of sides.

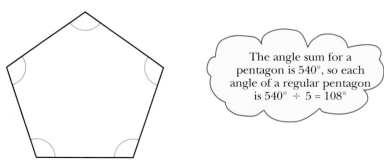

The angle sum for a pentagon is 540°, so each angle of a regular pentagon is 540° ÷ 5 = 108°

# 27: Angles and shapes

**1**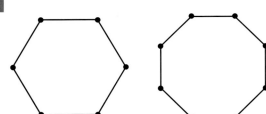

a) Copy and split each polygon into triangles.

b) Work out the angle sum of each polygon by multiplying the number of triangles by 180°.

c) Use the formula on the opposite page to work out the angle sum for each polygon. Check you get the same answers as in b).

**2** Work out the interior angle of

  a) a regular hexagon     b) a regular octagon

  c) a regular decagon (10 sides)   d) a regular dodecagon (12 sides).

**3** Here is a regular pentagon which has been split into 5 congruent (equal) triangles.

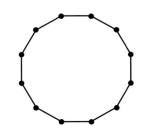

a) Work out the size of each blue angle.

b) Use your answer to a) to work out the size of each red angle.

c) Use your answer to b) to work out the interior angle of the pentagon.

**4**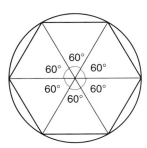

a) Draw a circle of radius 5 cm.

b) Draw angles of 60° at its centre, as shown in the diagram.

c) Now complete the hexagon.

d) Measure the sides of the hexagon. What do you notice? Explain it.

Stars are often used for decoration. The diagram shows how you can use a regular pentagon to make a 5-pointed star. Make one for yourself out of cardboard.

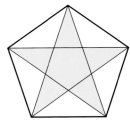

Now use a regular hexagon to make a 6-pointed star and a regular octagon to make an 8-pointed star.

## 27: Angles and shapes

# Exterior angles of polygons

Sue is using a robot to draw a regular octagon.

She needs to know what angle the robot has to turn through each time.

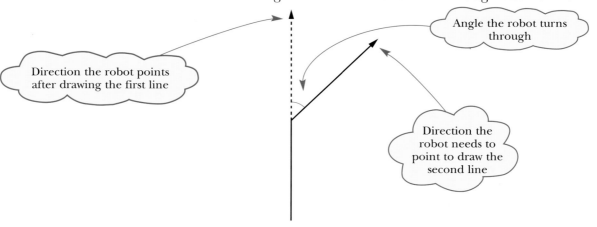

This diagram shows all the angles the robot has to turn through.

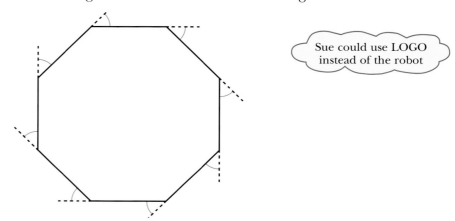

Sue could use LOGO instead of the robot

These are called the **exterior angles** of the octagon.

When it draws the octagon, the robot makes one complete turn. So it turns through 360° altogether.

 *What instructions should Sue give to the robot?*

Sue works out that there are 8 angles, so each exterior angle must be 360° ÷ 8 = 45°.

**Exterior angle of a regular polygon = 360° ÷ Number of sides**

 *What happens if the robot turns too much each time?*

*What happens if it turns too little each time?*

# 27: Angles and shapes

**1** Use the rule on the opposite page to work out the exterior angle of each of these polygons.

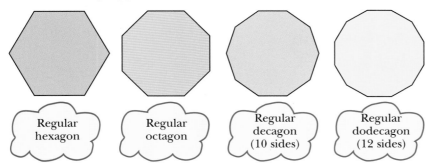

Regular hexagon   Regular octagon   Regular decagon (10 sides)   Regular dodecagon (12 sides)

**2** a) For each of the polygons in question 1, add together the interior angle and the exterior angle.
(You worked out the interior angles in question 2 on page 289.)
What do you notice?

b) Explain why this happens.

**3** a) Work out the exterior angle for each of these regular polygons.
  (i) 9 sides
  (ii) 25 sides
  (iii) 100 sides.

b) Now work out the interior angle of each of them.
(You will need to use your answer to question 2b).)

**4** In the diagram, *all* the sides of the polygon are equal.

The angles at A and E are both 90°.
a) Describe the polygon.
b) What sort of triangle is BCD?
c) What is the size of the interior angle at C?
d) Work out the size of the interior angles at B and D.
e) Copy the diagram and mark the exterior angles.

What do the exterior angles add up to?

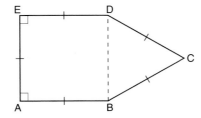

The more sides a polygon has, the more it looks like a circle. So if you want to program a robot to draw a circle, you have to program it to draw a polygon with lots of sides.

Write a set of instructions for a robot (or LOGO) to draw a circle.

## 27: Angles and shapes

# Tessellations

Look at these designs for floor tilings:

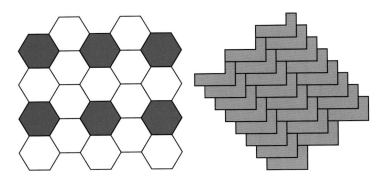

These patterns are examples of **tessellations**.

A tessellation is a repeating pattern without any gaps, made up of **congruent** shapes fitted together.

(Congruent means exactly the same size and shape.)

Tessellations can be made with very complicated shapes, as well as simple ones like those above. The Dutch artist M.C. Escher used fascinating tessellations in his work. Here are two examples:

'Reptiles'

'Day and Night'

# 27: Angles and shapes

**1** Use squared paper to draw tessellations using the following shapes. Colour each tessellation to make a pattern.

a)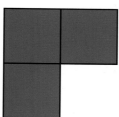

b)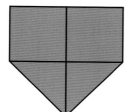

c)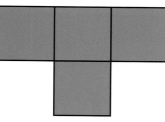

**2** Pentominoes are shapes made up of five squares which touch edge to edge.

These are two of the pentominoes.

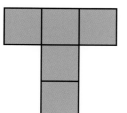 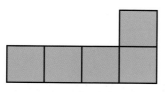

There are twelve pentominoes altogether.

a) Draw the other ten pentominoes.
b) Which pentominoes tessellate?

Draw and colour tessellations of these pentominoes.

---

### Exercise

1. Start with a simple shape that tessellates such as a square or a hexagon.

2. Draw a shape on to one side. Use tracing paper to remove the same shape from the opposite side.

3. Do the same on any other sides that you want to.

4. Cut out your design to make a template. Use your imagination to add any details you like.

5. Use your template to draw your tessellation. The shapes should fit together perfectly.

   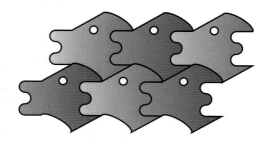

## 27: Angles and shapes

# Finishing off

**Now that you have finished this chapter you should**

- ★ be able to find pairs of equal angles where two lines cross and where a line intersects parallel lines
- ★ know that the angles in a triangle add up to 180°
- ★ know that the angles in a quadrilateral add up to 360°
- ★ be able to find the sum of the angles of any polygon
- ★ be able to find the interior and exterior angles of any regular polygon
- ★ be able to draw tessellations with simple shapes

Use the questions in the next exercise to check that you understand everything.

**Mixed exercise**

**1** Find the angles marked with letters in the diagrams below.

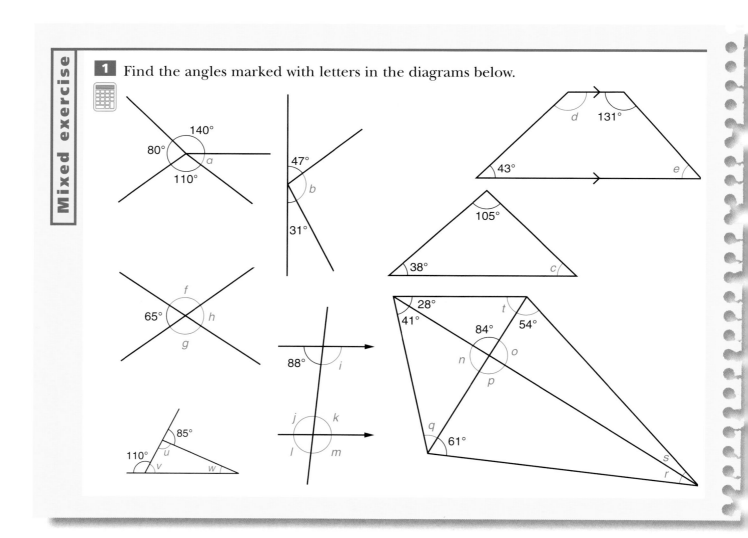

# 27: Angles and shapes

**Mixed exercise**

**2** Find the sum of the interior angles of the polygon shown below.

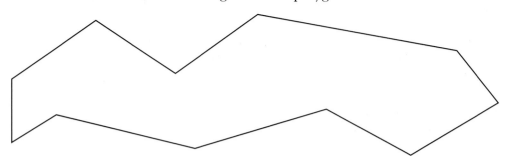

**3** For each of these shapes, find

a) the angle sum

b) the interior angle

c) the exterior angle.

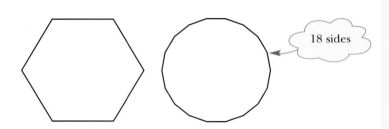

18 sides

**4** Draw tessellations on squared paper of each of the shapes below.

a)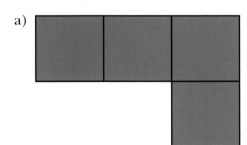

b)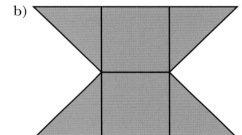

Test how good you are at judging distances and angles. You need plenty of space and a friend to tell you if you are about to bump into anything. Mark a starting point on the floor and close your eyes. Walk five metres forward and then turn through 90° to your right. Repeat until you have done it four times.
Open your eyes and see how far you are from where you started.

Do this several times.

How close can you get?

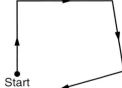

Start

# Twenty eight
# Using decimals

**Before you start this chapter you should be able to**

- ★ change $\frac{1}{4}$, $\frac{1}{2}$ and $\frac{3}{4}$ into decimals
- ★ change tenths and hundredths into decimals

## Tenths and hundredths

This number line is split into tenths.

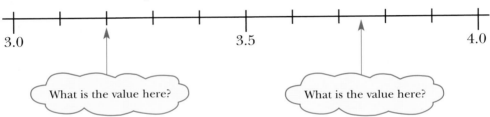

What is the value here?

What is the value here?

## Adding and subtracting

This is the design of Sophie's garden:

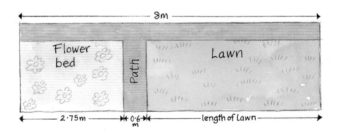

She wants to know how long her lawn is.

She works it out like this:

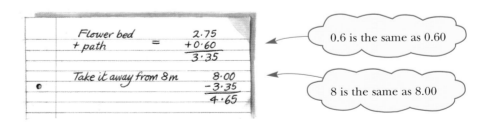

0.6 is the same as 0.60

8 is the same as 8.00

 *How long is the lawn?*

## 28: Using decimals

**1** Write down the tape measure readings.

a)   b)

**2** This bar chart shows the population, in millions, of 4 cities.

Write down the population, in millions, of each city.

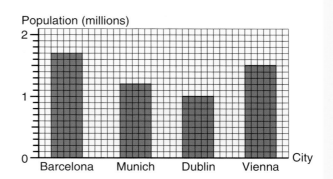

**3** Write down a fraction equal to

a) 2.3    b) 4.91    c) 3.07    d) 5.237.

**4** Write down the decimal equal to

a) $3\frac{9}{10}$    b) $2\frac{19}{100}$    c) $5\frac{3}{100}$    d) $\frac{109}{1000}$.

**5** Draw a number line between 8 and 9 and divide it into tenths.

Mark the points

a) 8.5    b) 8.25    c) 8.75.

**6** Gemma buys a birthday card for £1.80, a newspaper for £0.45 and some crisps for £0.36.

How much change does she get from £5?

**7** The school discus record is 39.24 metres.

On his first throw Mick throws 38.16 metres.

a) How far short of the record is he?

b) With his second throw Mick throws 41.02 metres.

How much did he beat the record by?

---

In metric units, 1 millimetre is $\frac{1}{1000}$ of a metre or 0.001 metres.
In the same way, 1 milligram is $\frac{1}{1000}$ of a gram or 0.001 grams.
You can say that milli- means 0.001.

There are other words like milli-, for example centi- and kilo-.

Make a list of all these words and what numbers they mean.

### 28: Using decimals

# Multiplication and division

Rachel works in a coffee shop.

She pours out 2.5 litres of lemonade for 10 children so that they all get the same.

How much does each child get?

$2.5 \div 10 = 0.25$

**So each child gets 0.25 litre.**

> ÷ 10
> 2.5 becomes 0.25
> To divide by 10 you move the decimal point one place left

What is $250 \div 10$?

What is $25 \div 10$?

What is the rule for dividing by 100?

You can check the amount each child gets by multiplying

$0.25 \times 10 = 2.5$

> × 10
> 0.25 becomes 2.5
> To multiply by 10 you move the decimal point one place right

What is $2.5 \times 10$?

What is $25 \times 10$?

What is the rule for multiplying by 100?

Rachel has a 3 litre pot of coffee.

How many mugs, each holding 0.2 litres, can she fill?

$$3 \div 0.2 = \frac{3}{0.2}$$
$$= \frac{30}{2}$$
$$= 15$$

> This needs to be a whole number before you divide, so multiply top and bottom by 10

**So Rachel can fill 15 mugs from the coffee pot.**

You can check this by multiplying

$15 \times 0.2$

> There is one decimal place

First work out $15 \times 2 = 30$.

Then add in the one decimal place to get 3.

> 3.0. gives 3.0 = 3

## 28: Using decimals

**1** Work these out.
   a) $1.7 \times 10$
   b) $1.7 \times 100$
   c) $1.7 \times 1000$
   d) $3.2 \div 10$
   e) $3.2 \div 100$
   f) $3.2 \div 1000$

**2** Work these out.
   a) $2 \times 1.5$
   b) $4 \times 0.2$
   c) $3 \times 5.4$
   d) $2.6 \times 1.2$

**3** Work these out.
   a) $4 \div 0.8$
   b) $3.2 \div 0.4$
   c) $3.9 \div 0.6$
   d) $5 \div 0.25$

**4** Caroline buys 10 metres of garden fencing at £3.25 per metre.
   a) How much does she pay?
   b) How much do 100 metres cost?

**5** A pen costs £0.18.

   How much do 100 pens cost?

**6** The temperature in Blackpool is 30 °C.
   a) Multiply 30 by 1.8 and then add 32 to get the temperature in °F.

   It is 20 °C in Scarborough.

   b) What is this in °F? (You need to do the same thing: multiply 20 by 1.8 and then add 32.)

**7** Rory buys 10 oranges, 0.6 kg of green peppers and 0.75 kg of mushrooms. How much change does he get from £10?

**8** It costs £120 to hire a boat.
The cost is shared between 10 people.

How much does each person pay?

**9** Alex is 150 cm tall. (100 cm = 1 m)

   a) Divide this by 100 to get her height in metres.

   b) Julie is 1.65 m tall.

   Multiply this by 100 to get her height in cm.

a) Find  i) $2^2$  ii) $20^2$  iii) $200^2$  iv) $2000^2$

How many zeros are there in $2\,000\,000^2$?
What about in $2\,000\,000^3$?

b) Find  i) $\sqrt{9}$  ii) $\sqrt{900}$  iii) $\sqrt{90\,000}$

How many zeros are there in $\sqrt{9\,000\,000}$?

What happens when you try $\sqrt{90}$ and $\sqrt{9000}$?

# 28: Using decimals

## Finishing off

**Now that you have finished this chapter you should be able to**

★ add and subtract decimals
★ multiply decimals (including multiplying by 10, 100, etc.)
★ divide decimals (including dividing by 10, 100, etc.)
★ work out squares and square roots of decimals

Use the questions in the next exercise to check that you understand everything.

### Mixed exercise

**1** How tall are these people?

a)    b)

**2** Change the decimal to a fraction:

a) 0.7    b) 5.4    c) 1.83    d) 6.371

**3** Change the fraction to a decimal:

a) $2\frac{3}{10}$    b) $3\frac{17}{100}$    c) $6\frac{7}{100}$    d) $\frac{141}{1000}$

**4** Kelly buys a newspaper for £0.60, a magazine for £1.75 and a bag of sweets for £0.49.

How much change does she get from £5?

**5** A pencil is 5 inches long. How many centimetres is it?

(Remember 1 inch = 2.54 cm.)

**6** This is Andrew's room.

a) What is the area of the room?

b) He wants a skirting board fitted all round the room but excluding the door.

The door is 0.9 m wide.

What length of skirting board is needed?

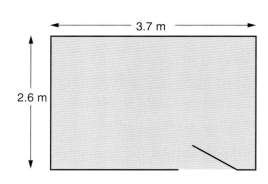

## 28: Using decimals

**7** Ceri buys 10 tins of cat food.

a) How much does she pay?

The supermarket pays £500 for 1000 tins of the cat food.

b) How much do they pay for 1 tin?

**8** Laura buys 0.5 kg of grapes, 1.2 kg of tomatoes and 3 peaches. How much change does she get from £10?

**9** A path is 9 metres long. How many paving slabs are needed to make the path if each slab is

a) 0.9 metres long?   b) 0.6 metres long?

**10** This is Shamir's bookcase. The wood is 1.6 cm thick.

a) Work out the inside width, $W$ cm, of the shelves.

b) The shelves are of equal height.

Work out the inside height, $H$ cm, of each shelf.

**11** Dale buys twenty litres of petrol at £0.84 per litre. What is the total cost?

**12** 50 000 people each pay £30 to attend a pop concert.

a) How much money is collected?

b) 4000 T-shirts are sold at the concert. They cost £52 000 altogether.

How much does one T-shirt cost?

**13** Find the value of

a) $5.3^2$   b) $\sqrt{8.41}$

c) $6.8^2$   d) $\sqrt{15.21}$

**14** Change these fractions into decimals.

a) $\frac{4}{5}$   b) $\frac{5}{8}$

c) $\frac{7}{20}$   d) $\frac{5}{6}$

**15** Arrange these in order of size. Start with the smallest.

5.1, 5, 5.02, 5.01

---

Find examples of measuring instruments with scales in

a) whole   b) 0.5

c) 0.2   d) 0.1   units.

Can you find examples of any other divisions?

# Twenty nine

# Sequences

**Before you start this chapter you should**

★ understand the word formula  ★ know that letters can be used to stand for numbers

## Patterns

Look at this pattern.

 *What is the next shape?*

A pattern like this forms a **sequence**.

This is the second **term** of the sequence.

Here is another pattern:

 *What are the next 2 terms in the sequence?*

When you count the dots, you get

    1, 4, 9, . . .

This is a sequence of numbers.

 *What are the next 2 terms?*

Now look again at the pattern at the top of the page.

 *Count the numbers of dots and make a sequence of numbers.*

# 29: Sequences

**1** Write down the next 3 terms in each of these sequences.

a) 2, 4, 6, . . .   b) 5, 10, 15, . . .   c) 1, 3, 5, . . .

**2** a) Draw the next 3 shapes in this sequence.

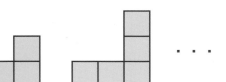

Write down the first 6 terms in the sequence of the numbers of small squares.

b) Draw the next 3 shapes in this sequence.

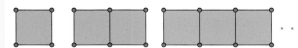

Write down the first 6 terms in the sequence of the numbers of

(i) small squares

(ii) dots.

c) Draw the next 3 shapes in this sequence.

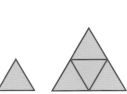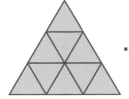

Write down the first 6 terms in the sequence of the numbers of small triangles.

**3** Rupesh wants a pond for his garden.

He looks at some designs in a brochure.

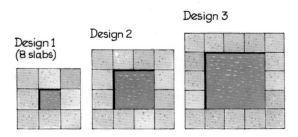

a) Write down the sequence of the numbers of slabs for each design.

b) What is the next term in the sequence?

c) Draw design 4 and count the number of slabs to check that you are right in b).

Make up four 'What comes next?' problems to try out on your friends. Some of them should involve numbers, but they don't all have to.

Here are two examples (don't use these!).

What comes next in

a) 20, 22, 24, 26, …, …?

b) J, F, M, A, …, …?

## 29: Sequences

# More sequences

During the day, trams go from Altrincham to Manchester every 12 minutes.

Trams leave Altrincham at this many minutes past every hour:

0, 12, 24, 36 . . .

This means o'clock

This means 12 minutes past, etc

The connection between the times is

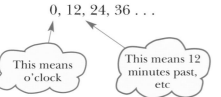

The rule is ADD 12 MINUTES.

 *What happens to this sequence when it gets to 60?*

Now look at this sequence:

20, 18, 16, 14, . . .

 *What are the next 3 terms?*

The connection between successive terms is

$$\boxed{20} \xrightarrow{-2} \boxed{18} \xrightarrow{-2} \boxed{16} \xrightarrow{-2} \boxed{14}$$

So the rule is SUBTRACT 2 and the sequence carries on like this:

20, 18, 16, 14, 12, 10, . . .

# 29: Sequences

**1** Write down the next three terms of each sequence.

Write down the rule that you use.

a) 2, 4, 6, 8, . . .

b) 1, 5, 9, 13, . . .

c) 10, 9, 8, 7, . . .

d) 128, 120, 112, 104, . . .

e) 0.5, 1, 1.5, 2, . . .

f) 9, 7, 5, 3, . . .

**2** Anna lives in a tower block.

The lift has broken down again.

She counts the steps and calls out the number as she reaches each floor.

The sequence of numbers is

15, 30, 45, . . .

a) What number does she call out next?

b) How many steps are there between each floor?

c) Anna lives on the 7th floor. How many steps does she have to climb?

**3** Ella is walking down the street.

a) She sees these house-number signs on the right.

| EVENS | EVENS | EVENS |
| 2 – 8 | 10 – 16 | 18 – 24 |

What are the next 3 signs in this sequence?

b) She sees these house-number signs on the left.

| ODDS | ODDS | ODDS |
| 1 – 15 | 17 – 31 | 33 – 47 |

What are the next 3 signs in this sequence?

c) Ella is visiting house number 71. Which sign is she looking for?

**4** John has 4 hospital appointments, one every two weeks beginning on 4 September.

Write down the sequence of dates for John's appointments.

**5** The Olympic Games are held every four years.

They were held in 2004.

Write down the years of the next three Olympic Games after this date.

Mandy and Phil always have a house party on the first Sunday of the year. In 2006, this was on 1 January.

Write down the sequence of dates of the party for 2007 to 2016.

# 29: Sequences

## Finding n

At a farm show, Mr Brown makes sheep pens with fences like this.

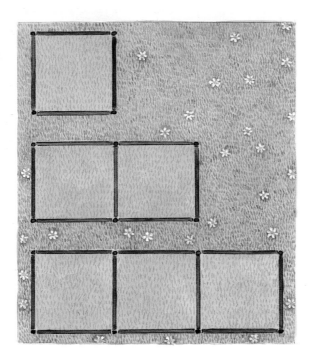

4 fences make 1 pen

7 fences make 2 pens

10 fences make 3 pens

 *How many fences does he need to make 4 pens?*

Look at the number of fences for each pen.

| Number of pens | Number of fences |
|---|---|
| 1 | 4 |
| 2 | 7 |
| 3 | 10 |

To find the number of fences, you multiply the number of pens by 3 and then add 1.

Number of fences = (Number of pens × 3) + 1

Call this $n$  Call this $p$

We can write it like this:

$n = (p \times 3) + 1$

If we want 4 pens, $p = 4$ so $n = (4 \times 3) + 1 = 13$.

# 29: Sequences

**1** Look at this sequence:

1, 6, 11, 16, . . .

The arrow diagram looks like this:

term number      term

1 ⟶ 1
2 ⟶ 6
3 ⟶ 11
4 ⟶ 16

a) What is the rule for getting the next term?

b) What is term number 5?

**2** Look at this sequence:

60, 55, 50, 45, . . .

Copy and complete this arrow diagram:

term number      term

1 ⟶ 60
2 ⟶
3 ⟶
4 ⟶

a) What is the rule for getting the next term?

b) What term number is 25?

**3** Michelle is given £50 for her birthday.

Each week after her birthday, she gets £10 pocket money.

She is saving up for a mountain bike, which costs £130.

a) Write down a sequence showing how much she has after 1 week, 2 weeks, etc.

b) How much does she have after 5 weeks?

c) Draw an arrow diagram to show how much she has saved each week.

Weeks after      Amount
birthday, $N$      saved, £$A$

1 ⟶ £60
2 ⟶

d) Work out a formula in words and letters for the amount Michelle saves after $N$ weeks.

e) How long does it take her to get the bike?

The diagram shows a house of cards with three layers. If you are careful you can build card houses with more layers. Find the numbers of cards in each layer as a sequence and find a formula for it.

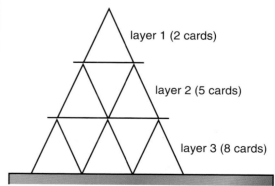

layer 1 (2 cards)

layer 2 (5 cards)

layer 3 (8 cards)

# More number sequences

Vicky is a scientist.

The number of bacteria in her experiment doubles every week.

At the start there are 2.

After 1 week there are 4.

After 2 weeks there are 8.

 *How many bacteria are there after 3 weeks?*

The numbers make this sequence:

   2, 4, 8, 16, . . .

The connection between them is

   ×2   ×2   ×2   ×2
   [2] → [4] → [8] → [16]

So the rule is MULTIPLY BY 2.

The Men's Singles Competition at Wimbledon has 128 players in round 1.

Half of the players (those who lose their matches) are knocked out at every round.

There are 64 players in round 2.

There are 32 players in round 3.

 *How many are there in round 4?*

The numbers make this sequence:

   128, 64, 32, 16, . . .

The connection between them is

   ÷2   ÷2   ÷2   ÷2
   [128] → [64] → [32] → [16]

So the rule is DIVIDE BY 2.

 *How many rounds are there before there is a winner?*

# 29: Sequences

**1** Write down the next 3 terms for each sequence.

Write down the rule that you use.

a) 1, 2, 4, . . .

b) 5, 10, 20, . . .

c) 1, 3, 9, . . .

d) 10, 100, 1000, . . .

**2** Write down the next 3 terms for each sequence.

Write down the rule that you use.

a) 800, 400, 200, . . .

b) 100 000, 10 000, 1000, . . .

c) 64, 32, 16, . . .

**3** Jane is folding a large sheet of thin card, 1 metre by 1 metre.

The card is 1 mm thick.

She folds it in half.

Now the card is 2 mm thick.

She folds the card in half again.

a) How thick is it now?

Jane carries on folding the card in half like this.

b) Write down the first 6 terms in the sequence of how thick the card is, starting with 1 mm.

Before she folds it, the area of the card is 1 $m^2$.

c) Write down the first 6 terms in the sequence of the area of the folded card, starting with 1 $m^2$.

d) What is the area of the card when it is 8 mm thick?

**4** Usma plants a creeper plant. It grows along her fence.

After 1 year it covers 1 fence panel.

Usma knows that this type of plant doubles every year.

a) How many panels does the plant cover after 2 years?

b) How many panels does it cover after 3 years?

c) Usma has 16 panels on her fence altogether.

How long is it before the plant covers the whole fence?

---

Paper comes in various sizes.

The sizes, starting large and getting smaller, are called

A0, A1, A2, A3, A4, A5, . . .

Find out how the paper sizes in the sequence are related.

# 29: Sequences

## Finishing off

**Now that you have finished this chapter you should**

★ understand the meaning of a sequence and a term

★ be able to continue a sequence

★ be able to find a rule for finding the next term of a sequence in words and symbols

★ be able to find the $n$th term

Use the questions in the next exercise to check that you understand everything.

**Mixed exercise**

**1** What is the next pattern in this sequence?

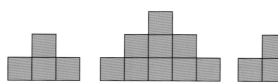

Write down the number of small squares in each of the patterns.

What do you notice?

**2** Write down the next 2 terms in each of these train time sequences.

0900, 0915, 0930, . . .

0800, 0930, 1100, . . .

0700, 0745, 0830, . . .

**3** A crocus bulb splits into 3 each year.

Write down the next 5 terms in the sequence of the number of new bulbs each year.

1, 3, 9, . . .

**4** Look at this number sequence:

2, 4, 6, 8, . . .

We can plot these points on a graph:

Copy the graph and plot the next 3 terms in the sequence.

Can you find a formula for term $n$?

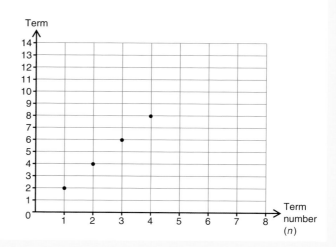

310

# 29: Sequences

**5** A small booklet is made by folding a sheet of A4 paper in half.

a) Now another sheet is folded and the two sheets are stapled like this:

How many pages does this booklet have?

b) Write down a sequence of the number of pages in a booklet made in this way.

c) Copy and complete the arrow diagram.

| Number of A4 sheets | Number of pages in booklet |
|---|---|
| 1 | ⟶ |
| 2 | ⟶ |
| 3 | ⟶ |
| 4 | ⟶ |

d) Copy and complete this sentence:

For each extra A4 sheet used the booklet has _____ more pages.

e) Find a formula for working out the number of pages in a booklet made with $n$ A4 sheets.

---

## Mixed exercise

In the activity on page 289 you made some stars.

In this activity you look at the angles of the points of the stars. You will find it helpful to draw the polygons in circles.

Here is a 9-sided polygon.

The angles at the centre are
$360° ÷ 9 = 40°$

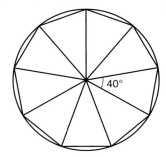

Draw stars with 5, 7, 9, … points and measure the angles. Find a formula connecting the angle, 180° and the number of points.

Do the same for stars with 6, 8, 10, … points.

Which stars do you think look best?

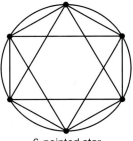

6-pointed star

7-pointed star

# Thirty

## Using percentages

**Before you start this chapter you should**

★ know the fraction and decimal equivalents of 50%, 25% and 75%

★ be able to calculate the percentage of a number

## Percentages, decimals and fractions

Ali is doing a survey for a travel agent.

He asks 50 people

Have you ever been to France?

20 people say 'yes'.

What is $\frac{20}{50}$ as a percentage and a decimal?

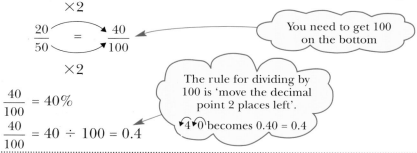

$$\frac{20}{50} \overset{\times 2}{\underset{\times 2}{=}} \frac{40}{100}$$

You need to get 100 on the bottom

$\frac{40}{100} = 40\%$

$\frac{40}{100} = 40 \div 100 = 0.4$

The rule for dividing by 100 is 'move the decimal point 2 places left'.
40 becomes 0.40 = 0.4

So $\frac{20}{50}$ **is the same as 40% or 0.4.**

Now Ali asks these 20 people

35% of these people say 'yes'.

How many people is this?

Did you visit Paris?

$35\%$ of 20

$= \frac{35}{100} \times 20$

$= \frac{700}{100} = 7$

**So 7 people have been to Paris.**

# 30: Using percentages

**1** Look at this pie chart.

It shows the amount of sales a company makes in different parts of the world.

a) Which of these is the correct answer?

(i) The amount of sales in the UK is

A less than 25%  B 25%  C more than 25%.

(ii) The amount of sales in the Rest of Europe is

A less than 50%  B 50%  C more than 50%.

(iii) The amount of sales in the USA is

A less than 25%  B 25%  C more than 25%.

b) Estimate the percentage of sales in other countries.

**2** A survey asks a group of teenagers what time they went to bed last night.

This pie chart shows the results.

What percentage went to bed

a) before 10 p.m.?

b) after 10 p.m. but before midnight?

c) after midnight?

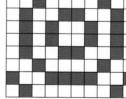

**3** Each of these floor designs is made of 100 tiles.

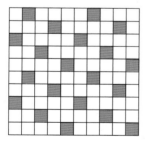 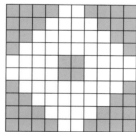 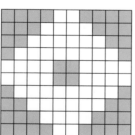

a) What percentage of each design is coloured?

b) Write each percentage as a fraction in its simplest form.

c) Write each percentage as a decimal.

---

Make two floor designs of your own where 20% of the tiles are white and the rest are black.

Look at real floors and tiles for ideas.

## 30: Using percentages

# Percentage calculations

Stephen runs a fashion business.

Last year he made a profit of £150 000.

This year he makes 20% more profit.

He works out his total profit for this year like this:

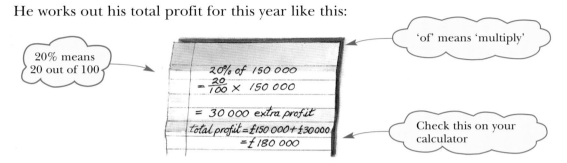

'of' means 'multiply'

20% means 20 out of 100

Check this on your calculator

**So Stephen's total profit this year is £180 000.**

Mark works for Stephen.

He earns £170 a week.

He gets a 3% pay rise.

Mark works out how much extra money he will get like this:

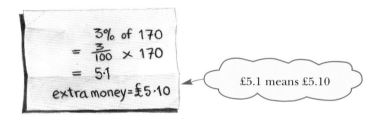

£5.1 means £5.10

*How much does Mark earn after the pay rise?*

Staff get 15% discount when they buy clothes from Stephen.

Mark buys these clothes, worth £125.60.

His discount is 15% of £125.60.

He works it out like this:

*How much does he pay for the clothes?*

*How much discount does he get off the polo-neck jumper?*

# 30: Using percentages

**1** Work these out.

a) 25% of 800  b) 80% of 300  c) 60% of 250  d) 50% of 133
e) 12% of 460  f) 75% of 1500  g) 40% of 320  h) 35% of 650

**2** These flights are on special offer.

a) How much do you save if you book a flight to San Francisco today?

b) How much do you save if you book a flight to Tokyo today?

c) How much do you pay if you book a flight to Mexico today?

| | |
|---|---|
| San Francisco | £240 |
| Egypt | £200 |
| Tokyo | £500 |
| Mexico | £320 |

Book today for 10% discount — SPECIAL OFFER

**3** Neil is given a 5% pay rise. He earns £8000 before the rise.

a) How much does he earn after the rise?

b) Neil goes out for a meal to celebrate.

It comes to £30 and he leaves a 15% tip.

How much tip does he leave?

**4** Danielle's heating bills are £900.

How much would she save with roof insulation?

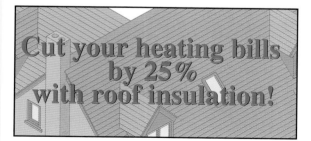

Cut your heating bills by 25% with roof insulation!

**5** Sam sells her house for £90 000.

She pays the estate agent 1.75% of this.

a) How much does she pay the estate agent?

b) Sam buys carpets worth £1600 for her new house.

She pays cash and gets a 5% discount.

How much does she pay for the carpets?

---

Watch a television programme (with adverts) that is scheduled to take $\frac{1}{2}$ an hour.

What percentage of the time is adverts and what percentage is the programme itself?

## 30: Using percentages

# Proportions

Dr Lee is trying out 2 new migraine treatments.

She gives 50 patients treatment A and 50 patients treatment B.

Next day she asks the patients how they feel.

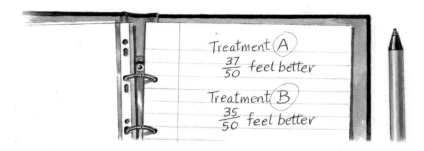

What percentage feel better after treatment A?

$$\frac{37}{50} = \frac{74}{100} = 74\%$$

**74% feel better after treatment A.**

*What percentage feel better after treatment B?*

*Which treatment works better?*

Dr Parker is also trying out the new treatments.

He gets these results:

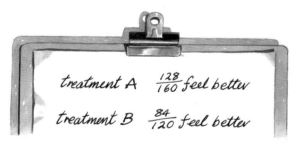

It's more difficult to work out which treatment works better for Dr Parker's patients.

He did not give the treatments to the same number of patients.

We can work out the **proportion** of patients who feel better.

With treatment A,

$\frac{128}{160} = 80\%$ or 0.8 feel better.

*What percentage feel better with treatment B?*

*Which treatment works better for Dr Parker's patients?*

## 30: Using percentages

**1** Write each fraction as a decimal and as a percentage.

a) $\dfrac{3}{4}$  b) $\dfrac{20}{80}$  c) $\dfrac{17}{50}$  d) $\dfrac{6}{10}$

e) $\dfrac{13}{15}$  f) $\dfrac{6}{25}$  g) $\dfrac{19}{20}$  h) $\dfrac{7}{9}$

**2** This table shows the members of two sports clubs.

|  | Female | Male |
|---|---|---|
| **Tennis** | 24 | 17 |
| **Badminton** | 31 | 23 |

a) How many members are there in the tennis club?

b) What proportion of the tennis club are female?

c) Which club has a higher proportion of female members?

d) Which club has a higher proportion of male members?

**3** Maureen manages a driving school.

| Instructor | Number taking test | Number passing test |
|---|---|---|
| Arthur | 45 | 26 |
| Barry | 53 | 17 |
| Caroline | 37 | 21 |

She makes this table to compare her instructors.

a) What proportion of Arthur's pupils pass the test?

b) What proportion of Barry's pupils pass the test?

c) What proportion of Caroline's pupils pass the test?

d) What do you think Maureen should do?

---

Ask 15 people you know to name the capital of Norway. (It's Oslo, if you have forgotten!)

What proportion get it right?

# 30: Using percentages

# Finishing off

**Now that you have finished this chapter you should be able to**

★ find the fraction and decimal equivalents of simple percentages

★ calculate the outcome of a percentage increase or decrease

★ change a fraction into a percentage

★ use percentages to calculate proportions

Use the questions in the next exercise to check that you understand everything.

## Mixed exercise

**1** What percentage of the costs is

a) labour?

b) overheads?

c) materials?

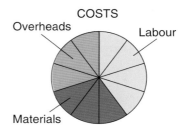

**2** 23% of marriages at a registry office are between partners who have both been married before.

31% are between partners where only one has been married before.

What percentage of couples have both never been married before?

**3** Work these out.

a) 75% of 600    b) 32% of 550    c) 7.5% of 140

**4** Jessica works in a clothes shop. She buys this coat.

She gets a 15% discount. How much does she pay?

**5** Lucy's sales figures last year were £53 000. Her target this year is an 8% increase.

What is her target figure? (Give your answer to the nearest hundred pounds.)

## 30: Using percentages

**6** Keith sees the same rucksack on sale in 2 shops.

Which shop is cheaper?

**7** Change the fraction to a percentage:

a) $\dfrac{4}{5}$   b) $\dfrac{7}{25}$   c) $\dfrac{5}{8}$   d) $\dfrac{161}{250}$

**8** In a spelling test, 30 people are asked to spell 'parallel'.

This is what they write. What percentage get it right?

| parallel | parallel | **pareloll** | parallel | **parallel** | parallel | **parallel** | **parallel** | **parallel** | parolell |
| parallel | **parrallell** | paralel | parallel | paralel | **parallel** | **parallel** | parelol | parelell | **parallel** |
| **paralell** | parallel | **parallel** | parralel | **parallel** | parolell | **parolell** | parelel | **parallel** | parolel |

**9** Becky's recommended daily allowance (RDA) of vitamin C is 60 milligrams.

She drinks a small glass of orange juice which has 20 milligrams in it.

a) Work out what percentage of her RDA this is.

b) How many small glasses of orange juice does she need to drink to obtain 100% of her RDA of vitamin C?

**10** A travel agent has four offices.

This bar chart shows the number of skiing holidays booked on Saturday at each office.

a) How many holidays are booked in total?

b) What proportion of these are booked at each office?

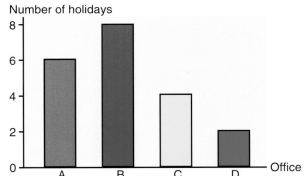

Describe 3 examples of when you have used percentages in other subjects.

# Thirty one

# Co-ordinates and graphs

**Before you start this chapter you should**

* understand what is meant by $x$ axis, $y$ axis and origin
* be able to write down the $x$ and $y$ co-ordinates of a point from a grid
* be able to plot a point $(x, y)$ on a grid

## More co-ordinates

M is the mid-point of AB.

Malik wants to find the co-ordinates of M.

3 is mid-way between 2 and 4.

*What is mid-way between 7 and 3?*

*Do you think that M is at $(3, 5)$?*

Draw this on squared paper and check whether $(3, 5)$ is the mid-point of AB.

Co-ordinates can also be used to describe a point in three dimensions.

The height above the $x$ and $y$ axes is measured on a $z$ axis.

The seal is at $(2, 0, 3)$.

*Where are the cat and the mouse?*

## 31: Co-ordinates and graphs

**1** Find the mid-point of the line AB in each of the following cases.
   a) A is (1, 4), B is (3, 8)
   b) A is (5, 2), B is (9, 8)
   c) A is (6, 0), B is (8, 4)
   d) A is (9, 3), B is (3, 5)
   e) A is (10, 5), B is (4, 7)
   f) A is (1, 7), B is (3, 9)

**2** Hang-glider R has position (3, 1) and hang-glider T has position (7, 3). Hang-glider S is mid-way between R and T.
Find the co-ordinates of S.

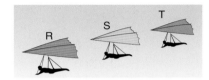

**3** The tops of three chimneys have co-ordinates A (1, 2) C (3, 6) and E (7, 10). Chimney B is mid-way between A and C, and chimney D is mid-way between C and E.
Find the co-ordinates of chimneys B and D.

**4** Five cable cars are travelling up a mountain. Car A is at (3, 8), car C is at (7, 14) and car E is at (11, 20). Car B is mid-way between A and C, and D is mid-way between C and E.
Find the co-ordinates of cars B and D.

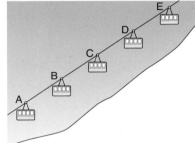

**5** An aircraft flying at a constant speed on a straight-line path is at A (2, 5). One minute later it is at B (6, 8). After one more minute it is at C.
Find the co-ordinates of C.

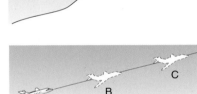

**6** Look again at the scaffolding on the previous page.
Give the positions in three dimensions of
   a) the fox
   b) the kangaroo
   c) the penguin
   d) the elephant
   e) the camel
   f) the footballer
   g) the blackbird
   h) the owl
   i) the dog.

**7** A stepped pyramid with a square base is shown in the diagram. Each step has a height of 3 units and each step has a width of 1 unit. So A is (1, 1, 6).
Given that B is (12, 0, 0), find the co-ordinates of C, D, E and F.

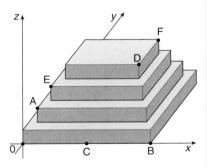

Choose one corner of the classroom as the origin.

Take measurements of your classroom and use them to give co-ordinates for the pen on your desk, the top and bottom of the legs on your chair, the chalkboard and the position of the light switch.

# 31: Co-ordinates and graphs

## Using co-ordinates

David and Elaine have found three corners P, Q and R, of Aemilia's house.

They have drawn a grid over the site.

*Aemilia's Summer House*
*The house is a parallelogram with a post at each corner.*

*Where is the fourth corner? Where do we dig?*

*M is the mid-point of PR. You can use it to find S.*

 What are the co-ordinates of S?

To find S you used a fact about parallelograms.

 Complete this statement: 'The diagonals of a parallelogram … each other.'

*There is another way to find S. From Q to R is 6 east and 2 north. So go to P and then walk 6 east and 2 north.*

 Where does this place S?

This time you used two more facts about parallelograms.

 What are the missing words in this statement?

'Opposite sides of a parallelogram are … and ….'

There are two other places S could be.

*These parallelograms are PSQR and PQSR*

 Where are they?

# 31: Co-ordinates and graphs

Use squared paper for this exercise.

**1** Three vertices of a parallelogram are at (1, 1), (–1, –2) and (–3, –1).

Find the possible positions of the fourth vertex.

**2** Three vertices of a parallelogram are at (4, 1), (–2, –3) and (–6, –1).

Find the possible values of the fourth vertex.

**3** Three vertices of a parallelogram ABCD are A (–2, 2), B(–1, 1) and C (6, 8).

a) Find the co-ordinates of D.

b) Measure the lengths AC and BD.

c) What is the special name of parallelogram ABCD?

**4** Three vertices of a parallelogram ABCD are A (3, –2), B (7, 6) and C (–1, 2).

a) Find the co-ordinates of D.

b) Measure the lengths AB and BC.

c) Measure the angle between AB and BC.

d) What is the special name of parallelogram ABCD?

**5** Three vertices of a parallelogram ABCD are A (–4, –3), B (1, –15) and C (13, –10).

a) Find the co-ordinates of D.

b) Measure the lengths AB and BC.

c) Measure the angle between AB and BC.

d) What is the special name of parallelogram ABCD?

**6** Three vertices of a parallelogram ABCD are A (4, 3), B (–1, –2) and C (1, –4).

a) Find the co-ordinates of D.

b) Measure the lengths AC and BD.

c) What is the special name of parallelogram ABCD?

---

Draw a parallelogram of your own on grid paper. Write down the co-ordinates of three of the points. Draw the other two parallelograms using these three points.

Exchange your three co-ordinates with a friend and find the missing co-ordinates for their three parallelograms. What shape do you get when you draw all three parallelograms on a single diagram?

## 31: Co-ordinates and graphs

# Equations and graphs

A farmer has 20 metres of fencing.

Here are some of the rectangular pens he can make:

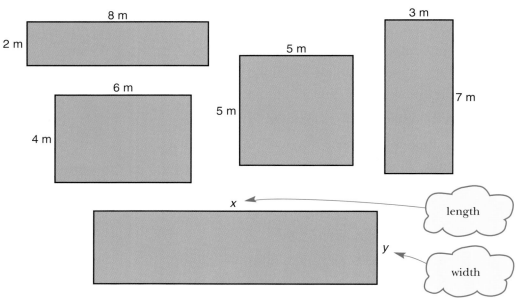

Call the length $x$ and the width $y$.

The length and the width add up to 10 metres

$x + y = 10$

$x$ and $y$ can be these numbers:

| $x$ | 0 | 1 | 2 | 3 | 4 | 5 | 6 | 7 | 8 | 9 | 10 |
|---|---|---|---|---|---|---|---|---|---|---|---|
| $y$ | 10 | 9 | 8 | 7 | 6 | 5 | 4 | 3 | 2 | 1 | 0 |

These points are plotted on the graph:

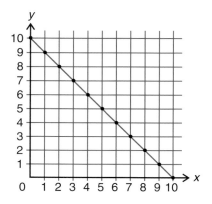

For any of the points on this line $x + y = 10$.

$x + y = 10$ is called the **equation** of the line.

This can also be written as $y = 10 - x$.

# 31: Co-ordinates and graphs

**1** a) Write down the co-ordinates of the red points on this grid.

b) Copy and complete this table to show the *x* and *y* co-ordinates of the points.

| x | 0 | 1 | 2 | 3 | 4 | 5 | 6 | 7 | 8 | 9 |
|---|---|---|---|---|---|---|---|---|---|---|
| y |   |   |   |   |   |   |   |   |   |   |

c) Look carefully at the values of *x* and *y*. Can you spot the connection between them? Write it down in words and symbols.

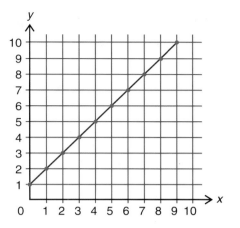

**2** a) (i) Write down the co-ordinates of the red points on this grid.

(ii) Make a table showing these values, like the one in question 1.

(iii) Write down in words and symbols a connection between *x* and *y*.

b) Repeat these three steps for the green points.

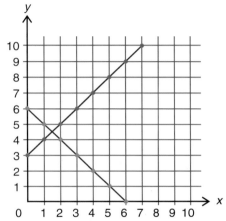

**3** A set of points has *y* co-ordinates which are always 2 more than the *x* co-ordinate. So when *x* is 0, *y* is 2, and so on.

a) Copy and complete this table.

| x | 0 | 1 | 2 | 3 | 4 | 5 | 6 | 7 | 8 |
|---|---|---|---|---|---|---|---|---|---|
| y | 2 |   |   |   |   |   |   |   |   |

b) Mark these points on a grid, as in the diagrams above.

Copy and complete this table for the graph of $y = 2x - 3$.

| x | 0 | 1 | 2 | 3 | 4 |
|---|---|---|---|---|---|
| 2x | 0 | 2 |   |   |   |
| −3 | −3 | −3 |   |   |   |
| y = 2x − 3 | −3 | −1 |   |   |   |

Draw the graph.

On the same graph paper draw the graph of $x + y = 6$.

Where do the two graphs meet?

325

## 31: Co-ordinates and graphs

# Curved graphs

The graphs on the previous pages have all been straight lines. Sometimes you need to draw curves, as in the two examples on this page.

**Example**

Draw the curve $y = x^2 - 4x + 3$ for values of $x$ between 0 and 4.

**Solution**

| $x$ | 0 | 1 | 2 | 3 | 4 |
|---|---|---|---|---|---|
| $x^2$ | 0 | 1 | 4 | 9 | 16 |
| $-4x$ | 0 | $-4$ | $-8$ | $-12$ | $-16$ |
| $+3$ | $+3$ | $+3$ | $+3$ | $+3$ | $+3$ |
| $y$ | $+3$ | 0 | $-1$ | 0 | $+3$ |

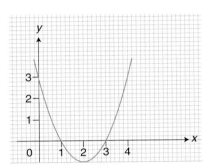

 Why is it wrong to join the points with straight lines?

What are the values of $x$ for which $y = 2$?

**Example**

Draw the curve $y = \dfrac{12}{x}$ for values of $x$ from 1 to 6.

**Solution**

| $x$ | 1 | 2 | 3 | 4 | 5 | 6 |
|---|---|---|---|---|---|---|
| $y = \dfrac{12}{x}$ | 12 | 6 | 4 | 3 | 2.4 | 2 |

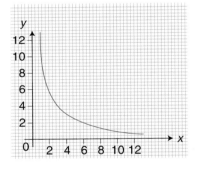

 In this graph $x$ cm and $y$ cm could be the length and width of a rectangle of area $12$ cm$^2$.

What is the length of the rectangle if the width is 2.5 cm?

What other meanings could $x$ and $y$ have?

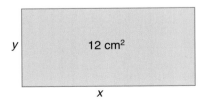

 What happens to $y = \dfrac{12}{x}$ when $x$ is 0?

# 31: Co-ordinates and graphs

**Exercise**

**1** Alka throws a tennis ball up in the air. Its height, $h$ metres above the ground, at time $t$ seconds is given by
$$h = 20t - 5t^2$$

a) Copy and complete this table of values of $h$.

| $t$ | 0 | 1 | 2 | 3 | 4 |
|---|---|---|---|---|---|
| $20t$ | | | | | 80 |
| $-5t^2$ | 0 | $-5$ | | | $-80$ |
| $h$ | | | | | 0 |

*For $-5t^2$ work out $t^2$ first and then multiply by $-5$: when $t = 4$, $4^2 = 16$ and $-5 \times 16 = -80$*

b) Choose suitable scales and draw the graph of $h$ against $t$.

c) Use your graph to estimate the times at which the ball is 10 m above the ground. Why do you get two answers?

d) Why is it not sensible to take values of $t$ greater than 4?

**2** A curve has equation $y = -4 + 5x - x^2$.

a) Copy and complete this table of values.

| $x$ | 0 | 1 | 2 | 3 | 4 | 5 |
|---|---|---|---|---|---|---|
| $-4$ | $-4$ | | | | | $-4$ |
| $+5x$ | 0 | | | | | 25 |
| $-x^2$ | 0 | | | | | $-25$ |
| $y$ | $-4$ | | | | | $-4$ |

*Notice that this is $-5^2$ not $(-5)^2$*

b) Choose suitable scales and draw the graph.
c) Estimate the greatest value of $y$ (i.e. the highest point of the curve).
d) Estimate the values of $x$ for which $y = 1$.

**3** A curve has equation $y = x^3 - 2x$.

a) Copy and complete this table of values.

| $x$ | $-2$ | $-1$ | 0 | 1 | 2 |
|---|---|---|---|---|---|
| $x^3$ | $-8$ | | | | 8 |
| $-2x$ | $+4$ | | | | $-4$ |
| $y$ | $-4$ | | | | 4 |

*Notice that $(-2)^3$ is $-8$*

b) Choose suitable scales and draw the graph.
c) Estimate the values of $x$ where the curve crosses the $x$ axis.
d) Describe the symmetry of the curve.

---

On graph paper draw $x$ and $y$ axes from $-5$ to 5. Use the same scale for $x$ and $y$. Plot the points (5, 0), (4, 3) and (3, 4). Now plot the images of these points when they are rotated anticlockwise, centre the origin, through 90°, 180° and 270°. You should now have 12 points on your graph paper.

What curve do they lie on?

Join them up by hand to draw the curve as well as you can.

# 31: Co-ordinates and graphs

## Finishing off

**Now that you have finished this chapter you should be able to**

- write down the *x* and *y* co-ordinates of a point from a grid
- make a table of *x* and *y* from a set of points and look for a simple connection between them
- plot a point (*x*, *y*) on a grid
- find the co-ordinates of the mid-point of a line
- make a table of *x* and *y* from a simple connection between them and draw the points on a grid
- use co-ordinates to describe a point in three dimensions

Use the questions in the next exercise to check that you understand everything.

## Mixed exercise

**1** Three vertices of a parallelogram ABCD are A (2, 1), B (6, 5) and C (2, 9).

a) Find the co-ordinates of D.

b) Measure the lengths AC and BD.

c) Measure the angle between AC and BD.

d) What is the special name of parallelogram ABCD?

**2** a) Copy and complete this table for

$y = 8 - x$

| x | 0 | 1 | 2 | 3 | 4 | 5 | 6 | 7 | 8 | 9 | 10 |
|---|---|---|---|---|---|---|---|---|---|---|----|
| y |   |   |   |   |   |   |   |   |   |   |    |

b) Draw a square grid with 0 to 10 on the *x* axis and −2 to 8 on the *y* axis.

c) Mark the points (*x*, *y*) shown in the table.

**3** Three vertices of a parallelogram ABCD are A (1, 2), B (−6, 1) and C (−1, −4).

a) Find the co-ordinates of D.

b) Measure the lengths AB and BC.

c) Measure the angle between AB and BC.

d) What is the special name of parallelogram ABCD?

# 31: Co-ordinates and graphs

**4** Jane jumps off the high diving board into the pool.
The graph shows the stages of her dive.

a) What is happening from
  (i) A to B?
  (ii) B to C?
  (iii) C to D?

b) How long after she jumps does Jane hit the water?

c) How far below the surface of the water does Jane dive?

d) How long is she under water?

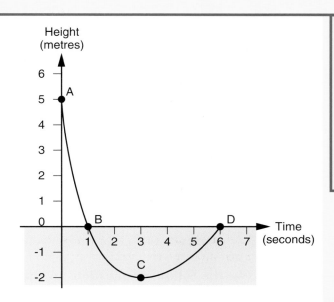

**5** This grid shows the position of a helicopter on a radar screen.

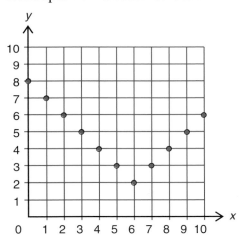

**6** A curve has equation $y = x^2 - 5$.

a) Copy and complete this table of values.

| $x$ | –3 | –2 | –1 | 0 | 1 | 2 | 3 |
|---|---|---|---|---|---|---|---|
| $x^2$ | +9 |  | +1 | 0 |  |  |  |
| –5 | –5 | –5 |  | –5 |  |  |  |
| $y$ | +4 |  |  | –5 |  |  |  |

b) Draw the graph. Use scales of 2 cm for 1 unit on each axis.

c) Find the value of $y$ when $x = 1.2$.

d) Estimate the values of $x$ for which $y = 2$.

a) What do you think happened at the point (6, 2)?

b) Make two tables to record the $x$ and $y$ co-ordinates, the first table to show the co-ordinates of the red points and the second to show the co-ordinates of the green points.

c) Write down in words and symbols a connection between $x$ and $y$ for each table.

---

At the end of a hockey match, Smruti finds that she has lost a gold ring somewhere on the pitch.

All 11 players on her team, and the coach, agree to look for it.

Work out a plan for them to do it systematically.

# Thirty two

# Using statistics

**Before you start this chapter you should be able to**

- ★ make and use a data collection sheet
- ★ make and use a tally chart
- ★ draw and use pictograms, bar charts, vertical line charts, line graphs
- ★ calculate the mean, median and mode of a set of data
- ★ calculate the range of a set of data
- ★ carry out a survey

Use the following questions to check that you still remember these topics.

## Revision exercise

**1** This pictogram shows the results of a football team one season.

a) Draw possible symbols for 1 match and 3 matches.

b) The team scores 3 points for a win, 1 for a draw and 0 for a loss. How many points did they get?

c) In the season, each team plays every other team in the league twice. How many teams are there in the league?

| Win | ✗✗✗✗✗ |
| Draw | ✗✗❯ |
| Lose | ✗✗ |

✗ means 4 matches

**2** Scott puts some bread on his bird table and keeps a record of the birds that eat it. His results are shown in the bar chart.

a) Make a frequency table.

b) How many birds does Scott see in total?

c) What percentage of the birds are starlings?

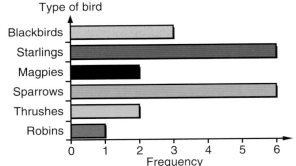

**3** Fran stands outside her house and counts the number of people in the first 20 cars that pass by.

1, 3, 4, 1, 4     5, 4, 1, 2, 1     2, 3, 1, 1, 2     3, 3, 2, 1, 6

a) Make a tally chart of the data.

b) Make a frequency table of the data.

c) Draw a vertical line chart to illustrate the data.

# 32: Using statistics

**4** Ten people work in an office. The numbers of days they are out of the office during one week are as follows.

John 2    Wendy 0    Jenny 0    Karen 0    Roger 2

Stuart 1    Ted 2    Ali 0    Sunil 3    Ray 5

Find

a) the mode    b) the mean

c) the median    d) the range

of the number of days people are absent.

**5** Joe is a caretaker for an office.

He must keep the office at a comfortable temperature, not too hot and not too cold.

One day he takes the temperature several times and draws this graph.

a) How many times does Joe take the temperature?

b) What is the hottest the office gets?

c) Chris is in the office from 9 am to 3 pm

What is the coldest the office gets while she is there?

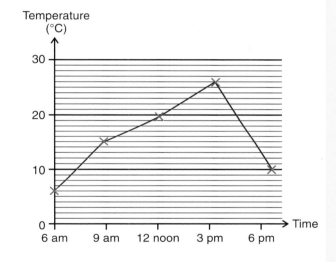

**6** Jenny has kept in touch with some of the girls in her class at school.

This vertical line chart shows the number of children they have had.

a) How many women does the chart represent?

b) What is the mode of their number of children?

c) What percentage have no children?

d) How many children do they have in total?

e) What is the mean number of children?

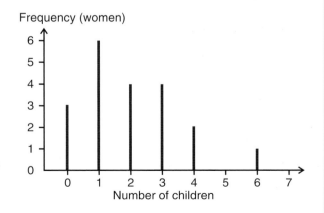

Carry out a survey into the sort of shoes your friends like for everyday wear.

## 32: Using statistics

# Pie charts

Leroy is asked to do a survey to find out how people in his company travel to work.

Here are his results:

| Car | Motor-bike | Bicycle | Bus | Train | Walk | Skate-board | Total |
|---|---|---|---|---|---|---|---|
| 90 | 5 | 8 | 15 | 15 | 45 | 2 | 180 |

He wants to show this information on a diagram and decides to use a pie chart.

There are 180 people in total and 360° in a full circle.

$$\frac{360°}{180} = 2°$$

So 1 person has 2° on the pie chart.

Leroy sets out his work like this:

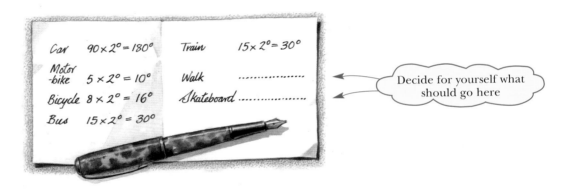

Car  90 × 2° = 180°    Train  15 × 2° = 30°
Motor-bike  5 × 2° = 10°    Walk  ...........
Bicycle  8 × 2° = 16°    Skateboard  ...........
Bus  15 × 2° = 30°

Decide for yourself what should go here

Then he draws the pie chart.

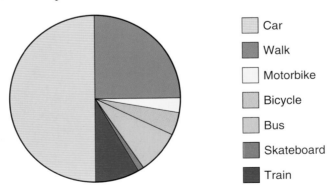

- Car
- Walk
- Motorbike
- Bicycle
- Bus
- Skateboard
- Train

What percentage come by car?

How can you tell this from the pie chart?

What percentage walk?

Why do you think the company wants to know this information?

# 32: Using statistics

**1** A clothing company makes £360 000 profit one year. The pie chart shows how this was divided between the children's, teenage and adult departments.

   a) Measure the angles in the pie chart.

   b) How much profit does each department make?

   c) What percentage of the profit comes from teenage clothes?

**2** Kevin times some traffic lights near his house.

| Colour | Red | Red and amber | Green | Amber |
|---|---|---|---|---|
| Time in seconds | 18 | 2 | 12 | 4 |

   a) Draw a pie chart to show these figures.

   b) What percentage of the time are the lights red?

**3** This pie chart shows the results for the 36 teams playing in a cricket league one day.

   a) How many teams win?

   b) How many teams lose?

   c) How many teams draw?

   d) Why are the angles for Win and Lose the same size?

**4** Anna says that on a typical day she spends her time like this:

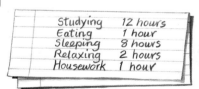

Studying   12 hours
Eating      1 hour
Sleeping    8 hours
Relaxing    2 hours
Housework   1 hour

She wants to show this on a pie chart.

   a) How many degrees represent 1 hour?

   b) Draw the pie chart.

   c) What percentage of her time is spent studying?

---

Do the popular colours for cars change from year to year?

Go round a large car park and record the colour of each car and the registration year letter.

Draw pie charts to compare the results, year by year.

## 32: Using statistics

# Mean, mode, median and range

20 people enter a fishing competition. The number of fish they catch is shown in the frequency table below. You can also see this information in the vertical line chart.

| Number of fish | Number of people (frequency) |
|---|---|
| 0 | 0 |
| 1 | 6 |
| 2 | 5 |
| 3 | 3 |
| 4 | 2 |
| 5 | 1 |
| 6 | 1 |
| 7 | 0 |
| 8 | 2 |

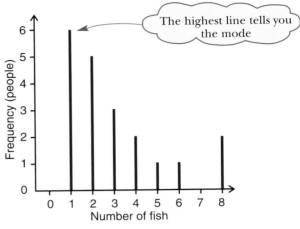

The highest line tells you the mode

*What is a typical number of fish?*

Here are three answers you could give.

The **mode** is the most common number of fish, the number with the greatest frequency.

*6 people caught 1 fish. The next most common is 2 fish*

The mode is 1.

The **mean** = $\dfrac{\text{the total number of fish}}{\text{the total number of people}}$

$= \dfrac{0\times0 + 1\times6 + 2\times5 + 3\times3 + 4\times2 + 5\times1 + 6\times1 + 7\times0 + 8\times2}{0 + 6 + 5 + 3 + 2 + 1 + 1 + 0 + 2}$

$= \dfrac{60}{20} = 3.$

*A total of 60 fish have been caught by the 20 people in the group*

The middle one when you write the data in order is called the **median**.

Number 10  Median  Number 11

1, 1, 1, 1, 1, 1, 2, 2, 2, 2, 2, 3, 3, 3, 4, 4, 5, 6, 8, 8

The median of 20 is half way between numbers 10 and 11. In this case these are both 2, so the median is 2.

How spread out are the answers? You can measure this by the **range**.

Range = Largest value − Smallest value

In this case Range = 8 − 1 = 7

# 32: Using statistics

**1** For each of these sets of data, find the mode, median, mean and range.

a) 8, 9, 10, 14, 14

b) 2, 3, 3, 3, 4, 4, 4, 4, 5, 8

c) 
| Value | 0 | 1 | 2 | 3 | 4 |
|---|---|---|---|---|---|
| Frequency | 2 | 1 | 4 | 1 | 2 |

d)
| Value | 21 | 22 | 23 | 24 | 25 |
|---|---|---|---|---|---|
| Frequency | 1 | 1 | 2 | 1 | 1 |

e) Frequency vs Value chart with bars: 0→4, 3→1, 4→3, 5→1, 6→1

**2** Alice keeps chickens. She keeps a record of how many eggs they lay each day during one month. Here is the frequency table.

| Number of eggs | 0 | 1 | 2 | 3 | 4 | 5 | 6 |
|---|---|---|---|---|---|---|---|
| Number of days (frequency) | 2 | 5 | 4 | 6 | 8 | 3 | 2 |

a) For how many days does she keep a record?

b) Find

    (i) the mode     (ii) the mean

    (iii) the median     (iv) the range    of the number of eggs.

**3** Some households buy a newspaper every day, others some evenings and others never. These vertical line charts refer to houses in two streets.

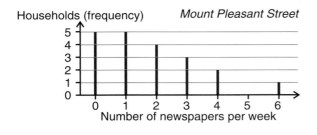
Mount Pleasant Street

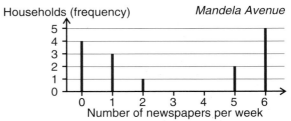
Mandela Avenue

a) Make a frequency table for each street.

b) Find the mean of the number of newspapers per week for each street.

c) Find the range of the number of newspapers per week for each street.

d) Compare the results for the two streets.

> Take a tabloid newspaper and count the number of words in each of 20 sentences. Find the mean and range.
>
> Now take a broadsheet newspaper and do the same.
>
> Compare the sentences in the two newspapers.

## 32: Using statistics

# Grouping data

Melissa's parents are angry with her. They say she is spending too long on the telephone. One week she makes 20 calls. These are their lengths, in minutes and seconds.

| 11m 22s | 34m 12s | 19m 16s | 22m 41s | 50m 00s |
| 48m 3s  | 11m 11s | 26m 12s | 16m 42s | 17m 36s |
| 51m 36s | 17m 11s | 20m 48s | 12m 33s | 10m 39s |
| 49m 29s | 18m 12s | 25m 47s | 44m 12s | 52m 18s |

It is quite hard to read data when they are given like this. It is better if they are grouped.

| Length (minutes) | $0 \leq x < 10$ | $10 \leq x < 20$ | $20 \leq x < 30$ | $30 \leq x < 40$ | $40 \leq x < 50$ | $50 \leq x < 60$ |
|---|---|---|---|---|---|---|
| Frequency (number of calls) | 0 | 9 | 4 | 1 | 3 | 3 |

Melissa says 'Look. Most of my calls are less than 20 minutes.'

*Is Melissa correct? What does the frequency table really show?*

She draws a frequency chart to make her point.

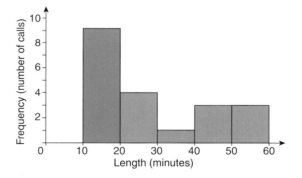

The highest bar is the highest frequency; it is called the **modal class**.

Another way of showing these data is the frequency polygon below.

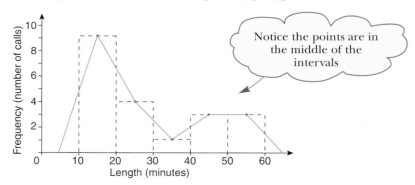

Notice the points are in the middle of the intervals

*Do you think Melissa's parents are right to be angry with her?*

# 32: Using statistics

**1** A company is testing a new type of seed. They plant twenty packets each containing 15 seeds and count how many grow.

| 9 | 15 | 1 | 2 | 3 | 10 | 12 | 13 | 1 | 6 |
| 6 | 11 | 15 | 0 | 1 | 14 | 13 | 14 | 14 | 1 |

a) The company group the data. Copy and complete this frequency table.

| Number of plants growing | $0 \leq x < 4$ | $4 \leq x < 8$ | $8 \leq x < 12$ | $12 \leq x < 16$ |
|---|---|---|---|---|
| Description | Very low | Low | Medium | High |
| Frequency | | | | |

b) Draw a bar chart.

c) What do you think will happen if the company sells the seeds?

**2** Look at the lengths of Melissa's telephone calls opposite.

a) Her total time on the telephone that week was 560 minutes.

How could you have worked this out?

What is the mean length of her 20 calls?

b) How long does she spend on the telephone each day, on average?

c) Melissa's parents make out a new table, with the data grouped like this:

$0 \leq x < 5, 5 \leq x < 10, \ldots, 55 \leq x < 60$

Draw a frequency chart, and draw a frequency polygon on it.

d) What is the modal class now?

---

To be done as a group.

Find out how good you are at estimating 1 minute. One person says 'Start' and starts timing you. When you think 1 minute is over you say 'Stop'. Group the data and show the results as a frequency chart.

---

**3** The sale prices for 30 cars at an auction are as follows.

£1430  £1750  £2430  £4560  £3480
£2520  £4160  £1995  £2460  £2840
£5100  £3275  £2160  £4050  £5120
£3500  £2750  £1850  £3520  £2650
£1200  £4200  £4000  £5800  £4950
£4510  £3840  £2380  £5750  £4800

Going, going, gone for £4800

a) Make out a tally chart to put the data into groups $0 \leq x < 1000$, $1000 \leq x < 2000$, $2000 \leq x < 3000$, and so on.

b) Make a frequency table.

c) Draw a frequency chart and a frequency polygon.

d) Which price range is the modal class?

## 32: Using statistics

# Scatter diagrams

Nina and her mother are entering a fun run. Nina says, 'You are older than me so you will take longer.'

*Is it true that older people usually take longer over a run?*

Here are scatter diagrams for three age groups: Time against Age.

You can see three different patterns.

*Notice the symbol* . *This shows a break in the scale on the axis. Why is it used on the graphs on this page?*

In the Juniors the line slopes down.

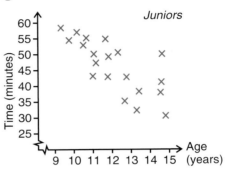

This is called **negative correlation**.

The greater the age, the less time they take.

In the Adults there is no pattern on the graph.

There is **no correlation**.

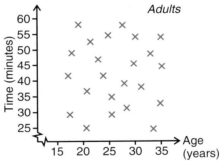

In the Seniors the line slopes up.

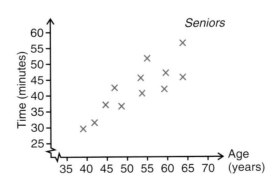

This is called **positive correlation**.

The greater the age, the greater the time they take.

When you look for correlation you are looking for an overall pattern. Not all of the points will fit into this pattern.

*These graphs are Time against Age. What can you say about Speed against Age?*

# 32: Using statistics

**1** For each of the scatter diagrams a) to f) answer 'positive correlation', 'no correlation' or 'negative correlation'.

a)   b)  c)

d)  e)   f)

**2** A company gives its secretaries a typing test. They all type for one hour. Here is how many pages they type and how many mistakes they make.

| Name | Pages | Mistakes | Name | Pages | Mistakes |
|---|---|---|---|---|---|
| Sue | 4 | 1 | Ali | 4 | 2 |
| Kevin | 6 | 4 | Mandy | 7 | 0 |
| Madeleine | 5 | 1 | Rus | 4 | 0 |
| Dolly | $3\frac{1}{2}$ | 0 | Savita | $6\frac{1}{2}$ | 3 |
| Lee | 6 | 2 | Linda | 5 | 2 |

a) Draw a scatter diagram.

b) Describe the correlation and say in simple English what it means.

c) Who is the best typist?

---

Look at the football league tables.

Draw scatter diagrams of

a) 'Goals for' against 'Points'.

b) 'Goals against' against 'Points'.

Describe any correlation your scatter diagrams show.

## 32: Using statistics

# Mean, median and mode of grouped data

Debbie organises coach tours. She keeps a computer record of the age-group of each person booked on a particular tour. When she is ready to make the detailed arrangements for a tour she prints out a frequency chart like this one. She can then try to suit the activities and timings to their ages.

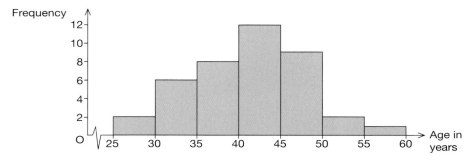

*How many people on this trip are aged 25–29?*

*What is the modal age group?*

*How can you estimate the mean age of the people booked on this tour?*

It is impossible to work out the mean of grouped data accurately. You need the raw data. However, you can make an estimate like this.

| Age | 20–24 | 25–29 | 30–34 | 35–39 | 40–44 | 45–49 | 50–54 | 55–59 | |
|---|---|---|---|---|---|---|---|---|---|
| **Mid-point of age group** | 22 | 27 | 32 | 37 | 42 | 47 | 52 | 57 | |
| **Frequency (number of people)** | 0 | 2 | 6 | 8 | 12 | 9 | 2 | 1 | Total 40 |
| **Mid-point × frequency** | 0 | 54 | 192 | 296 | 504 | 423 | 104 | 57 | Total 1630 |

*(This is the total number of people)*

*(The median 40 ages is between the 20th and 21st ages. The median age group is 40–44)*

Mean age of people = $\dfrac{\text{total age}}{\text{total number of people}}$

$\approx \dfrac{1630}{40}$

Mean age $\approx 40.75$

*(This is an estimate of the total age of this group - it assumes they are all 47)*

*(This is an estimate of the total age of all the people)*

*Estimate the new mean when Frank Heys (aged 54) drops out of the tour.*

# 32: Using statistics

**1** Each passenger on a flight to Cairo is allowed one item of hand-luggage. The weights, to the nearest kilogram, of the hand-luggage items are given in the table.

| Weight (kg) | 1–5 | 6–10 | 11–15 | 16–20 |
|---|---|---|---|---|
| Frequency (no. of items) | 12 | 24 | 10 | 4 |

a) Which is the modal class?
b) What is the mid-point of the 6–10 kg class?
c) Estimate the total weight of hand-luggage on the plane.
d) Estimate the mean weight of the items.

**2** Ben works for an estate agency which earns a percentage of the house price for each house it sells. He is working out the mean and mode of the house prices for the last year's sales, in order to prepare a cashflow forecast for the next year. He uses this table.

| Price (£ thousands) | 31–40 | 41–50 | 51–60 | 61–70 | 71–140 |
|---|---|---|---|---|---|
| Frequency (no. of houses) | 35 | 42 | 56 | 7 | 3 |

a) Which is the modal class?
b) What is the mid-point of the class '71–140'?
c) Estimate the mean of these data.

**3** In a recent cross-country skiing championship, the times for the first hundred competitors were recorded as follows.

| Time (minutes) | 80–84 | 85–89 | 90–94 | 95–99 | 100–104 | 105–109 |
|---|---|---|---|---|---|---|
| Frequency (number of skiers) | 8 | 27 | 33 | 20 | 8 | 4 |

a) Which is the modal class?
b) Estimate the mean time taken by these hundred skiers.

The fastest skier in this table was actually disqualified, so another skier whose time was 107 minutes entered the top hundred skiers.

c) Estimate the new mean time for the top hundred.
d) What is the new modal class for the top hundred?

> Find the mean height of the people in your class or maths group.
>
> Do it first using the individual heights, then from a grouped frequency table (if possible using a spreadsheet). Compare the answers you get by these two methods.

# 32: Using statistics

# Line of best fit

If the points on a scatter diagram lie in a narrow band there is a strong correlation. The stronger the correlation, the narrower the band. Provided there is some correlation, you can add a line of best fit to your diagram as in the example below.

Adding a line of best fit to your scatter diagram helps you to estimate the value of one variable, given the value of the other.

This table shows the mean heights of parents and the mean heights of their adult offspring.

| Surname | Mean height of parents (cm) | Mean height of adult offspring (cm) |
|---|---|---|
| Wilson | 161 | 163 |
| Allan | 164 | 167 |
| Gupta | 166 | 168 |
| Smith | 169 | 170 |
| Lipton | 174 | 173 |
| Morris | 179 | 177 |
| Spicer | 183 | 179 |
| Gibson | 183 | 183 |

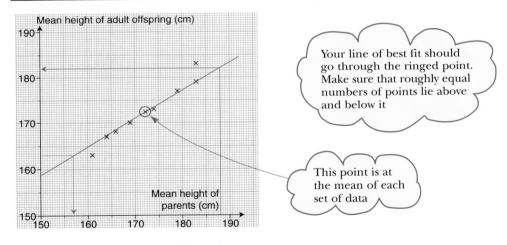

What does the diagram tell you?

A group of brothers and sisters found that their mean height was 163 cm. Use the scatter diagram to estimate the mean height of their parents. (The red line will help.)

A couple whose mean height is 188 cm are worried their children will be giants. They draw the green line on the diagram.

Do you think their children will all be 182 cm tall?

# 32: Using statistics

**1** This table records the force $P$ newtons required to move an object of mass $M$ kg on a rough surface.

| $M$ | 1 | 2 | 3 | 4 | 5 | 6 | 7 | 8 |
|---|---|---|---|---|---|---|---|---|
| $P$ | 7.5 | 14.4 | 22.5 | 30.5 | 37.4 | 45.4 | 53.5 | 61.4 |

a) Draw a scatter diagram and add the line of best fit.

b) Estimate the force required to move a mass of 2.6 kg.

c) Estimate the mass you could move with a force of 40 newtons.

**2**

The people in Gil's chemistry class have done an experiment in which an acid reacts with a solid. They recorded the volume of gas given off during the reaction, and the loss in mass of the reactants. Here are their results.

| Name | Alan | Bridie | Callie | Donna | Ed | Flick | Gil | Helen |
|---|---|---|---|---|---|---|---|---|
| Loss in mass (g) | 0.060 | 0.032 | 0.120 | 0.083 | 0.090 | 0.160 | 0.140 | 0.107 |
| Volume of gas (cm³) | 45 | 24 | 90 | 62 | 85 | 120 | 105 | 80 |

a) Draw a scatter diagram of their results.

b) Comment on what your diagram shows.

c) One student made a mistake during the measuring. Who do you think it was?

d) Ignoring the data from the person who made a mistake, draw a line of best fit (by eye) for the data.

e) Use your line to estimate the mass of 100 cm$^3$ of the gas.

**3** The table shows the number of hours of sunshine at a seaside resort one week in June, together with the number of people visiting the aquarium there.

| Day | Hours of sunshine | No. of visitors |
|---|---|---|
| Mon | 2.5 | 90 |
| Tue | 5.5 | 70 |
| Wed | 16 | 14 |
| Thurs | 13 | 28 |
| Fri | 10 | 44 |
| Sat | 0 | 96 |
| Sun | 7.5 | 64 |

a) Draw a scatter diagram.

b) Work out the mean of the number of hours of sunshine and the mean of the number of visitors. Plot the mean point on your diagram. Draw a line of best fit passing through this point.

c) How many visitors can the aquarium expect on a day when about 8 hours of sunshine are forecast?

Sometimes it does not make sense to draw a line of best fit. Draw two scatter diagrams for which this is the case.

## 32: Using statistics

# Finishing off

**Now that you have finished this chapter you should be able to**

- ★ draw pie charts and frequency polygons
- ★ work with mode, median, mean and range
- ★ group data
- ★ draw scatter diagrams and identify any correlation
- ★ draw a line of best fit

Use the questions in the next exercise to check that you understand everything.

## Mixed exercise

**1** This table shows the height in metres above sea level and the temperature in Celsius, on one day at 9 different places in Europe.

| Height (m) | 1400 | 400 | 280 | 790 | 370 | 590 | 540 | 1250 | 680 |
|---|---|---|---|---|---|---|---|---|---|
| Temperature (°C) | 6 | 15 | 16 | 10 | 14 | 14 | 13 | 7 | 13 |

a) What is the mean height?

b) What is the mean temperature?

c) Plot a scatter diagram and describe the correlation.

d) Add a line of best fit.

e) Use your diagram to estimate the temperature at a height of 500 m.

f) Use your diagram to estimate the height of a place with a temperature of 8 °C.

**2** Kevin does a survey of how many nights his friends went out in the last week.

The data are given below as two frequency tables, one for boys and one for girls.

| Boys | Nights out | 0 | 1 | 2 | 3 | 4 | 5 | 6 | 7 |
|---|---|---|---|---|---|---|---|---|---|
| | No of boys (frequency) | 4 | 2 | 1 | 0 | 2 | 2 | 5 | 4 |
| Girls | Nights out | 0 | 1 | 2 | 3 | 4 | 5 | 6 | 7 |
| | No of girls (frequency) | 0 | 0 | 2 | 4 | 9 | 2 | 3 | 0 |

a) For both boys and girls, find the mode and range of the number of nights out.

b) What do the mode and range tell you about the two groups?

c) Draw vertical line charts to illustrate the two groups.

# 32: Using statistics

**Mixed exercise**

**3** Denny wants to know if girls pass their driving test sooner than boys. She asks some of her friends how many tests they had to take to pass.

Here are the results.

Girls

| Number of tests | 1 | 2 | 3 | 4 | 5 | 6 |
|---|---|---|---|---|---|---|
| No of girls (frequency) | 9 | 6 | 3 | 1 | 0 | 1 |

Boys

| Number of tests | 1 | 2 | 3 | 4 | 5 | 6 |
|---|---|---|---|---|---|---|
| No of boys (frequency) | 1 | 3 | 6 | 0 | 0 | 0 |

a) Find the mean and range for the girls.

b) Find the mean and range for the boys.

c) What do these answers tell Denny?

**4** An athletics club holds a trial for local students. They are asked to run 400 metres. These are their times in seconds.

| 68.2 | 72.1 | 83.0 | 65.5 | 76.3 | 74.8 | 89.0 | 76.7 | 78.9 | 71.2 |
| 75.4 | 69.6 | 75.8 | 76.6 | 81.2 | 89.9 | 83.0 | 85.8 | 71.2 | 76.4 |
| 83.6 | 79.2 | 69.1 | 75.0 | 74.9 | 86.7 | 88.0 | 72.2 | 73.2 | 75.6 |

a) How many students took part?

b) Are the data discrete or continuous?

c) Copy and complete this table for their times.

| Time $t$ (seconds) | $65 \leq t < 70$ | $70 \leq t < 75$ | $75 \leq t < 80$ | $80 \leq t < 85$ | $85 \leq t < 90$ |
|---|---|---|---|---|---|
| Frequency (No of runners) | | | | | |

d) Draw a frequency chart and frequency polygon.

e) What is the modal group?

**5** A new fertiliser is being tried out on some apple trees. The data give the amount of fertiliser used and the fruit yield of each tree.

| Fertiliser (g) | Yield (kg) | Fertiliser (g) | Yield (kg) |
|---|---|---|---|
| 5 | 52 | 12 | 32 |
| 8 | 40 | 8 | 44 |
| 6 | 48 | 6 | 46 |
| 10 | 38 | 5 | 50 |
| 4 | 50 | 10 | 40 |

a) Plot these data on a scatter diagram. Use the horizontal axis (→) for the Amount of fertiliser. Use the vertical axis (↑) for the Yield.

b) What sort of correlation is there?

c) Do you think it is a good thing to use a lot of fertiliser?

Carry out a survey among your friends to discover whether they think there should be a death penalty.

If yes, for what crimes?

Present your findings as a report.

# Thirty three

# Using formulae

**Before you start this chapter you should be able to**

★ put numbers into a formula to solve a problem
★ write down a formula in words and symbols

## Using brackets

The perimeter of this rectangle is
$5 + 3 + 5 + 3 = 16$

You can write this as

*The lengths* → $2 \times 5 + 2 \times 3$ ← *The widths*

You can write this neatly using brackets.

$2 \times (5 + 3)$

$2 \times 8 = 16$

You can do the same thing for any rectangle.

The perimeter is $a + b + a + b = 2a + 2b$

You can write this using brackets as $2(a + b)$.

*$2(a+b)$ means $2 \times (a+b)$*

What is the value of $2(a + b)$ when $a = 6$ and $b = 2$?

Always work out what is inside brackets first.

When $a = 6$ and $b = 2$, $2(a + b)$ is $2 \times (6 + 2) = 2 \times 8 = 16$

 What is the value of $2(a + b)$ when $a = 30$ and $b = 20$?

**Example**

a) Write $3p + 6q + 9r$ using brackets.
b) Find the value when $p = 4$, $q = 5$ and $r = 2$.

**Solution**

a) Notice that 3 divides into 3, 6 and 9. So 3 can be taken outside a bracket.

*$3 \times 2 = 6$*    $3p + 6q + 9r = 3(p + 2q + 3r)$    *$3 \times 3 = 9$*

b) Putting $p = 4$, $q = 5$ and $r = 2$ gives

$3 \times (4 + 2 \times 5 + 3 \times 2)$ so $3 \times 20 = 60$

 *Write $5(s + 2t + 4u)$ without brackets.*

# 33: Using formulae

**1** Use brackets to write the following formulae more simply.

a) $p = 2l + 2w$   b) $t = 3a + 3b$

**2** a) Find the perimeter of this square:

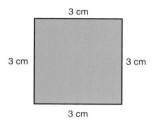

b) (i) Find a formula for the perimeter, $P$ cm, of this square:

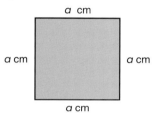

(ii) Use your formula to check your answer to a).

**3** a) Find the perimeter of this kite:

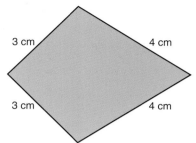

b) (i) Find a formula for the perimeter, $P$ cm, of this kite:

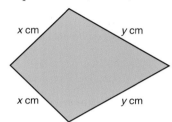

(ii) Use your formula to check your answer to a).

**4** Write down a formula for the perimeter, $P$, of each of these shapes:

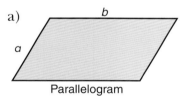

a) Parallelogram    b) Isosceles triangle    c) Regular hexagon

**5** The formula for the total length, $t$, of all the edges of this cuboid is

$$t = 4(l + w + h)$$

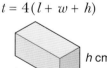

Find the value of $t$ when

a) $l = 5$, $w = 4$, $h = 3$

b) $l = 4$, $w = 4$, $h = 4$

What is the name of the shape in this case?

---

A printing company quotes this price for making personalised Christmas cards: 'Setting up costs £20, after that it's 5p per card.'

Write this as a formula.

Use a spreadsheet to make a list of the total cost and the cost per card for different numbers of cards.

# 33: Using formulae

## Multiplying out brackets

 *How is $n(3n + 5)$ multiplied out?*

You use the same method as before. The term outside the bracket, $n$, multiplies each term inside the bracket.

$$n(3n + 5) = 3n^2 + 5n$$

*($n \times 3n = 3n^2$ and $n \times 5 = 5n$)*

You can show this by looking at these rectangles.

3n + 5 | n

3n | 5 | n | $n \times 3n = 3n^2$ | $n \times 5 = 5n$

The whole area is $n(3n + 5)$.   The smaller areas added together are $3n^2 + 5n$.

These must be the same so $n(3n + 5) = 3n^2 + 5n$

Here are some more examples of multiplying out brackets.

- $n(4n + 1) = 4n^2 + n$
- $2n(n - 3) = 2n^2 - 6n$
- $x(x - 6) = x^2 - 6x$
- $-5x(3 + 2x^2) = -15x - 10x^3$

 *How is $(n + 1) \times (n + 2)$ multiplied out?*

Look at these rectangles.

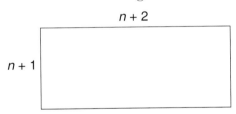
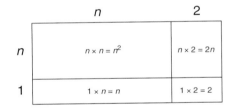

The whole area is $(n + 1)(n + 2)$   The smaller areas are $n^2 + 2n + n + 2$

The whole area is the same as the four smaller areas added together so

$$(n + 1)(n + 2) = n^2 + 2n + n + 2$$
$$= n^2 + 3n + 2$$

*Simplify by collecting like terms $2n + n = 3n$*

Here are two more examples

- $(n + 5)(n - 2) = n^2 - 2n + 5n - 10$
  $= n^2 + 3n - 10$
- $(n - 4)(2n - 3) = 2n^2 - 3n - 8n + 12$
  $= 2n^2 - 11n + 12$

## 33: Using formulae

**1** Simplify

a) $n(n+6)$    b) $n(n-2)$    c) $n(n+1)$
d) $n(n-1)$    e) $2n(n+5)$    f) $-4(2n+1)$
g) $x(2x-3)$    h) $-p(7p-1)$    i) $3a(5a+2)$
j) $-c(2c+5)$    k) $4k(3+5k)$    l) $5x(5x^2-1)$
m) $-a^2(a-4)$    n) $-4b(b^3+7)$    o) $3x^2(x^2-6)$

**2** Write each of these as simply as possible.

a) $(n+2)(n+4)$    b) $(n+1)(n+6)$    c) $(n+5)(n-3)$
d) $(n-4)(n+7)$    e) $(n-8)(n-2)$    f) $(2n+3)(n+4)$
g) $(x+6)(3x+1)$    h) $(4x-5)(x+2)$    i) $(3x+7)(x-4)$
j) $(a+3)(5a-1)$    k) $(b-2)(4b-3)$    l) $(3c+4)(2c+7)$
m) $(8d+3)(2d-1)$    n) $(2t-9)(3t-4)$    o) $(5y-11)(4y+7)$
p) $(1+2y)(1+3y)$    q) $(4-s)(3+2s)$    r) $(7+u)(2-3u)$
s) $(3+4b)(b-2)$    t) $(4d-5)(4-3d)$    u) $(5+3f)(5-3f)$

**3** Work out $(x+4)^2$ by multiplying out $(x+4)(x+4)$.

**4** Work out

a) $(x+5)^2$    b) $(x-6)^2$    c) $(3x+1)^2$    d) $(4x-3)^2$

**5** Write each of these as simply as possible.

a) $n(n+2)+4n$    b) $12n^2+n(n-3)$
c) $(3x+5)^2+x^2$    d) $2p(p+2)+p(4p+1)$
e) $(2x-7)(x+3)+3(4x+5)$    f) $(4a+3)(4a-3)-16a^2$
g) $3c(c+4)-2(6c+1)$    h) $2n(n+1)-3n-2(n^2-5)$
i) $(7x-2)^2-6(x-3)$    j) $(2x+3)^2-(2x-3)^2$

**6** For each of these sequences write down a formula for the $n$th term.

a) $1\times 2, 2\times 3, 3\times 4 \ldots$    b) $4\times 1, 5\times 2, 6\times 3, \ldots$
c) $2\times 3, 3\times 4, 4\times 5, \ldots$    d) $2\times 0, 3\times 1, 4\times 2, \ldots$
e) $1\times 2\times 3, 2\times 3\times 4, 3\times 4\times 5, \ldots$    f) $1\times 3+2, 2\times 4+3, 3\times 5+4, \ldots$

## 33: Using formulae

# More formulae

To find the area of this rectangle you multiply the length by the width.

$A = 4 \times 3 = 12$

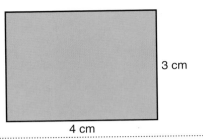

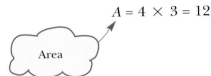

**So the area is 12 cm².**

You can do this for any rectangle.

$A = a \times b$

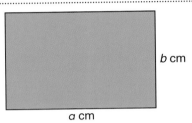

You can write $a \times b$ as $ab$ without the $\times$ sign.

For a square the formula is

$A = a \times a$

You normally write this as $a^2$.

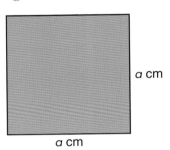

This is a power, you say '$a$ squared'

Look at these two rectangles:

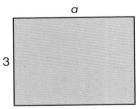
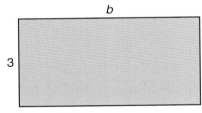

Area = $3a$        Area = $3b$

So the total area is $3a + 3b$.

But you can think of these as one big rectangle, like this:

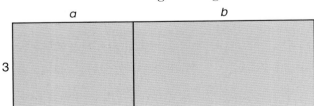

and write the area as $3(a + b)$.

# 33: Using formulae

**1** Find the value of $2x + 2y$ when.
  a) $x = 3$ and $y = 4$
  b) $x = 5$ and $y = 1$
  c) $x = 10$ and $y = 5$

**2** Find the value of $xy$ when
  a) $x = 4$ and $y = 4$
  b) $x = 3$ and $y = 1$

**3** Find the value of $2(x + y)$ when
  a) $x = 3$ and $y = 4$   b) $x = 5$ and $y = 1$
  c) $x = 10$ and $y = 5$   d) $x = 9$ and $y = -4$

**4** a) Find a formula for the area of this rectangle:

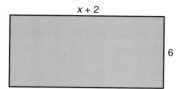

  b) Find the area when $x = 10$.

**5** This cube has edges which are $a$ cm long.
  a) Find formulae for
    (i) the volume
    (ii) the surface area.

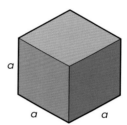

  b) Find the volume and surface area when $a = 2$.

**6** a) Find a formula for the area of this letter L.

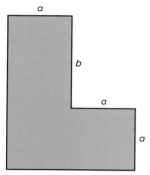

  b) Find the area when $a = 2$ and $b = 3$.

**7** The formula for the area of a triangle, $A$, is
$$A = \frac{1}{2} \times \text{base} \times \text{height}$$
  a) Find a formula for the area of this triangle:

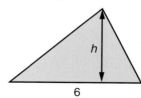

  b) Find the area when $h = 4$.

**8** a) Enter the following key strokes in your calculator.

( − 1 ) × ( − 1 ) + 3 × ( − 1 ) =

Did you get −2?
You have just found $x^2 + 3x$ when $x = -1$.
Find $x^2 + 3x$ when
  a) $x = 0$
  b) $x = -2$
  c) $x = -3$
  d) $x = 1$.

---

Find out how a taxi fare is worked out for short journeys with no delays.

Explain it in words and write it as a formula.

# 33: Using formulae

# Changing the subject of a formula

Look at this formula for finding the temperature, $F$, in degrees Fahrenheit from the temperature, $C$, in degrees Celsius.

$$F = 1.8C + 32$$

The Fahrenheit temperature $F$ is on its own on the left. It is the **subject** of the formula. It is easy to find the value of $F$ when you know the value of $C$.

*Find the value of $F$ when $C$ is 20.*

You probably found the formula very easy to use in this way. But what if you know the value of $F$ and you want to find $C$?

*Find the value of $C$ when $F$ is 77.*

This would be easier with a formula for $C$ in terms of $F$, in other words to have $C$ as the subject of the formula. You can make $C$ the subject by rearranging the formula as follows.

$$F = 1.8C + 32$$

Subtract 32 from both sides $\quad F - 32 = 1.8C$

Divide both sides by 1.8 $\quad \dfrac{F - 32}{1.8} = C$

or $\quad C = \dfrac{F - 32}{1.8}$

> Rearranging the formula is a bit like solving an equation. You want to get $C$ by itself on one side, but at each step you have to do the same thing to each side of the formula

> This means the same as $(F - 32) \div 1.8$

> You might have seen this formula in another form:
> $C = \dfrac{5}{9}(F - 32)$
> They are both the same

The formula for the volume of a cylinder is $V = \pi r^2 h$. You can make $r$ the subject by rearranging the formula.

$$V = \pi r^2 h$$

Divide both sides by $\pi h$ $\quad \dfrac{V}{\pi h} = r^2$

Square root both sides $\quad \sqrt{\dfrac{V}{\pi h}} = r$

$$r = \sqrt{\dfrac{V}{\pi h}}$$

> Notice that $\sqrt{r^2} = r$
> In the same way,
> $(\sqrt{x})^2 = x$

# 33: Using formulae

**1** Make $x$ the subject of each of these.
  a) $y = x + 4$
  b) $y = x + 20$
  c) $y = x + a$
  d) $y = 3 + x$
  e) $y = 13 + x$
  f) $y = c + x$
  g) $y = x - 5$
  h) $y = x - 11$
  i) $y = x - b$
  j) $y = 6 - x$
  k) $y = 1 - x$
  l) $y = d - x$

**2** Make $x$ the subject in each of these.
  a) $y = 2x$
  b) $y = 0.1x$
  c) $y = ax$
  d) $y = \dfrac{x}{4}$
  e) $y = \dfrac{x}{10}$
  f) $y = \dfrac{x}{b}$
  g) $y = \dfrac{3}{4}x$
  h) $y = \dfrac{5}{3}x$
  i) $y = \dfrac{a}{b}x$
  j) $y = \dfrac{4x}{5}$
  k) $y = \dfrac{11x}{2}$
  l) $y = \dfrac{ax}{b}$
  m) $p = x^2$
  n) $q = 2x^2$
  o) $l = x^2 - m$
  p) $s =$

**3** Make $t$ the subject of each of these.
  a) $x = 2t - 3$
  b) $y = 3t + 4$
  c) $p = 6 + 2t$
  d) $c = 4 - t$
  e) $z = 6 - 2t$
  f) $s = 2t + a$
  g) $x = 5t - c$
  h) $n = 7t - 3x$
  i) $p = t^3$

**4** In each of these, make the given letter the subject.
  a) $v = u + at$, $u$
  b) $p = 2l + 2b$, $l$
  c) $V = 4x - 9y$, $x$
  d) $v = u + at$, $t$
  e) $a = b + x^2$, $x$
  f) $a = b + \sqrt{x}$, $x$

**5** In each of these, expand the bracket and then make $x$ the subject.
  a) $p = 2(x + y)$
  b) $V = 12(r + x)$
  c) $s = 4(2 - x)$
  d) $y = 4(a - x)$

**6** In each of these, make the given letter the subject.
  a) $A = lb$, $l$
  b) $V = lbh$, $h$
  c) $V = IR$, $R$
  d) $c = \pi d$, $d$
  e) $c = 2\pi r$, $r$
  f) $I = \dfrac{r}{100} \times P$, $P$
  g) $I = \dfrac{PRT}{100}$, $T$
  h) $I = \dfrac{PRT}{100}$, $R$
  i) $A = \dfrac{\pi d^2}{4}$, $d$

All of the formulae in question 6 are real.

What do they refer to?

Write down six more formulae that you can use in mathematics or elsewhere.

# 33: Using formulae

## Expanding two brackets

You have already expanded expressions like $2(x + 3)$ and $x(x + 3)$. You know that you have to multiply each term inside the bracket by the term outside it.

How would you do $(x + 2)(x + 3)$?

Start like this $\qquad x(x + 3) + 2(x + 3)$

$\qquad\qquad\qquad\qquad x^2 + 3x + 2x + 6$

Now collect like terms $\qquad x^2 + 5x + 6$

*Check your answer by putting in $x = 10$ and working out $12 \times 13$*

You need to multiply each number in the first bracket by each number in the second.

One order for doing this is **F**irsts **O**utsides **I**nsides **L**asts (remember it as **FOIL**).

If you join the numbers as you multiply, the lines look like a smiley face.

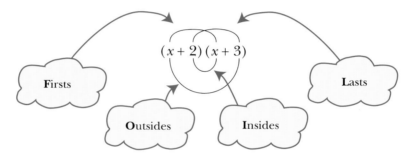

$$x^2 + 3x + 2x + 6$$
$$x^2 + 5x + 6.$$

*Use this method to work out $102 \times 104$ without a calculator.*

### Example

Expand $(2x - 5)(x - 3)$.

### Solution

$\qquad\quad$ **F** $\qquad$ **O** $\qquad\quad$ **I** $\qquad\quad$ **L**

$2x \times x + 2x \times (-3) + (-5) \times x + (-5) \times (-3)$

$\qquad\qquad 2x^2 - 6x - 5x + 15$

$\qquad\qquad\quad \mathbf{2x^2 - 11x + 15}$

## 33: Using formulae

**1** Write these as briefly as possible.
   a) $2a \times 3$
   b) $2c \times 3c$
   c) $4y \times 5y$
   d) $-3 \times 2x$
   e) $3 \times (-4x)$
   f) $-3x \times (-5)$

**2** Expand and then simplify each of these.
   a) $(x+1)(x+3)$
   b) $(y+2)(y+5)$
   c) $(4+x)(3+x)$
   d) $(5+y)(6+y)$
   e) $(x-1)(x+2)$
   f) $(y+3)(y-2)$
   g) $(x-6)(x+2)$
   h) $(y+1)(y-5)$
   i) $(x-7)(x-2)$
   j) $(y-3)(y-4)$
   k) $(x-3)(x-5)$
   l) $(y-6)(y-5)$

**3**
   a) Calculate $21^2$ by expanding $(20+1)(20+1)$.
   b) Calculate $31^2$ by the same method.
   c) Expand $(x+1)^2 = (x+1)(x+1)$.
   d) Use your answers to part c) to find $41^2$ without a calculator.

**4** Expand each of these and then simplify your answer.
   a) $(a+3)(a+3)$
   b) $(x+5)^2$
   c) $(a-3)(a-3)$
   d) $(x-5)^2$
   e) $(2y+1)^2$
   f) $(3x+2)^2$

**5** Expand and then simplify these.
   a) $(2x+5)(3x+2)$
   b) $(2x-5)(3x-2)$
   c) $(2x+5)(3x-2)$
   d) $(2x-5)(3x+2)$

**6** Expand each of these into 4 terms.
   a) $(x+1)(y+1)$
   b) $(a+3)(d+2)$
   c) $(x-1)(y-1)$
   d) $(c+4)(k-10)$
   e) $(a-3)(x+4)$
   f) $(x+10)(y-1)$

**7** The rectangle in the diagram has sides of length $(x+1)$ and $(x+2)$.
   a) Write down an expression for its area.
   b) Write down the area of each smaller part: A, B, C and D.
   c) Expand $(x+1)(x+2)$.
   d) Which rectangles correspond to the middle term of c)?

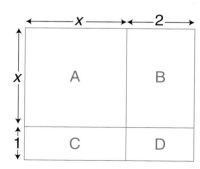

## 33: Using formulae

# Finishing off

**Now that you have finished this chapter you should be able to**

- ★ write down a formula in words and symbols
- ★ put numbers into a formula to solve a problem
- ★ collect like terms to simplify a formula
- ★ use brackets

Use the questions in the next exercise to check that you understand everything.

**Mixed exercise**

**1** Work out $x + y$ when

a) $x = 7$ and $y = 4$   b) $x = 8$ and $y = 2$   c) $x = 12$ and $y = 0$.

**2** Work out $ab$ when

a) $a = 5$ and $b = 7$   b) $a = 2$ and $b = 6$   c) $a = 3$ and $b = 9$.

**3** Work out $x^2$ when

a) $x = 1$   b) $x = 3$   c) $x = 5$.

**4** Work out $x^3 - 2x$ when

a) $x = 3$   b) $x = -3$   c) $x = 1$   d) $x = -1$.

**5** The mean of two numbers is found by adding them together and dividing the answer by two.

a) Find the means of

(i) 2 and 4

(ii) 6 and 10.

b) Write a formula to find the mean, M, of two numbers P and Q.

**6**

**CAR HIRE**
Hire charge £30
+
30p for each mile

What is the cost if I travel 100 miles?

Write a formula for the cost, £C, for M miles.

# 33: Using formulae

**7** Write the following without brackets:
 a) $5(a+b)$
 b) $2(x-y)$
 c) $3(x+3y)$

**8** Write the following with brackets:
 a) $3a+3b$
 b) $4x-4y$
 c) $2x+6y$

**9** The cost £$C$ of having $T$ of these cards printed is given by the formula

$C = T + 5$

How much does it cost for

 a) 2 cards?
 b) 10 cards?
 c) 20 cards?

**10** Four members of a tennis club share out the tennis balls there like this:

There are $B$ balls in each box

Each of them has one full box containing $B$ balls, plus 3 spare balls.

a) Which of these formulae gives the total number of balls, $N$?

  A  $N = 4B + 3$

  B  $N = 4B + 12$

  C  $N = (4 + 3)B$

b) At the end of the day, the friends count the balls.

 There are 36 of them.

 Someone says 'This means that $B = 6$'.

 Is she right?

 How do you know?

---

Different currency exchange bureaux work out the total cost in different ways. Choose a currency and an amount (such as £500 of Euros or Hong Kong dollars). Get information from a number of bureaux.

In each case write the total cost in words and as a formula.

Use your formulae to work out the cheapest way to change your money.

# Thirty four

# Equations and inequalities

> **Before you start this chapter you should be able to**
> - ★ expand brackets
> - ★ collect together like terms
> - ★ solve simple equations

Use the questions in the next exercise to check that you still remember these topics.

**Revision exercise**

**1** Find the value of $x$ to make these scale pans balance.

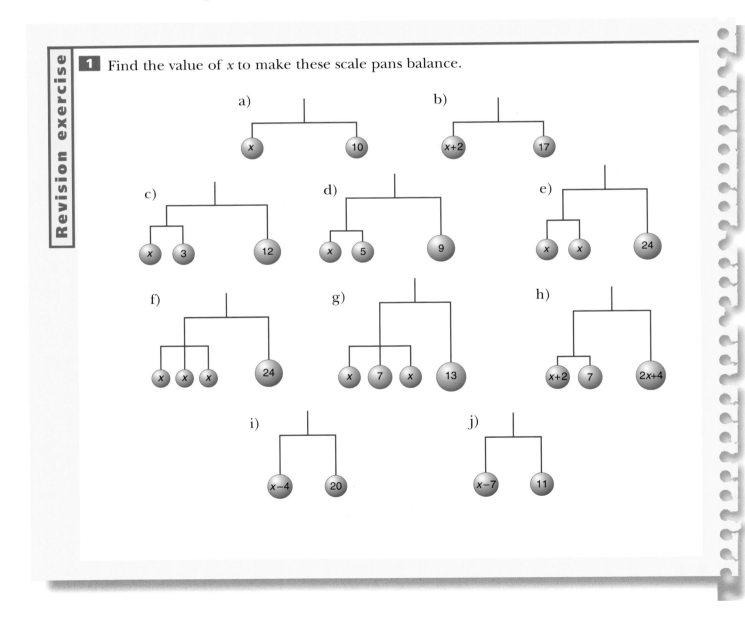

## 34: Equations and inequalities

**Revision exercise**

**2** Simplify these expressions.
  a) $3 + 25$
  b) $4 - 12$
  c) $5 + 2 - 10$
  d) $1 - 5 + 3$
  e) $7x - 4x$
  f) $3x + 2x - 8x$
  g) $4y + 2y - 9y$
  h) $5c + 4c - 15c$

**3** Write these expressions as briefly as possible.
  a) $3 \times a + 6 \times b$
  b) $4 \times m - 6 \times n + 3 \times p$
  c) $2 \times b + 7 \times c - 5 \times d$
  d) $2 \times a + 5 \times b - 3 \times c$
  e) $8 \times f + 7 \times g - 4 \times h$
  f) $4 \times r + 5 \times s - 6 \times t$

**4** Solve each of these equations.
  a) $x + 5 = 37$
  b) $x - 5 = 37$
  c) $7x = 21$
  d) $7x + 3 = 24$
  e) $5x + 4 = 24$
  f) $7 - x = 4$
  g) $15 - 3x = 6$
  h) $30 - 7x = 2$

**5** Expand these brackets.
  a) $5(x + 2)$
  b) $6(x - 3)$
  c) $2(5 - x)$
  d) $9(5 - 2x)$
  e) $4(3x + 2)$
  f) $5(7x - 3)$
  g) $4(5x - 2y)$
  h) $3(3x - 4y)$

**6** Simplify these expressions.
  a) $3 \times a + 6 \times a$
  b) $4 \times m - 6 \times m + 3 \times m$
  c) $2 \times b + 7 \times b - 5 \times b$
  d) $2 \times a + 5 \times b - 3 \times a$
  e) $8 \times f + 7 \times g - 4 \times f$
  f) $4 \times r + 5 \times r - 6 \times r$

**7** Collect together like terms and simplify.
  a) $2x + 3 + 3x - 4$
  b) $6m - 5 + 10 - 6m$
  c) $4x + 3 - 2x - 4$
  d) $5y + 2 - 4y - 3$
  e) $3r - 4 + 7 + 2y$
  f) $3b + 4 + 5 - 2b$
  g) $7s + 2 - 4s + 5$
  h) $4t + 12 - 3t - 7$

**8** Expand the brackets and simplify.
  a) $4(x + 3) - 3x - 6$
  b) $5(b - 3) + 7b - 8$
  c) $7(a - 3) + 5a - 2$
  d) $9(2x + 3) - 4x + 20$
  e) $4(3b - 3) + 4b + 12$
  f) $5(4s - 2) + 3s - 12$
  g) $5(4x - 1) + 5 - 20x$
  h) $2(5x + 2) - 10x$

Design a mobile of your own to be made out of cardboard.

Be careful to check that every part of it balances.

Make the mobile.

## 34: Equations and inequalities

# Solving equations

*Can you think of a number which, if you multiply it by 7 and then subtract 12, gives you the original number again?*

You can write an equation to help you do this.

Call the number $x$. Then the equation is $7x - 12 = x$

> Multiplying $x$ by 7 and subtracting 12 gives $x$

You can solve this using the same method as before.

| | |
|---|---|
| Start with | $7x - 12 = x$ |
| Add 12 to both sides | $7x = x + 12$ |
| Subtract $x$ from both sides | $7x - x = 12$ |
| (Tidy up) | $6x = 12$ |
| Divide both sides by 6 | $x = 12 \div 6$ |
| (Tidy up) | $x = 2$ |

> Add and subtract to get all the $x$ terms on the left and the numbers on the right

> Multiply or divide once you have separated the $x$ terms from the numbers

**Check** by substituting $x = 2$ in both sides of the original equation.

Left-hand side $= 7 \times 2 - 12 = 2$

Right-hand side $= 2$  ✓ Both sides are equal so $x = 2$ is correct.

> This is called **back-substitution**

**The solution is $x = 2$.**

Sometimes you need to solve equations with brackets in. To do this, you just expand the brackets and continue as before.

### Example

Solve the equation $\quad 3(5 + x) = 33$

### Solution

| | |
|---|---|
| | $3(5 + x) = 33$ |
| (Expand the brackets) | $15 + 3x = 33$ |
| Subtract 15 from both sides | $3x = 33 - 15$ |
| (Tidy up) | $3x = 18$ |
| Divide both sides by 3 | $x = 18 \div 3$ |
| | $x = 6$ |

> In expanding the brackets we have just written the left-hand side differently: the right-hand side isn't affected

*Check by back-substitution that $x = 6$ satisfies the original equation.*

## 34: Equations and inequalities

**1** Solve these equations. Write them out carefully and say what you have done at each step. Check your answers by back-substituting.

a) $h + 5 = 11$
b) $n + 3n = 16$
c) $3 + 2x = 14 + x$
d) $3y = 2 - y$
e) $k + 24 = 36 - 11k$
f) $2x - 7 = 17 - 2x$
g) $4x - 7 = 10 + 2x$
h) $5y + 12 = 72 + 3y$
i) $8x - 20 = 6x + 12$

**2** Solve these equations. Check each answer by back-substituting.

a) $10 - 2x = 2$
b) $12 - 2a = 8$
c) $26 - 3x = 10x$
d) $22 - 5y = 14 + 3y$
e) $d + 21 = 4d$
f) $10 - 2x = 0$
g) $104 - 10k = 4 + 10k$
h) $4 - 10k = 104 + 10k$

**3** Solve these equations. Check each answer by back-substituting.

a) $5(a - 2) = 4a$
b) $4(t - 3) = 8$
c) $3y + 6 = 2(y + 2)$
d) $11(c + 1) = 35 - c$
e) $3(x + 2) = 6 - x$
f) $19 - d = 7(1 - d)$

**4** Solve these equations and check your answers.

a) $x + 3 = 2x + 11$
b) $x - 3 = 2x + 11$
c) $4(x - 3) = 5x + 1$
d) $3(y + 2) = 4y - 9$
e) $6(x + 1) = 5(x + 2)$
f) $7(x + 2) = 4(x + 5)$
g) $3(x - 1) = 4x$
h) $2x + 7 = 4x - 3$

**5** Solve these equations and check your answers.

a) $3x = -21$
b) $x + 7 = -10$
c) $2x - 5 = -7$
d) $4 - x = 6 + x$
e) $3x + 1 = -5$
f) $5(2x + 3) - 3(x + 1) = 26$
g) $22x = 7x - 15$
h) $3(x - 2) = 5x$

### Investigation

Solve the equations in order, one by one.

$3a + 2 = a + 4$

$3(b - 2) = 12(1 - a)$ — Use the value of $a$ you have just found

$c = a + b$

$3d + 2 = 2(d + c)$

$e = \dfrac{a + b + c + d}{2}$ — Use the value of $a$ and $b$ you have just found

## 34: Equations and inequalities

# Trial and improvement

Ann thinks of a number between 1 and 100. Rachel tries to guess it.

Ann says 'too big' or 'too small' until Rachel gets it right.

This method is called **trial and improvement**. You often use it in everyday life.

When you run a bath, you use the hot and cold taps. You turn each one, a bit more hot or a bit more cold, until the water is the right temperature.

 *Think of some more everyday trial and improvement situations.*

### Example

A cube has volume 200 m³.

Without using the cube root on your calculator, find the length, $x$ m, of the sides, to one decimal place.

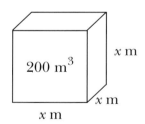

### Solution

The volume is $x \times x \times x$ or $x^3$. So you need to solve the equation
$$x^3 = 200$$

Try $x = 5$: $5^3 = 125$ so 5 is too small.

Try $x = 6$: $6^3 = 216$ so 6 is too big but much closer.

*You know $x$ is between 5 and 6*

Try $x = 5.9$: $5.9^3 = 205.79$ so 5.9 is still too big but getting closer.

Try $x = 5.8$: $5.8^3 = 195.112$ so 5.8 is too small.

Try $x = 5.85$: $5.85^3 = 200.2$ so 5.85 is very close but is still too large.

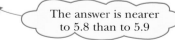

*The answer is nearer to 5.8 than to 5.9*

The sides of the cube are 5.8 m long (correct to one decimal place).

 *How can you use your calculator to check your answer?*

# 34: Equations and inequalities

**1** For each of these equations, find two consecutive whole numbers between which the solution must lie. Then find the solutions of the equations correct to 1 decimal place.

a) $x^3 = 4$  
b) $x^3 - x - 80 = 0$  
c) $4x^3 - 157 = 0$  
d) $2x^3 + 3x - 1500 = 0$  
e) $x + \dfrac{1}{x} = 5$  
f) $x^3 - 7x^2 + 8x - 3 = 0$

**2** Use trial and improvement to solve $x^3 - 7x - 9 = 0$ to 3 decimal places.

**3** Use trial and improvement to solve $x^2 - \dfrac{1}{x} = 1$ to 3 decimal places.

**4** Matt is designing a carton for a special party fruit drink. The manufacturer wants the carton to hold 4 litres and Matt decides to make it the shape in the diagram.

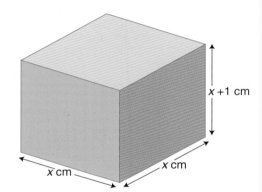

a) Show that the volume is $(x^3 + x^2)$ cm$^3$.

b) Explain why $x^3 + x^2 = 4000$.

c) Solve the equation $x^3 + x^2 - 4000 = 0$, and so find the lengths of the sides to the nearest millimetre.

**5** Majid throws a tennis ball over his parents' garage. The graph shows its trajectory. As you can see it just clears the roof on the far side. The equation of the path of the ball is $y = 1 + 2x - 0.2x^2$ and the height of the garage is 3 m.

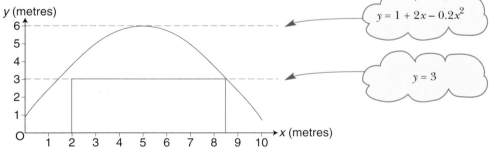

Use a trial and improvement method to find the length of the garage to the nearest centimetre.

Find out how using a spreadsheet can help you to solve an equation by trial and improvement. Choose an equation and solve it by this method. The spreadsheet printouts should form part of your answer.

# 34: Equations and inequalities

## Using inequalities

Look at these notices.

They all describe restrictions on ages, heights and prices.

These restrictions can be written as **inequalities** using special symbols.

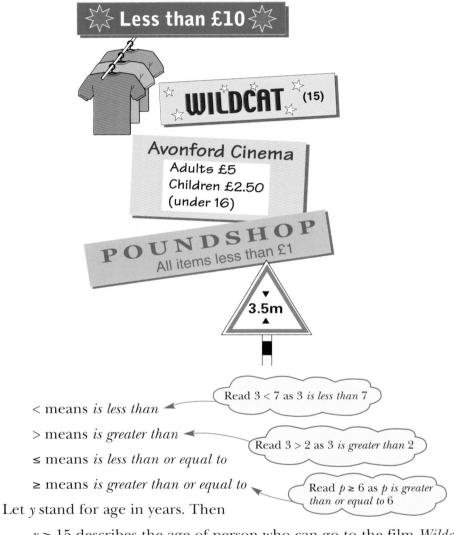

< means *is less than*  (Read 3 < 7 as 3 *is less than* 7)

> means *is greater than*  (Read 3 > 2 as 3 *is greater than* 2)

≤ means *is less than or equal to*

≥ means *is greater than or equal to*  (Read $p \geq 6$ as $p$ *is greater than or equal to* 6)

Let $y$ stand for age in years. Then

$y \geq 15$ describes the age of person who can go to the film *Wildcat*.

$y < 16$ describes the age of a person whose cinema ticket will cost £2.50.

 *Write inequalities, using the symbols, for the restrictions in the other notices.*

*What other restrictions are commonly shown in notices and signs?*

You have already met the inequality symbols in Chapter 10, when you were grouping data. In this chapter you will learn about other ways to use them.

 *What is the other way of writing 7 > 3 ?*

## 34: Equations and inequalities

**1** Write these statements as inequalities.
   a) $x$ is greater than 6
   b) $x$ is greater than or equal to zero
   c) $x$ is less than 3
   d) $x$ is less than or equal to $-4$

**2** Write these inequalities in words.
   a) $x < 10$
   b) $p > 20$
   c) $q \leq 10$
   d) $y < 11$
   e) $x \leq 24$
   f) $b \geq -5$

**3** For each of these, copy the numbers down in the same order, replacing the comma by an inequality sign ($<$ or $>$).
   a) 2, 7
   b) 13, 4
   c) $-3, -10$
   d) 13, $-2$
   e) 5, $-4$
   f) $-2, 2$

**4** Look at each of these signs and write down the restriction shown using the inequality symbols. (Choose a suitable letter, in each case, to stand for the quantity that is being restricted.)

a)

b)

c) **Old Man Theatre** Party rates for groups of **10** or more

d)

e) *Holidays in Spain* 7 nights less than £350

f)

g)

h) **RAINBOW Building Society** HIGH INTEREST ACCOUNT Minimum deposit £200

**5** You can write an inequality in two ways.

For example, $9 > 6$ could be written as $6 < 9$.

Write down each of your inequalities from question 3 in another way.

## 34: Equations and inequalities

# Number lines

It is often helpful to show an inequality on a number line.
This is how you show the inequality $x > 5$

*The empty circle shows that x cannot be equal to 5*

*Any number to the right of 5 is greater than 5*

This is how you show $x \leq 3$.

*The solid circle shows that x can be equal to 3*

*Any number to the left of 3 is less than 3*

 *Think of some possible values for x for each inequality and check that they come on the marked part of the number line.*

*Is −3 greater than 2 or less than 2?*

## *Combining inequalities*

'Roxys' is a club for people in their late twenties and early thirties.

The inequalities which describe the ages of its members are

$y \geq 25$ and $y < 35$.

As $y$ is *between* 25 and 35 you can combine these and write $25 \leq y < 35$.

You can show it on a number line like this.

*People who have had their 25th birthday are included*

*People who have had their 35th birthday are not included*

 *Write a similar inequality for the age of a student at your school or college.*

*Jason is 1.8 m tall (correct to 1 decimal place). Write an inequality to show exactly what this means, and show it on a number line.*

To qualify for cheap rail tickets you must be under 25, or else 60 or over. Your age has to satisfy the inequality

$x < 25$  or   $x \geq 60$.

You can show it on a number line like this.

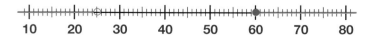

# 34: Equations and inequalities

**1** Show each of these inequalities on a number line.

a) $x \geq 4$   b) $x > -2$   c) $x < 6$   d) $x \leq 3$   e) $x < 0$

f) $x$ is positive   g) $x$ is not more than 7   h) $x$ is at least 5   i) $x$ is not positive

**2** A second class stamp can be used to send a letter weighing no more than 60 g.

a) Write this as an inequality for $w$, the weight of the letter in grams.

b) Show your answer on a number line.

**3** Show these measurements on a number line.

a) 2.8 m (correct to 1 d.p.)   b) 3.0 kg (correct to 1 d.p.)

c) 4 m (to the nearest metre)   d) 7000 m (to the nearest 1000m)

e) 2000 m (to the nearest 100 m)   f) 0.12 kg (correct to 2 d.p.)

**4** Combine the inequalities and show the results on a number line

a) $x \geq 2$ and $\leq 4$   b) $x \leq 3$ and $x \geq -1$   c) $x < 0$ and $x \geq -2$

d) $y \geq 65$ and $y < 75$   e) $a > -3$ and $a \leq 0$   f) $b < -2$ and $b > -4$

g) $x \geq -2$ and $x \geq 3$ (Be careful!)   h) $c \leq 5$ and $c \leq 3$

**5** You join the queue at the supermarket checkout with this sign.

Write down an inequality for the number of items, $x$, in your basket.

Write down all the possible values of $x$ and mark these points on a number line.

**6** State whether the following children would be allowed into the ball pool. Give a reason if the answer is no.

a) Martin is 5 years old and 1.3 m tall.

b) Betty is 3 years old and 0.9 m tall.

c) Jason is 2 years old and 1.1 m tall.

d) Geri is $2\frac{1}{2}$ years old and 0.95 m tall.

---

Inequality signs have been used to show the class limits in the table on page 334.

Draw a number line to show the range of possible values for an item in each class. Use a different colour for each one.

What do you notice?

## 34: Equations and inequalities

# Solving inequalities

 *For what values of x is $4x - 5 > 23$?*

You can solve this inequality in just the same way as you solve an equation.

|  | An equation | An inequality |
|---|---|---|
|  | $4x - 5 = 23$? | $4x - 5 > 23$ |
| Add 5 to both sides | $4x = 28$ | $4x > 28$ |
| Divide both sides by 4 | $x = 7$ | $x > 7$ |

*The equation has just one solution*

*Check using a number just greater than 7*

*This is the solution of the inequality. It means that x can have any value greater than 7*

Check:       When $x = 7$           When $x = 8$

$4x - 5 = 28 - 5 = 23$ ✓    $4x - 5 = 32 - 5 = 27$ ✓

*The inequality holds because $27 > 23$*

 *Check for yourself that when $x = 6$ the inequality is not true.*

 *Find the value of the expression $4x - 5$ when $x = 10$ and when $x = 7.5$.*

*What do you think will happen when $x = 2$?*

You can do what you like to an equation, so long as you do the same to both sides. You can add or subtract any number, and you can multiply or divide by any number.

> **You solve inequalities just as you solve equations.**
> **Instead of one answer you get a range of answers.**
> **You can show these on the number line.**

Sometimes you may be subtracting $x$.

**Example**

Solve $17 - x < 9$

**Solution**

Add $x$ to both sides         $17 < 9 + x$

*Always put $x$ on the side where it is positive*

Subtract 9 from both sides    $8 < x$

 *Check for yourself that when $x$ is 9 the inequality is true.*

*Check for yourself that when $x$ is 5 the inequality is not true.*

## 34: Equations and inequalities

**1** Solve these inequalities.
a) $x + 1 < 8$  b) $2x + 1 \leq 15$
c) $x - 3 \geq 7$  d) $3x - 2 \geq 13$
e) $x - 4 < -2$  f) $2x + 17 > 27$
g) $5x - 2 \leq 18$  h) $2x + 11 \leq 21$
i) $3 + 4x > 15$  j) $5 - x \geq 3$
k) $7 - 2x \leq 5$  l) $2 \leq 5 - 3x$

**2** Find all the possible values of $y$ when
a) $11 < y < 30$ and $y$ is a prime number
b) $30 \leq y \leq 50$ and $y$ is a square number
c) $1 < y < 10$ and $y$ is a factor of 12
d) $21 > y > 3$ and $y$ is a multiple of 3.

**3** Last Saturday Grandad went to the races. He placed a bet of £$x$ on the first race and his horse came in first at 3 to 1. Grandad won £$3x$. After that he lost £100 of his winnings.

a) Write an expression for the winnings he had left.

b) This amount was still more than he had bet on the first horse.

   Write this as an inequality and solve it for $x$.

**4** Nat looks at the weight card his mother has kept from when he was a baby. He finds that he now weighs 56 kg more than he did when he was born. This is more than 14 times his birthweight.

Write this as an inequality for $w$, his birthweight, and solve it.

**5** Solve these inequalities.
a) $2x + 3 \geq x + 18$  b) $4x - 6 \geq x + 3$  c) $5 < x + 1$
d) $2(x + 4) < 24$  e) $7 \leq x + 3 \leq 13$  f) $18 \geq 2x > 8$
g) $10 < 2x < 18$  h) $0 \leq x - 2 \leq 12$  i) $7 < 2x + 1 \leq 21$
j) $3(x - 1) > 2x + 4$  k) $10(x - 2) \geq 5(x + 2)$  l) $2 - x > 5 - 2x$

---

Choose an inequality for $x$, such as $x < 5$ or $x \geq 1$.

Multiply both sides by the same number.

Add or subtract the same number on both sides.

You now have an inequality for a partner to solve.

Write it on another piece of paper so your working is not visible.

See if your partner can solve the inequality and get back to your starting point.

## 34: Equations and inequalities

# Finishing off

**Now that you have finished this chapter you should**

★ be able to solve simple equations using algebra

★ be able to solve equations involving brackets and fractions

★ be able to use trial and improvement methods to solve equations

★ understand the symbols <, >, ≤ and ≥

★ be able to represent an inequality on a number line

★ be able to solve simple inequalities using algebra

Use the questions in the next exercise to check that you understand everything.

## Mixed exercise

**1** Solve these equations.

a) $5a = 35$  
b) $15b = 30$  
c) $70 = 35c$  
d) $2d + 11 = 17$  
e) $19 = 16e - 13$  
f) $22 - 3f = 7$  
g) $14 - 5g = 9$  
h) $22 - 4h = 10$

**2** Solve these equations by collecting the unknown terms together on one side.

a) $5z - 4 = 2z + 2$  
b) $6y + 1 = 7y - 4$  
c) $8x - 9 = 15 - 4x$  
d) $3w + 4 = 19 - 2w$  
e) $12 + 6v = 2v + 20$  
f) $10 + 5u = 26 - 3u$  
g) $62 - 9t = 2t - 4$  
h) $999 - 50s = 49s + 900$

**3** Multiply out the brackets and solve these equations.

a) $2(8 + x) = 22$  
b) $3(8 - x) = 5x$  
c) $21x = 7(x + 4)$  
d) $7 + x = 4(x - 2)$  
e) $3(x + 1) = 2(x + 3)$  
f) $5(2x + 1) - 3(x + 3) = 10$

**4** Lloyd thinks of a number $x$.

He multiples it by 7 and subtracts 12 and makes the answer 23.

Make an equation and solve it to find Lloyd's number.

**5** All these equations have negative solutions. Solve them.

a) $3x = -18$  
b) $-2x = 10$  
c) $x + 7 = -10$  
d) $2x + 5 = 3$  
e) $2x - 3 = -11$  
f) $4 - x = 8 + x$

## 34: Equations and inequalities

**Mixed exercise**

**6** Write each of these inequalities in words and show it on a number line.

a) $x > 3$  b) $x \leq 2$  c) $x \geq 5$

d) $x < -1$  e) $-1 < x < 3$  f) $2 \leq x \leq 5$

**7** Madeleine has inherited some money from her aunt.

She puts it in one of these accounts.

It is earning 6.50% annual interest.

Write an inequality for the sum, £$m$, that Madeleine inherited.

| RAINBOW Building Society | SAVINGS RATES |
|---|---|
| Balance | Annual Interest |
| £1 – £4,999 | 5.35% |
| £5,000 – £9,999 | 5.70% |
| £10,000 – £24,999 | 6.20% |
| £25,000 – £49,999 | 6.50% |
| £50,000 – £99,999 | 6.70% |
| £100,000 + | 6.80% |

**8** Solve these inequalities.

a) $x + 3 \leq 10$  b) $2x - 5 \geq 15$  c) $2x + 1 > x + 7$

d) $5(x + 3) < 40$  e) $2(2x - 3) \leq 3x + 1$  f) $5 < x + 1$

g) $4(x + 2) - 3(x + 1) > 14$

**9** 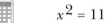 Sally is using trial and improvement to solve

$x^2 = 11$

This is how she starts:

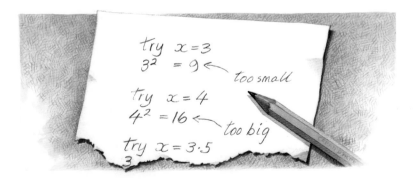

Finish off her working to find the solution to 1 decimal place.

# Thirty five

# Ratio and proportion

**Before you start this chapter you should be able to**

★ change simple fractions into decimals

★ change simple fractions into percentages

★ understand simple ratio and proportion

## Simplifying ratios

Two families are going for a day trip to Blackpool.

There are 6 people in the Green family and 4 in the Brown family.

They rent a minibus for the day. It costs £60.

How much does each family pay?

There are 10 people altogether. So each person pays $\frac{£60}{10} = £6$.

*Find the cost of each person first*

**The Green family pays 6 × £6 = £36**
**The Brown family pays 4 × £6 = £24**

*Notice that £36 + £24 = £60. It works out*

The £60 is split in the **ratio** 36:24.

You can simplify this: Divide by 6 to get 6:4

and further: Divide by 2 to get 3:2

(You could also divide by 12 straight away to get 3:2.)

*This is the ratio of the numbers in the two families*

*Simplest form*

Ratios are usually written in their simplest whole number form.

In this case, it is 3:2.

Another way to work out the cost is to say

Greens' share
$\frac{6}{10} \times £60 = £36$

Browns' share
$\frac{4}{10} \times £60 = £24$

The fractions $\frac{6}{10}$ and $\frac{4}{10}$ are called the **proportions** of the cost that the two families have to pay.

Notice that $\frac{6}{10}$ is $\frac{\text{the number of Greens}}{\text{the number of people}}$.

*Greens and Browns together*

# 35: Ratio and proportion

**1** Write each ratio in its simplest form.

a) 6:3    b) 20:100    c) 24:6    d) 10:15

e) 21:7   f) 75:25     g) 12:30   h) 250:400

**2** Write each ratio in its simplest form. (You need to put both sides in the same units first.)

a) 20 minutes : 1 hour

b) 1 kilogram : 5 grams

c) 2 cm : 5 mm

d) 80 pence : £2

e) $2\frac{1}{2}$ seconds : 15 seconds

f) 10 hours : $7\frac{1}{2}$ hours

g) 2.4 kg : 6 kg

h) 5.6 kg : 2.1 kg

**3** 450 grams of brass is made using the metals copper and zinc in the ratio 2:1.

How much of each metal is used?

**4** Milton and Spencer share a £60 bonus in the ratio 3:2.

How much do they each get?

**5** Ceri and Andrea are business partners who share profits in the ratio 3:5. How much will each get from a profit of £20 000?

**6** This bar chart shows the number of employees a company has at three factories.

a) What is the ratio of female to male employees at

(i) A?   (ii) B?   (iii) C?

b) What is the ratio of female to male employees in the company?

c) What fraction of the company's employees are male?

---

Look at the Nutrition information on a cereal packet. For each item, two figures are given. The first is the amount per 100 g. The second is the amount in a typical serving. Some packets include the milk in the second figure, others do not.

Find an example of each, and in each case explain how the second figure is worked out.

# 35: Ratio and proportion

## Unitary method

Sunil is at the supermarket.

He needs some milk.

Sunil compares the prices by working out the price of 1 pint from each container.

From the 4 pint container: $\frac{100}{4} = 25$ ← This is £1 in pence

So 1 pint is 25p.

From the 6 pint container: $\frac{144}{6} = 24$ ← This is £1.44 in pence

So 1 pint is 24p. ← So this is the better buy

 *Why doesn't everyone buy this size container?*

Sunil wants to buy enough butter to bake 15 cakes.

He has a recipe for 6 cakes which uses 200 g butter.

He writes down

This is called the unitary method.

*Why?*

 Complete Sunil's calculation.

 Sunil wrote '1 cake needs $\frac{200}{6}$ g butter,' but he did not work it out. Why not?

## 35: Ratio and proportion

**Exercise**

In questions 1 to 4, which is the better buy?

**1** Ice cream: 2 litres £1·29 or 4 litres £2·49

**2** Fizzy drinks: £1·72 or £2·52

**3** Toilet rolls: £1 or £4

**4** Choco Bran: 500g £1·35 or 750g £1·95

**5** Sophie sees French marigold plants for sale at 6 for £2.25 in the garden centre and at 10 for £4.00 at the market.

Which place offers the better deal?

**6** Roshan sees a 1 litre tin of paint priced at £2.19 and a 2.5 litre tin of paint priced at £4.99. Which is the better buy?

**7** A train travel 520 m in 24 seconds. How far does it travel in one minute?

**8** John's watch loses 20 seconds in 4 days. How much does it lose in a week?

**9** Mary cycles 42 km. Her faulty cyclometer records 36 km. What will her cyclometer record if Mary cycles for 63 km?

**10** A market gardener can grow 6200 plants in 250 m$^2$. How many plants can he grow in 475 m$^2$.

**11** Mark is tiling an area 120 cm by 120 cm.

He has to choose which of these tiles to use.

Small tile: 15 cm × 15 cm, £0.80
Big tile: 20 cm × 20 cm, £1.49

a) (i) How many small tiles does he need to tile the area?
(ii) How much do they cost altogether?

b) (i) How many big tiles does he need to tile the area?
(ii) How much do they cost altogether?

c) Which size is cheaper?

Find 3 different size jars of the same type of coffee.
What is the cost per 100 g in each jar?
Which is the best buy?

## 35: Ratio and proportion

# Changing money

Each country has its own money.

The UK uses pounds and pence.

*What does the USA use?*

*How do you get foreign money?*

Tina is on holiday in New York.

She wants to change £200 into dollars.

$ means dollars

The bank changes pounds into dollars by multiplying by 1.60.

200 × 1.60 = 320

**Tina gets $320 for her £200.**

*How much does she get for £300?*

*How much for £150?*

Tina buys a guide book for $20.

She works out the cost in pounds by dividing by 1.60.

20 ÷ 1.60 = 12.5

**So the book costs her £12.50.**

*What is $50 in pounds?*

Tina writes down some conversions so that she knows how much she is spending.

*What should the missing entries be?*

*How can Tina use this to work out $70 in pounds?*

# 35: Ratio and proportion

Look at these exchange rates.

Use them to answer the questions on this page.

| Exchange rates £1= |
|---|
| 15 Swedish kroner (Sweden) |
| 185 yen (Japan) |
| 2.28 Swiss francs (Switzerland) |
| 1.61 euros (EU) |

**1** How many Swedish kroner do you get for

   a) £1?      b) £10?      c) £60?

**2** How many Japanese yen do you get for

   a) £1?      b) £5?      c) £25?

**3** How many Swiss francs do you get for

   a) £1?      b) £10?      c) £15?

**4** Reni is on holiday in Japan.

   a) He buys a meal for 4440 yen. What is this in pounds?

   b) He spends 500 yen on postcards. What is this in pounds and pence?

**5** Copy and complete this Swedish kroner conversion chart rounding down to a whole number of pence.

| 1K | 2K | 3K | 4K | 5K | 10K | 20K | 50K | 75K | 100K |
|---|---|---|---|---|---|---|---|---|---|
| £0.06 | £0.13 | | | | £0.66 | | | | |

**6** Paula is in Paris for the weekend.

She wants to buy her sister a present, but can't afford to spend more than £5.

Which of these can she afford?

Mug €9, Keyring €7, Glass €8, Eiffel Tower €10

---

Exchange rates are changing all the time.

Go to a foreign exchange or look in a newspaper or the internet and find out the current exchange rates for Russia, South Africa, Mexico, Australia, Japan, United States of America, India and EU.

How much do you get in each of these countries for £100?

## 35: Ratio and proportion

# Distance, speed and time

What is the speed limit on motorways?

How fast does an aeroplane fly?

How many metres per second can you run?

Speed is usually measured in miles per hour (m.p.h.), kilometres per hour (km/h) or metres per second (m/s).

Tim drives at 100 km/h on the motorway.

*Do you think that he drives at exactly 100 km/h all the time?*

Tim drives 90 km on country roads in $1\frac{1}{2}$ hours.

What is his average speed in km/h?

**Average speed** = $\dfrac{\text{distance covered}}{\text{time taken}}$

= $\dfrac{90}{1\frac{1}{2}}$ = 60

**Tim's average speed is 60 km/h.**

*Kanwal does the same journey in $1\frac{1}{4}$ hours.*

*What is her average speed?*

Tim's average speed on a motorway is 100 km/h.

How long does a motorway journey of 225 km take?

**Time taken** = $\dfrac{\text{distance covered}}{\text{average speed}}$

= $\dfrac{225}{100}$ = 2.25 or $2\dfrac{1}{4}$

Be careful here!
.25 hours is not 25 minutes.
To change 0.25 hours into
minutes you multiply by 60
$0.25 \times 60 = 15$ so
0.25 hours = 15 minutes

**The time taken is $2\frac{1}{4}$ hours (or 2 hours 15 minutes).**

*How long does it take Tim to travel 60 km?*

*How far does Tim travel in $1\frac{3}{4}$ hours?*

**Distance covered = average speed × time taken**

= $100 \times 1\dfrac{3}{4}$ = 175

**Tim travels 175 km in $1\frac{3}{4}$ hours.**

*How far does Tim travel in 45 minutes?*

# 35: Ratio and proportion

**1** What distance is covered by

  a) Linton cycling at 50 km/h for 2 hours?

  b) Jovanka driving at 70 km/h for $1\frac{1}{2}$ hours?

  c) Philip flying at 880 km/h for $2\frac{1}{4}$ hours?

  d) Liz running at 15 km/h for 1 hour 20 minutes?

**2** Hana has four meetings today. This is her schedule.

  Work out the average speed

  a) between London and Milton Keynes (84 km apart)

  b) between Milton Keynes and Leicester (81 km apart)

  c) between Leicester and Sheffield (105 km apart).

  TUESDAY 6
  MEETING
  Leave London 10.00
  Arrive Milton Keynes 11.30
  MEETING
  Leave Milton Keynes 12.30
  Arrive Leicester 13.45
  MEETING
  Leave Leicester 14.45
  Arrive Sheffield 16.00
  MEETING

**3** Work out the time it takes to

  a) cycle 75 km at 30 km/h

  b) fly 2250 km at 600 km/h

  c) drive 100 km at 60 km/h

  d) run a marathon (42.2 km) at 20 km/h.

**4** Darren from Liverpool and Vicky from Hull drive to Manchester to meet for lunch.

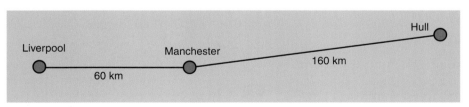

  a) Darren leaves home at 1140 and his average speed is 60 km/h. What time does Darren arrive in Manchester?

  b) Vicky leaves at 1045 and expects to take 2 hours. What will her average speed be?

  c) Vicky's journey takes 30 minutes longer than planned. What is her average speed?

  d) How long did Darren have to wait on his own?

# 35: Ratio and proportion

## Finishing off

**Now that you have finished this chapter you should be able to**

★ write a ratio as a fraction, decimal or percentage

★ write a ratio in its simplest form

★ solve problems with distance, speed and time

★ solve simple problems using ratio

★ compare prices and work out the 'best buy'

★ change money

Use the questions in the next exercise to check that you understand everything.

### Mixed exercise

**1** Helene caught a plane to Paris. There were 20 English passengers. The ratio of English passengers to French passengers was 1 to 5. How many French passengers were there on the plane?

**2** Concentrated orange juice must be diluted with 5 times as much water as juice.

a) How much water must be added to 10 ml of orange juice?

b) How much orange juice is in a diluted drink of 300 ml?

**3** This bar chart shows how a group of students travel to college.

Work out the ratio of

a) bus to walk

b) bus to train.

Give each answer in its simplest form.

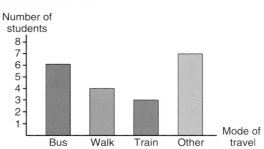

**4** Write each ratio in its simplest form.

a) 5 : 10   b) 12 : 8

c) 40 : 100   d) 225 : 75

e) 1 hour : 15 minutes

f) 5 cm : 1 m

g) $4\frac{1}{2}$ miles : 6 miles

h) 3.2 kg : 4 kg

**5** A house plan uses a scale of 2 centimetres to 1 metre.

a) Write this ratio in its simplest form.

b) The lounge is 4 m long.

How long is it on the plan?

c) The kitchen is 7 cm long on the plan.

How long is it in real life?

## 35: Ratio and proportion

**6** Helen and Clare share a house.

a) The rent is £100 and they pay it in the ratio 2:3.
How much does each of them pay?

b) The phone bill is £140 and they pay it in the ratio 5:2.
How much does each of them pay?

**7** For each of these items work out which is the better buy.

a) Margarine 500g £1.05, Margarine 750g £1.59

b) 60 Vitamin Tablets £2.79, 120 Vitamin Tablets £5.25

c) 5 Kg Potatoes £3, 10 Kg Potatoes £5

**8** Nicky wants to learn to drive. She sees these adverts.

MOTOR SCHOOL — 12 lessons for £150

Learn to drive — 9 lessons for £120, 10th lesson free

How much does one lesson cost at each school?

**9** Kerry and Liam are on holiday in the USA.
The exchange rate is £1 = $1.60.

a) How many dollars does Liam get for £80?

b) How many dollars does Kerry get for £140?

c) Liam pays $20 for a coach ticket.
How much is this in pounds and pence?

d) Kerry pays $110 for a hotel room. How much is this to the nearest pound?

**10** a) Lucy cycles for 45 minutes at an average speed of 30 km/h.
How far does she travel?

b) Anant is a airline pilot. He has 900 km to cover in $1\frac{1}{4}$ hours.
What average speed does he need to attain?

c) Tracey has 50 km to drive. She thinks her average speed will be 65 km/h.
How long, to the nearest 5 minutes, will it take her?

Describe 4 ways in which you have used ratio and proportion in other subjects.

# Thirty six

# Area and volume

**Before you start this chapter you should know how to work out**

- ★ the area of a rectangle
- ★ the area of a triangle from base and height measurements
- ★ the area of a shape made up of rectangles and triangles
- ★ the volume of a cuboid from the lengths of its sides

## Parallelograms and trapezia

### Area of a parallelogram

A **parallelogram** is a shape with two pairs of parallel sides.

The rule for working out the area of a parallelogram is

**Area of a parallelogram = base × vertical height**

For this parallelogram,
Area (in cm²) = 8 × 5 = 40

*Explain why the rule works. This diagram may help.*

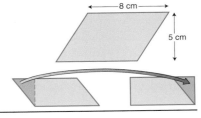

### Area of a trapezium

A **trapezium** is a shape with one pair of parallel sides.

The rule for working out the area of a trapezium is

**Area of a trapezium = $\frac{1}{2}(a+b)h$**

For this trapezium,
Area (in cm²) = $\frac{1}{2} \times (4+6) \times 3$
$= \frac{1}{2} \times 10 \times 3$
$= \frac{1}{2} \times 30$
$= 15$

*Explain why the rule works. This diagram may help. It shows two trapezia fitted together to make a parallelogram.*

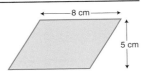
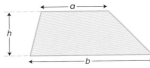
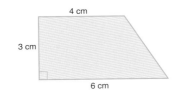
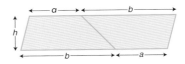

# 36: Area and volume

**1** Find the areas of these shapes. Find the perimeters of a), b) and c).

a) 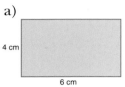   b) 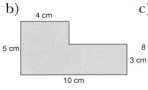   c)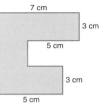

d) 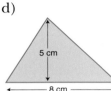   e) 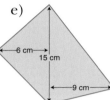   f)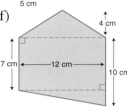

**2** Find the areas of the shapes below.

a) 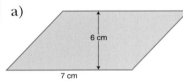   b)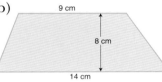

c) 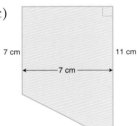   d)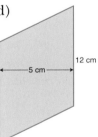

**3** You are making the roof for a dolls' house out of two trapezia and two triangles, as shown.

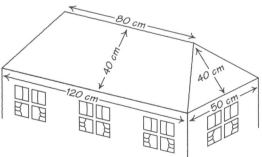

Work out the area of wood you need to make the roof.

a) Find the area of this rectangle (i) in $cm^2$ (ii) in $m^2$.

b) Use your answers to show that $1\,m^2 = 100^2\,cm^2$.

## 36: Area and volume

# Cuboids

The diagram shows a small box of cereal.

*How much room is there in the box?*

The amount of space taken up by the box is called the **volume**.

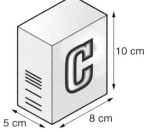

It is measured in cubic units. The volume of the box of cereal is measured in cubic centimetres ($cm^3$).

To find the volume of the box, you need to know how many cubic centimetres it takes up.

So imagine you have a lot of cubes, 1 cm long, 1 cm wide and 1 cm high.

Each cube has a volume of 1 $cm^3$.

You are finding out how many of these cubes fit in the box.

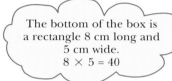

The bottom of the box is a rectangle 8 cm long and 5 cm wide.
$8 \times 5 = 40$

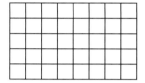

So 40 cubes can be fitted in the bottom of the box.

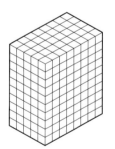

The box is 10 cm high.
Each layer has 40 cubes.
There are 10 layers.
$40 \times 10 = 400$

There are 400 cubes altogether. So the box has a volume of 400 $cm^3$.

The number of cubes in the bottom of the box was found by multiplying the length by the width. To find the total number of cubes in the box, this was then multiplied by the height of the box.

**Volume of a cuboid = Length × Width × Height  V = L × W × H**

The density of a material is found by dividing its mass by its volume.

$$\text{Density} = \frac{\text{Mass}}{\text{Volume}}$$

*How can you find the density of a cuboid?*

## 36: Area and volume

**1** Find the volume of each of the following cuboids.

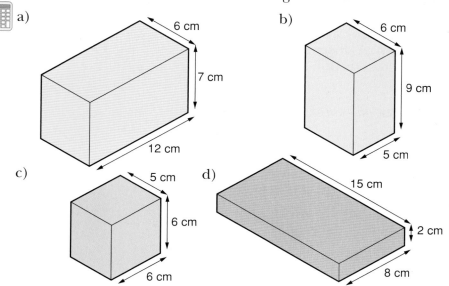

**2** The masses of the cuboids in question 1 are given below.

Find their densities in grams per cubic centimetre ($g\ cm^{-3}$).

a) 1008 g   b) 810 g   c) 900 g   d) 2760 g

**3** A fish tank is 60 cm long and 40 cm wide.

It is filled with water to a depth of 30 cm.

Find the volume of water in the tank.

**4** A box containing 350 $cm^3$ of icing sugar is 7 cm long and 4 cm wide.

What depth of icing sugar is in the box?

**5** An open topped box is made by cutting out a rectangle 15 cm by 12 cm, and cutting a square 3 cm by 3 cm off each corner.

Find the volume of the box.

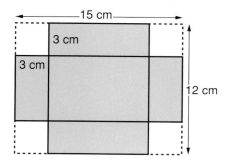

**6** A swimming pool has 3750 $m^3$ of water in it.

It is 3 metres deep, and the length of the pool is twice its width.

Find the length of the pool.

Look at packets of your favourite cereal in the supermarket.

Work out the volume of the different sized packets it comes in.

Open the packets out. Work out the area of cardboard each one uses.

## 36: Area and volume

# Volume of a prism

When slices are cut (as shown) from this wedge of cheese, each piece is the same size and shape. The wedge of cheese is an example of a **prism**. A prism is a solid which has the same cross section all the way along its length.

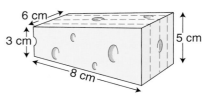

 *All prisms have at least one plane of symmetry. Where is it?*

 *Think of 3 other familar objects whose shapes are prisms.*

To find the volume of a prism, the first step is to find the area of the cross section. In the case of the wedge of cheese, this is a trapezium.

Area of trapezium (in cm²) = $\frac{1}{2} \times (3 + 5) \times 8$
= 32

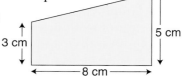

An end view without perspective is sometimes called an **elevation**

Now multiply the area of cross section by the length.

**The volume of the wedge of cheese = 32 cm² × 6 cm = 192 cm³.**

The volume of any prism can be found in the same way, by multiplying the area of the cross section by the length of the prism.

**Volume of a prism = area of cross section × length**

This Swiss roll is a prism with circular cross section. This shape is usually called a **cylinder**.

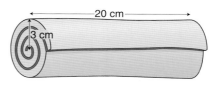

Using the above rule,

Area of cross section (in cm²) of Swiss roll = $\pi \times 3^2 = 9\pi$

Volume = $9\pi$ cm² × 20 cm = 565 cm³ (to the nearest cm³).

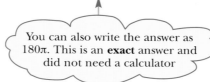

You can also write the answer as $180\pi$. This is an **exact** answer and did not need a calculator

 *The formula for the volume of a cylinder is $V = \pi r^2 h$. What do V, r and h stand for?*

# 36: Area and volume

**1** A child's toy consists of five plastic prisms (shown below) with different cross sections, and a box with matching holes through which to post them.

Each prism is 5 cm long. Find the volume of each prism.

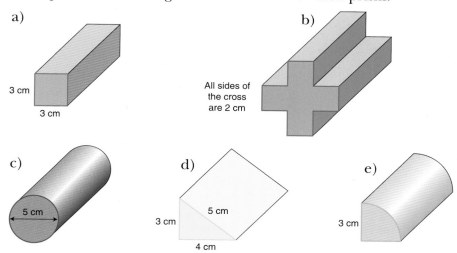

**2** The cross section of a swimming pool is shown in this diagram. The pool is 10 m wide, and it is filled to the brim.

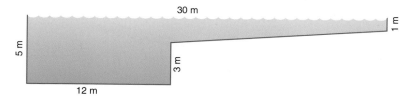

Find the volume of water in the swimming pool.

**3** Metal discs for pet collars have a diameter of 4 cm. Each disc is 0.1 cm thick. How many discs can be made from 1000 cm³ of metal?

a) Find the volume of this cuboid
  (i) in cm³ (ii) in m³.

b) Use your answers to show that $1\,m^3 = 100^3\,cm^3$.

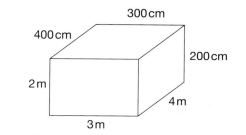

Measure the diameter and thickness of a 1p coin and a 2p coin and work out the volume of each coin. Does a 2p coin contain twice as much metal as a 1p coin?

Compare the amounts of metals in other coins, and make a table of your results.

## 36: Area and volume

# More about prisms

Claire designs food packaging. She is working on a new design for a packet of ground coffee.

The existing packet is a cuboid as shown. The volume of this packet is

13 cm × 4.5 cm × 4 cm = 234 cm³

The new design must have the same volume so that it holds the same amount of coffee.

Claire's first idea is a triangular prism like this:

Claire needs to work out how long the packet must be to have a volume of 234 cm³.

Here is Claire's calculation.

- First she works out the area of the cross section
- Then she uses the volume to work out the length needed

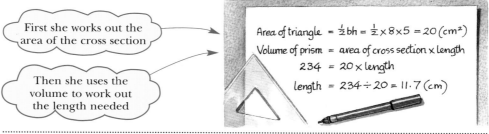

**The packet needs to be 11.7 cm long.**

Claire's next idea is a packet in the shape of a cylinder. She wants a cylinder 12 cm high. She needs to work out the radius of the cylinder.

Here is Claire's calculation.

- First she uses the volume to work out the area of the cross section
- Then she works out the radius of the circle

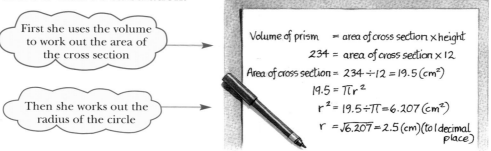

**The radius of the cylinder should be 2.5 cm.**

*What are the advantages and disadvantages of Claire's designs?*

# 36: Area and volume

**1** Find the lengths marked with letters on these prisms, whose volumes are marked.

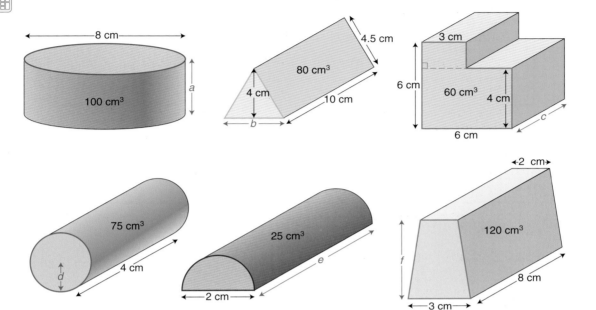

**2** Find the surface area of the first five prisms above.

**3** A drinks company is trying out different sizes for cans of fizzy drink.

a) One size is to hold 250 ml. The radius of the base of the can is 3 cm. Find the height of the can.

b) Another size is to hold 400 ml. The height of the can is 12 cm. Find the radius of the base.

**4** A skip has a capacity of 6 m³. It is in the shape of a prism with the cross section shown. How wide is the skip?

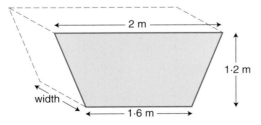

Find 4 containers, each in the shape of a prism.

For each one, draw a diagram and mark on its dimensions. Work out the capacity of each container.

## 36: Area and volume

# Finishing off

**Now that you have finished this chapter you should be able to**

★ find the volume of a cuboid from the length of its sides

★ find the other dimensions of a prism from the volume

★ find the volume and surface area of a prism

Use the questions in the next exercise to check that you understand everything.

**Mixed exercise**

**1** Find the volume of each of these boxes.

a)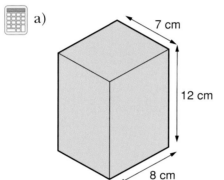

b)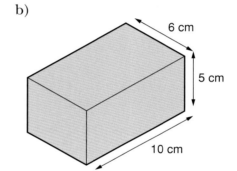

**2** A box of cereal is 20 cm long, 8 cm wide and 30 cm high.

a) Find the volume of the box.

b) Another box is to be made which has $\frac{3}{4}$ of the volume of the first one. It is to be 18 cm long and 8 cm wide.
Find the height of the second box.

**3** A cylinder has radius r, length l, area of cross section A and volume V.

a) Write down the formula for A and V.

b) If the lengths r and l are measured in cm, what are the units for A and V?

c) Find the area of cross section and the volume of a cylinder of radius 5 metres and length 40 metres. Give your answers in exact form in terms of π and state the units.

# 36: Area and volume

**4** A baby's bottle is approximately the shape of a cylinder. The diameter of its base is 6 cm. The bottle can be filled to a maximum depth of 9 cm.

a) What is the maximum amount of milk that the bottle can hold?

b) Baby Sam drinks 150 ml of milk at each feed (1 ml is the same as 1 cm$^3$). To what depth should his bottle be filled?

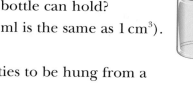

**5** The diagram shows designs of two chocolate novelties to be hung from a Christmas tree.

a) The 'Christmas tree' shape is 6 mm thick.
Find the volume of chocolate in the 'Christmas tree' shape.

b) The 'Santa' shape is 5 mm thick.
It must have the same volume of chocolate as the 'Christmas tree'.
Find the diameter of the 'Santa' shape.

**6** A hollow metal pipe 10 m long has external diameter 50 cm and internal diameter 40 cm.

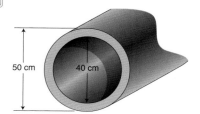

Find the volume of metal needed to make the pipe.

(Hint: find the volume of a solid cylinder and then subtract the volume of the hole.)

**7** A carton of fruit juice contains one litre of juice.

The carton is 16 cm long and 5 cm wide.

How deep is the juice in the carton?

*Remember that 1 litre = 1000 cm$^3$*

*Mixed exercise*

# Thirty seven

# Transformations

### Before you start this chapter you should be able to

* recognise different types of symmetry
* recognise congruent shapes
* decide whether a shape or pattern has rotational symmetry, and if so, its order
* draw the reflection of a shape in the *x* axis or the *y* axis
* rotate a shape about its centre or the origin through $\frac{1}{4}$, $\frac{1}{2}$ or $\frac{3}{4}$ turn

## Reflection, rotation and translation

Shape A has been **reflected** in the mirror line to give shape P.

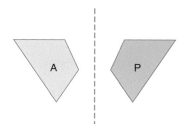

Shape A has been **rotated** through $\frac{1}{4}$ turn clockwise about the point O to give shape Q.

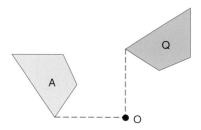

Shape A has been **translated** 3 units to the right and 1 unit down to give shape R.

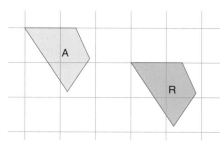

> Shapes A, P, Q and R are all **congruent**. The lengths of lines and the sizes of angles are not changed by these transformations

# 37: Transformations

**1** a) Copy the diagram.

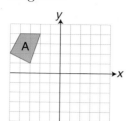

Reflect shape A in the y axis. Label the reflection B.
Reflect shape A in the x axis. Label the reflection C.

b) Make a second copy of the diagram. Rotate shape A clockwise about the origin through $\frac{1}{4}$ turn, $\frac{1}{2}$ turn and $\frac{3}{4}$ turn. Label the rotations D, E and F respectively.

**2** a) Copy the diagram.

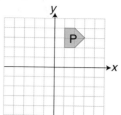

b) Translate shape P 5 squares to the left and 4 squares down. Label the translation Q.

c) Translate shape Q 3 squares to the right and 1 square up. Label this translation R.

d) What translation is needed to move shape R back to the position of shape P?

**3** a) Which of the triangles in this diagram are congruent to A?

b) Which of the triangles are congruent to B?

c) Describe the transformations which map

(i) I → A  (ii) C → D  (iii) E → F

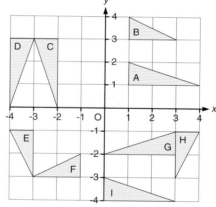

**4** The shape P is formed by joining the points
(0, 0), (5, 0), (5, 2), $(3\frac{1}{2}, 3\frac{1}{2})$

a) Draw P on graph paper. Use the same scales for both x and y. You will need values between −5 and 5 on both axes.

Now draw the following transformations of P.

b) R, T and V are rotations of P with centre O through 90°, 180° and 270° anticlockwise respectively.

c) W and S are reflections of P in the x and y axes.

d) Q and U are reflections of P in the lines $y = x$ and $y = -x$.

e) You have now drawn 8 congruent shapes which together make a polygon.

Describe this polygon, and state what symmetry it has.

## 37: Transformations

# Translations using column vectors

Describing a translation by saying how many squares it has moved in each direction is rather long-winded.

A quicker way of describing a translation is to use a **column vector**.

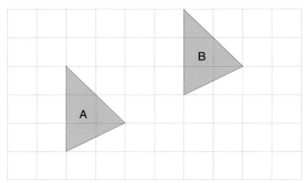

Here is the column vector that describes the translation from A to B:

The top number is the number of units moved to the right

$\binom{4}{2}$

The bottom number is the number of units moved up

- If the translation is to the left instead of the right, the top number is negative.
- If the translation is down instead of up, the bottom number is negative.

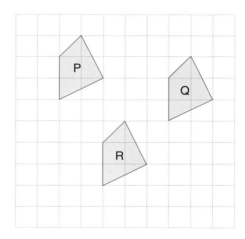

The translation from P to Q is $\binom{5}{-1}$.

*What is R to P?*

*What is Q to R?*

# 37: Transformations

**1** Write down the column vectors for the following translations.

a) A to B   b) D to E   c) B to C

d) E to B   e) C to A   f) B to D

g) A to D   h) E to C

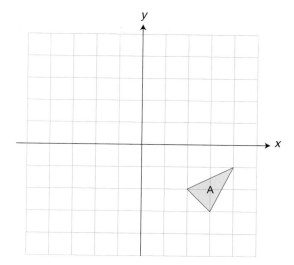

**2** Copy the diagram.

a) Translate shape A using the vector $\begin{pmatrix} -5 \\ 3 \end{pmatrix}$. Label the translated shape B.

b) What vector translates shape B back to shape A?

c) Translate shape B using the vector $\begin{pmatrix} -2 \\ -4 \end{pmatrix}$. Label the translated shape C.

d) What vector translates shape C back to shape B?

e) What vector translates shape A to shape C?

f) What vector translates shape C to shape A?

**3** Look back at your answers to question 2.

a) What is the rule connecting a vector with its reverse (e.g. A→B and B→A)?

b) Compare the vector A→C with the vectors A→B and B→C. What do you notice?

---

For chess players only.
A knight is on one of the centre squares of an empty chess board. There are 8 squares that it can move to; one of these needs the translation $\begin{pmatrix} 2 \\ -1 \end{pmatrix}$.

Describe the other moves.
Now the knight is placed in one of the corners. Which square of the board takes the greatest number of moves for the knight to reach it?

# 37: Transformations

# Reflection

## Reflection in a vertical or horizontal line

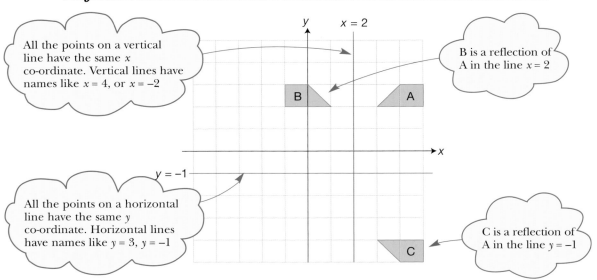

All the points on a vertical line have the same $x$ co-ordinate. Vertical lines have names like $x = 4$, or $x = -2$

B is a reflection of A in the line $x = 2$

All the points on a horizontal line have the same $y$ co-ordinate. Horizontal lines have names like $y = 3$, $y = -1$

C is a reflection of A in the line $y = -1$

## Reflection in a diagonal line

When you draw a reflection in a diagonal line, it is best to reflect its corners one at a time. The diagram shows you how to do this.

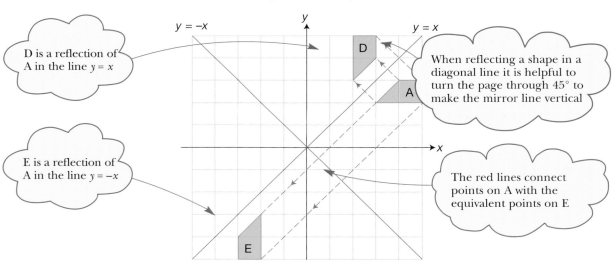

D is a reflection of A in the line $y = x$

When reflecting a shape in a diagonal line it is helpful to turn the page through 45° to make the mirror line vertical

E is a reflection of A in the line $y = -x$

The red lines connect points on A with the equivalent points on E

Why are the names of the two diagonal lines $y = x$ and $y = -x$?
(Think about the co-ordinates of points on the lines.)

# 37: Transformations

**1** Copy these diagrams onto squared paper and draw the reflection of each shape in the mirror line shown.

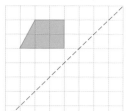 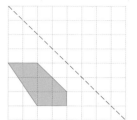 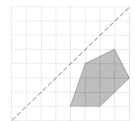

**2** Make 4 copies of this diagram on squared paper. Reflect the triangle in each of these mirror lines.

a) $y = 1$   b) $x = 2$
c) $y = -2$   d) $x = -1$

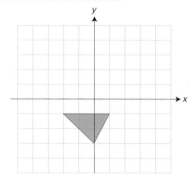

**3** a) Write down the co-ordinates of each corner of triangle A.

b) Copy the diagram and reflect triangle A in the line $y = x$. Label this triangle B.

c) Write down the co-ordinates of each corner of triangle B. What is the rule connecting the co-ordinates of A with the co-ordinates of B?

d) Copy the diagram again and reflect triangle A in the line $y = -x$. Label this triangle C.

e) Write down the co-ordinates of each corner of triangle C. What is the rule connecting the co-ordinates of A with the co-ordinates of C?

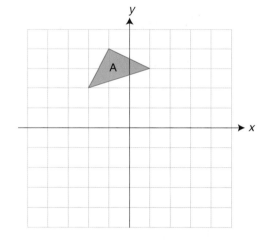

Which letters of the alphabet have mirror symmetry
a) about the *x* axis?
b) about the *y* axis?
State a word which has each type of symmetry.
Find out the meaning of the word palindromic.

## 37: Transformations

# Rotation

You already know how to rotate a shape about the origin. The diagram shows how to rotate a shape about any other point.

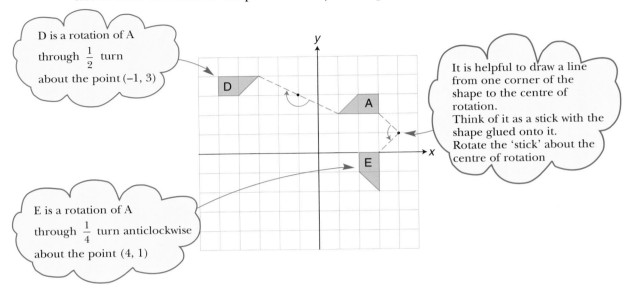

D is a rotation of A through $\frac{1}{2}$ turn about the point (−1, 3)

It is helpful to draw a line from one corner of the shape to the centre of rotation.
Think of it as a stick with the shape glued onto it.
Rotate the 'stick' about the centre of rotation

E is a rotation of A through $\frac{1}{4}$ turn anticlockwise about the point (4, 1)

## Recognising reflections, rotations and translations

Sometimes it is easy to recognise a reflection, rotation or translation and to spot the mirror line or centre of rotation. In cases where you are not sure, try joining up pairs of corresponding points.

For a translation, the lines will be parallel and the shapes will be the same way round.

For a reflection, the lines will be parallel and the shape will be 'flipped over'. The mirror line goes down the middle.

For a rotation through $\frac{1}{2}$ turn, the lines all meet at a point.

 *What is special about this point?*

If the lines are not parallel and do not meet at a point, you need to check for $\frac{1}{4}$ or $\frac{3}{4}$ turn. (In some cases it is hard to find the centre of rotation, but not in the examples in this book.)

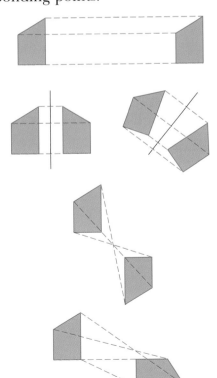

# 37: Transformations

**1** Copy this diagram.

For each of these rotations, draw the rotation and label it clearly.

a) $\frac{1}{4}$ turn clockwise about (−4, 1)

b) $\frac{3}{4}$ turn clockwise about (1, 2)

c) $\frac{1}{2}$ turn about (1, 0)

d) $\frac{1}{4}$ turn anticlockwise about (−1, 1)

e) $\frac{1}{2}$ turn about (1, −1)

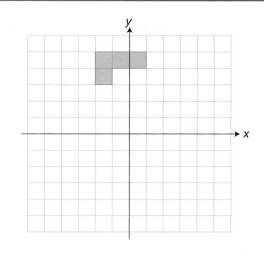

**2** For each of the following transformations, state whether it is a reflection, a rotation or a translation. For a reflection, give the equation of the mirror line, for a rotation, give the co-ordinates of the centre of rotation, and for a translation, give the translation vector.

a) A to B  b) B to C  c) C to D

d) D to E  e) E to F  f) F to G

g) G to H  h) H to I  i) I to A

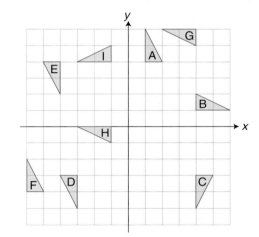

Triangle A is transformed into triangle B, using two transformations, first one and then the other.

How many such pairs of transformations can you find?

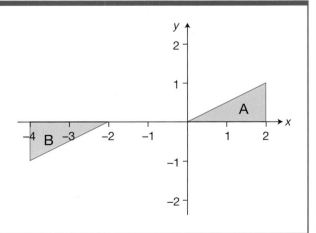

## 37: Transformations

# Enlargement

Jason and Sharon are buying a new fitted kitchen. Jason makes a scale drawing of the kitchen on squared paper so that they can plan the kitchen.

This is Jason's drawing.

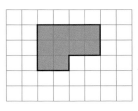

Sharon thinks that Jason's drawing is too small. She makes an enlargement of the drawing.

This is Sharon's drawing.

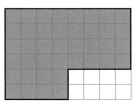

These figures are the same shape but different sizes. They are **similar**. They are not congruent

All Sharon's lines are twice as long as Jason's lines.

Sharon's drawing is an enlargement of Jason's with **scale factor 2**.

*Do the angles stay the same in an enlargement? Do the lengths stay the same?*

## Using a centre of enlargement

Another way to draw an enlargement is to use a **centre of enlargement**. The diagrams below show how to make an enlargement with scale factor 2 using a centre of enlargement C.

1. Draw lines from the centre of enlargement C to each corner of the shape.

2. Make each line twice as long.

3. Join up the ends of the lines to make the enlargement.

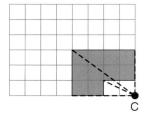 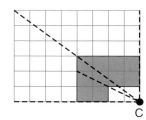 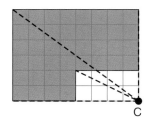

*Draw two squares, one bigger than the other.*

*Can you always find a centre of enlargement?*

*What about two rectangles? ... two circles?*

400

# 37: Transformations

**1** Which of these triangles are enlargements of triangle A?

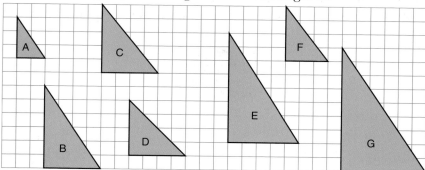

**2** a) Copy this picture on to squared paper.

b) Draw an enlargement of the picture with scale factor 2.

c) Draw an enlargement of the original picture with scale factor 3.

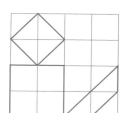

**3** Copy each diagram and draw an enlargement, centre C, with the scale factor given.

a)
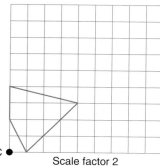
Scale factor 2

b)
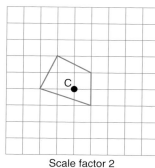
Scale factor 2

c)
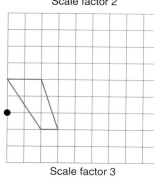
Scale factor 3

d)
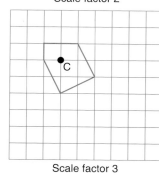
Scale factor 3

Draw a diagram showing a film projector working. Mark in the centre of enlargement, the object and the image.

How can you work out the scale factor?

## 37: Transformations

# Scale factors less than 1

This photograph has been reduced in size to fit into a magazine column. Each side is half as long as it was in the original.

Even though this is a reduction in size, in mathematics it is still called an enlargement. It is an enlargement with scale factor $\frac{1}{2}$.

**An enlargement with scale factor greater than 1 makes things larger.
An enlargement with positive scale factor less than 1 makes things smaller.**

 What do you think a scale factor of $\frac{3}{2}$ means?

You can draw enlargements with fractional scale factors using a centre of enlargement. This example shows how to draw an enlargement with scale factor $\frac{1}{2}$. You can see that the method is the same as before.

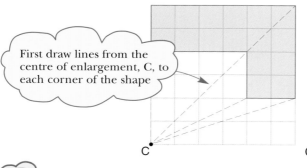

First draw lines from the centre of enlargement, C, to each corner of the shape

Then make all the lines half as long

Join up the ends of the lines to make the enlargement

 The examples on this page are all two-dimensional.

You often meet enlargements of three-dimensional objects too.

One such enlargement is shown in the illustration. Estimate its scale factor.

*What happens to the area when a 2-dimensional shape is enlarged with scale factor 3?*

*What about the volume when a 3-dimensional shape is enlarged with scale factor 3?*

# 37: Transformations

**1** A drawing is 18 cm wide.

Find the width of each of these photocopied enlargements.

a) Scale factor 3

b) Scale factor $\frac{1}{4}$

c) Scale factor $\frac{3}{2}$

d) Scale factor $\frac{2}{3}$

**2** For each of these, copy the diagram and draw an enlargement with the given scale factor and centre of enlargement (C).

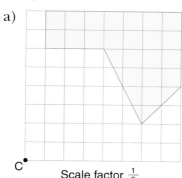

a) Scale factor $\frac{1}{2}$   b) Scale factor $\frac{3}{2}$   c) Scale factor $\frac{3}{4}$

**3** A photograph is 12 cm wide. The widths of some enlargements of the photograph are given below.

Find the scale factor of each enlargement.

a) 24 cm    b) 4 cm    c) 36 cm    d) 6 cm

**4** In each of these, shape A has been enlarged to create shape B. Find the scale factor and the co-ordinates of the centre of enlargement.

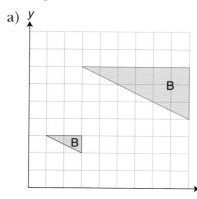
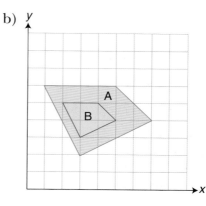

Find 3 examples of enlargements of solid objects.

For each one, work out its scale factor.

## 37: Transformations

# Similar shapes

In mathematics, the word similar does not just mean that two things are rather alike.

**Similar shapes** are shapes that are enlargements of each other.

Look at this poster that is displayed outside a photographic studio.

There is a mistake in one of the print sizes. They should all be similar shapes.

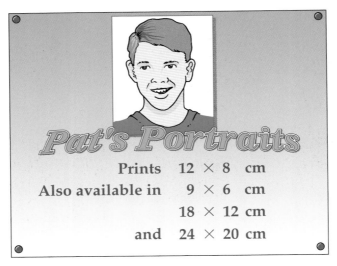

 *Can you see which size is incorrect?*

To check that two shapes are similar you need to make sure that the height and width have been enlarged by the same scale factor.

You find the scale factor for the width by dividing the new width by the old width.

For the 9 × 6 cm print this gives

$$\text{width scale factor} = \frac{6}{8} = 0.75$$

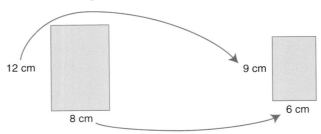

In the same way,

$$\text{height scale factor} = \frac{9}{12} = 0.75$$

The 9 × 6 cm print is a true enlargement because the height and width have been enlarged by the same scale factor.

 *Use this method to check which print size is incorrect on the poster.*

# 37: Transformations

**1** Find the scale factor of these enlargements.

a) A photograph 15 cm long is enlarged to 20 cm long.
b) A drawing 8 cm wide is enlarged to 18 cm wide
c) A diagram 24 cm long is reduced to 9 cm long.

**2** a) Which of the photographs B – F are similar to photograph A? Explain your answers.

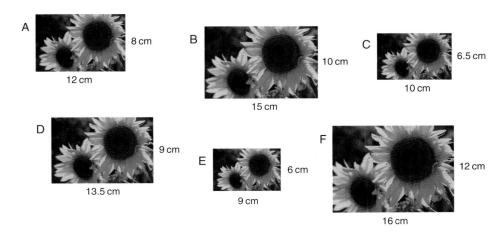

b) I need an enlargement of photograph A, 20 cm high.

How wide will the enlarged photograph be?

**3** Each of the triangles D, E and F is similar to one of the triangles A, B and C.

Match up the pairs of similar triangles, colouring the equal angles.

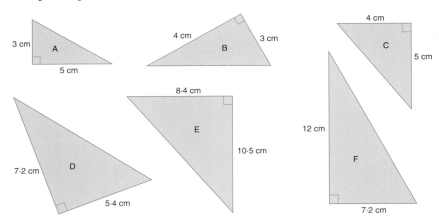

Measure the width and height of at least 3 different television screens. Are they similar to each other?

# 37: Transformations

# Finishing off

**Now that you have finished this chapter you should be able to**

- ★ find out how many lines of reflection symmetry a shape or pattern has
- ★ find out whether a shape or pattern has rotational symmetry, and if it has, give the order of rotational symmetry
- ★ draw the reflection of a shape in the *x* axis or the *y* axis
- ★ rotate a shape about its centre or the origin through $\frac{1}{4}$ turn or $\frac{1}{2}$ turn or $\frac{3}{4}$ turn
- ★ carry out and describe a translation of a shape
- ★ recognise and draw an enlargement of a simple shape using a whole number scale factor
- ★ recognise and draw an enlargement of a simple shape using a whole number scale factor and a centre of enlargement

Use the questions in the next exercise to check that you understand everything.

## Mixed exercise

**1** For each of these patterns, say

  (i) how many lines of reflection symmetry, if any, it has

  (ii) whether it has rotational symmetry, and if it has, what the order of rotational symmetry is.

a)

b)

c)

d)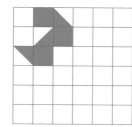

**2** a) Copy the pattern below and complete it so that the dotted line is a line of reflection symmetry.

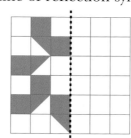

b) Copy the pattern below and complete it so that it has rotational symmetry of order 4.

# 37: Transformations

**Mixed exercise**

**3** Copy this diagram.

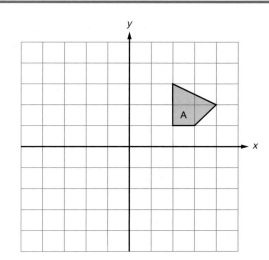

a) Reflect shape A in the *x* axis. Label this shape B.

b) Rotate shape B through $\frac{1}{4}$ turn clockwise about the origin. Label this shape C.

c) Translate shape C 2 squares to the left and 7 squares up. Label this shape D.

d) Reflect shape D in the *y* axis. Label this shape E.

e) Rotate shape E through $\frac{1}{2}$ turn about the origin. Label this shape F.

**4** Marie has a photograph 10 cm by 6 cm. She wants to have it enlarged to fit one of the frames below. For each frame, say whether it is possible to enlarge the photograph to fit the frame, and if it is, give the scale factor of the enlargement.

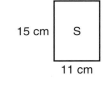

P: 16 cm × 12 cm  
Q: 20 cm × 12 cm  
R: 30 cm × 18 cm  
S: 15 cm × 11 cm

**5** a) Copy shape A in question 3 on to squared paper.

b) Draw an enlargement of shape A with scale factor 2.

c) Draw an enlargement of shape A with scale factor 3.

d) Draw an enlargement of shape A with scale factor 4.

**6** Make two copies of the diagram below and draw enlargements with scale factors 2 and 3, using centre of enlargement C.

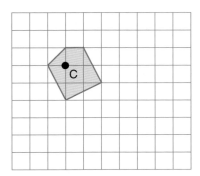

Solid objects may have planes of symmetry instead of lines of symmetry. Where is the plane of symmetry of a person?
Describe the symmetries of a cube, a cuboid and a sphere.

# Thirty eight

# Locus

### Before you start this chapter you should be able to

★ make accurate drawings, full size or to scale

## Simple loci

The **locus** of a point means all the possible positions for that point.

Ainsley and Sarah are going on a camping holiday. They need to decide where to pitch their tent. Ainsley wants to be no more than 200 m from the shop. The map of the campsite shows the area where he would like to be.

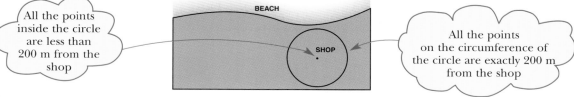

 *How could you describe the points outside the circle?*

**The locus of a point a fixed distance, *d*, from a fixed point, O, forms a circle, centre O and radius *d*.**

Sarah wants to be less than 100 m from the beach. The map shows the area where she would like to be.

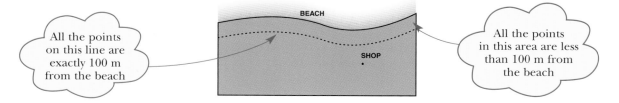

**The locus of a point a fixed distance, *d*, from a line, AB, forms a line parallel to AB and distance *d* from AB.**

 *Why is the line in this diagram dotted and not solid?*

All the points in the shaded area are both no more than 200 m from the shop and less than 100 m from the beach.

Ainsley and Sarah should camp somewhere in this area.

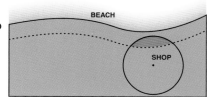

# 38: Locus

**1** For each part, draw this rectangle full size and shade in the required locus. Only shade points inside the rectangle.

a) The locus of all points less than 3 cm from A.

b) The locus of all points more than 2 cm from BC.

c) The locus of all points at least 1 cm from the centre of the rectangle.

d) The locus of all points more than 6 cm from C and more than 5 cm from D.

e) The locus of all points less than 3 cm from AB and at least 2 cm from B.

f) The locus of all points no more than 1 cm from the perimeter of the rectangle.

**2** A goat is tied to a point (marked G on the diagram) on the outside of a barn, 3 metres from the corner. The width of the barn is 5 metres. The rope is 5 metres long. Make a scale drawing of the diagram and shade in the locus of the points where the goat can go.

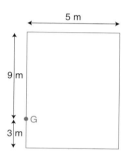

**3** Molly wants to plant a tree in her garden. She must not plant the tree within 2 metres of the house. Molly's house is 12 m long and 10 m wide. Make a scale drawing of the house and shade in the locus of the points where Molly must *not* plant the tree.

**4** Keith is looking for somewhere to live. He wants to be no more than 3 miles away from the station as he catches a train to work every morning. He is also a keen cinema-goer and would like to be no more than 5 miles away from the local cinema. The station and the cinema are 6.5 miles apart.

Make a scale drawing and show the area where Keith would like to live.

**5** Simon and his brother Mark are playing in a rubber dinghy in the sea. The coastguard has told them not to go more than 50 m from the shore. There is a rock 40 m from the beach and he has also told them not to go within 5 m of the rock.

Make a scale drawing and shade the area where they are allowed to go. (Assume that the shoreline is straight.)

Investigate the locus of a point on the circumference of a bicycle wheel as the bicycle moves along.

# 38: Locus

## A point equidistant from two fixed points

A new road is being built between two villages A and B. So that the new road makes as little disturbance as possible in the two villages, the road is being built so that it is always the same distance from A as from B.

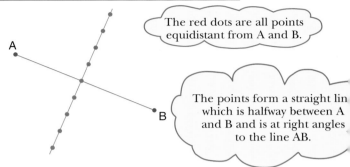

The red dots are all points equidistant from A and B.

The points form a straight line which is halfway between A and B and is at right angles to the line AB.

The position of the road will be the locus of a point **equidistant** (the same distance) from A and B.

This is called the **perpendicular bisector** of AB.

> **The locus of a point equidistant from two points is the perpendicular bisector of the two points.**

Where are the points which are nearer to A than to B?

### Drawing a perpendicular bisector accurately

These instructions explain how to draw the perpendicular bisector of the line AB.

1. Place your compass point on A. Open the compass to a radius more than half the distance from A to B. Draw an arc each side of AB.

2. Leave the compass at the same radius. Put the compass point on B and make another arc each side of AB, so that they cross the other arcs.

3. Draw a line joining the two intersections. This is the perpendicular bisector.

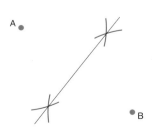

## 38: Locus

**1** a) Mark two points P and Q (not in a horizontal or vertical line). Construct the perpendicular bisector of PQ.

b) Mark 3 points on the perpendicular bisector. Measure the distance of each point from P and from Q to check that it is the same.

**2** a) Draw a circle and mark three points, A, B and C on the circumference.

b) Construct the perpendicular bisectors of AB, BC and CA.

c) You should find that they meet at a point.

What is special about the point where they meet?

**3** Sophie works in town S and her parents live in town T, 50 miles from town S. She is looking for a place to live somewhere between the two towns. She wants to be nearer to S than to T but would like to be within 30 miles of her parents.

Make a scale drawing and shade the area where Sophie would like to live.

**4** The diagram below shows three schools A, B and C in a large town. Mr and Mrs Hammond and their son Michael are moving into the area, and Michael wants to go to school C. The Hammonds need to live somewhere which is nearer to school C than to school A, and nearer to school C than to school B.

Trace the diagram and shade the area where the Hammonds should live.

A •

• B

C •

Find a map of your area with some of the local schools marked on it. Use perpendicular bisectors to show the areas which are nearer to each school than to any other school.

# 38: Locus

## A point equidistant from two lines

The designer of a new housing estate is putting two houses at the end of a cul-de-sac. She wants the boundary fence between the two houses to be the same distance from the wall of each house. The position of the fence will be the locus of a point equidistant from two lines.

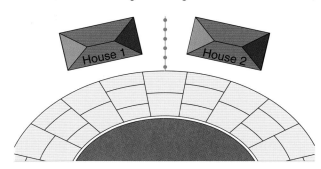

*The red dots are all points which are the same distance from each house*

*The dots form a straight line which will be the position of the boundary fence*

In this diagram, the lines showing the wall of each house have been extended until they meet, forming an angle. The red line showing the position of the fence has also been extended.

You can see that the fence line cuts the angle formed by the other two lines in half. It is called the **angle bisector**.

**The locus of a point equidistant from two lines is the angle bisector of the two lines.**

*Where are the points nearer to house 1 than to house 2?*

### Drawing an angle bisector accurately

1. Put the compass point on the point of the angle and mark off two points as shown.

2. Put the compass point on each of the points you have marked off and draw two arcs which meet each other.

3. Draw a line through the point where the arcs intersect to the point of the angle. This line is the angle bisector.

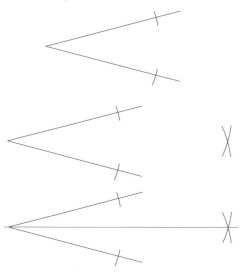

# 38: Locus

**1** a) Draw two lines to make an angle. Construct the angle bisector.

b) Measure the angle and check that the angle bisector cuts the angle in half.

**2** a) Draw a triangle. Construct the angle bisector of each of the three angles of the triangle.

b) The three angle bisectors should meet at a point. What is special about this point?

**3** Trace this triangle four times.

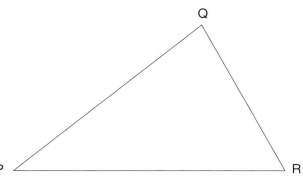

a) On the first diagram, shade the locus of all points nearer to PR than to PQ.

b) On the second diagram, shade the locus of all points nearer to QR than to PQ and within 3 cm of Q.

c) On the third diagram, shade the locus of all points nearer to PR than to QR and nearer to PQ than QR.

d) On the fourth diagram, shade the locus of all points nearer to PQ than to PR and nearer to Q than to P.

**4** This diagram shows a field in which there are some rabbits. The field is surrounded by hedges on sides AB, BC and CD. A fox appears at point F and all the rabbits run to the nearest hedge. Trace the field and divide it up, showing the areas which are nearest to each of the three hedges.

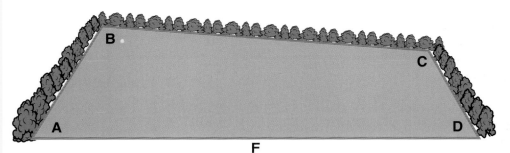

# 38: Locus

## Finishing off

**Now that you have finished this chapter you should be able to**

- ★ construct the perpendicular bisector of a line
- ★ construct the bisector of an angle
- ★ solve problems involving loci, including intersecting loci
- ★ find the locus of a point which is a fixed distance from a point, a fixed distance from a line, equidistant from two points, or equidistant from two lines

**Use the questions in the next exercise to check that you understand everything.**

### Mixed exercise

**1** Mark a point X. Draw the locus of all points which are less than 4 cm from X.

**2** Mark points S and T 5 cm apart (not in a horizontal or vertical line). Draw the locus of all points equidistant from S and T.

**3** Draw an angle ABC of size 63°. Draw the locus of all points equidistant from the lines AB and BC.

**4** Draw a line XY 3 cm long. Draw the locus of all points exactly 2 cm from XY. (Hint: think carefully what happens to points at each end of the line.)

**5** Draw points M and N 8 cm apart (not in a horizontal or vertical line). Draw the locus of all points nearer to M than to N.

*For questions 6 to 11, trace the triangle XYZ and shade the locus of the point P. In all cases, P is inside the triangle XYZ.*

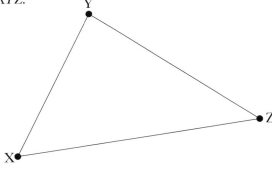

**6** P is less than 4 cm from X and less than 5 cm from Z.

**7** P is nearer to X than to Z and is no more than 2 cm from Y.

# 38: Locus

**Mixed exercise**

**8** P is less than 3 cm from XZ and less than 2 cm from Y.

**9** P is no more than 3 cm from X, less than 4 cm from Y and at least 5 cm from Z.

**10** P is nearer to XY than to XZ and is more than 1 cm from YZ.

**11** P is nearer to Y than to Z and is nearer to YZ than to XY.

**12** Jed is at a rock concert. He wants to be equidistant from the two speakers to get the best stereo effect. He also wants to be less than 10 m from the stage.

Make a scale drawing and show the possible places where Jed would like to be.

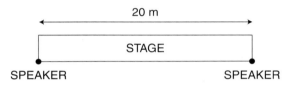

**13** This diagram is the plan of a church which is going to be fitted with a burglar alarm inside. Two motion sensors, each with a range of 8 m in all directions, are shown. Make a scale drawing of the church and shade the areas which are not covered by the sensors. (Remember that the sensors do not work round corners!)

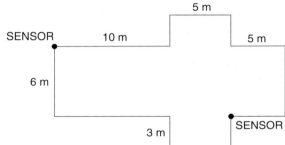

**14** Alex needs to draw a line through A, perpendicular to the given line in these two cases.

He starts by drawing two arcs of a circle centre A.

Copy and complete the constructions.

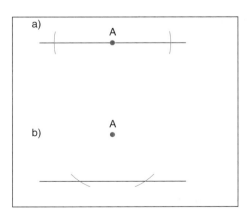

# Thirty nine

# Pythagoras' rule

## Finding the hypotenuse

*Measure the triangles below in millimetres and find the areas of squares A, B and C in each diagram. What do you notice?*

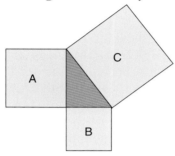
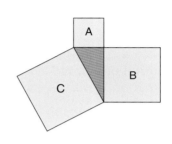

You should have found that the areas of the two smaller squares add up to the area of the largest square in each case.

This rule is called Pythagoras' rule (or theorem). It is true for all right-angled triangles. It is usually written like this:

$$a^2 + b^2 = c^2$$

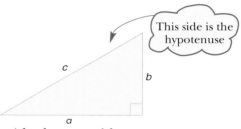

This side is the hypotenuse

The side labelled $c$ must be the hypotenuse (the longest side, always the one opposite the right angle).

Robert is a farmer and is building a gate for one of his fields. He wants to work out how long the diagonal piece of wood needs to be.

The diagonal bar on Robert's gate can be drawn as the hypotenuse of a right-angled triangle like this:

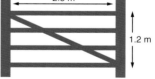

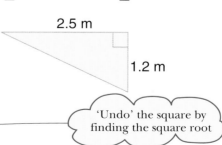

Pythagoras' rule is: $a^2 + b^2 = c^2$

In this case:
$2.5^2 + 1.2^2 = c^2$
$6.25 + 1.44 = c^2$
$7.69 = c^2$
$c = 2.77$

'Undo' the square by finding the square root

**So the diagonal piece on Robert's gate needs to be 2.77 m long.**

*Why must the value of $c$ be greater than 2.5 metres?*

# 39: Pythagoras' rule

**1** Use Pythagoras' rule to find the length of the hypotenuse in the triangles below.

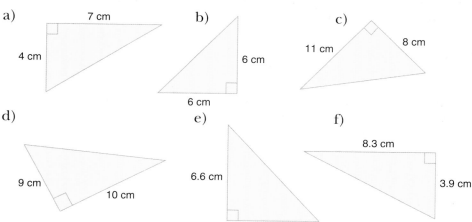

**2** A field is 150 metres long and 120 metres wide. A footpath goes diagonally across the field. How long is the footpath?

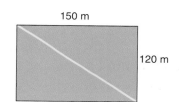

**3** A ship sails 23 km due north and then 17 km due east. It then sails back to its starting point in a straight line. How far is the distance back to the starting point?

**4** The diagram shows the two points A (1, 2) and B (4, 4).

a) What is the distance from A to the point P?

b) What is the distance from B to the point P?

c) Use Pythagoras' rule to find the distance from A to B.

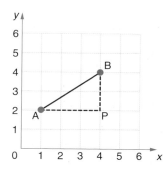

**5** Use the same method as in question 4 to find the distances between each of the following pairs of points.
(You will find it helpful to draw diagrams showing the points.)

a) (3, 1) and (5, 6)     b) (4, 2) and (1, 5)

c) (−2, 4) and (3, 5)    d) (0, 3) and (−2, −1)

## 39: Pythagoras' rule

# Finding one of the shorter sides

Terry is a window-cleaner. His ladder is 8 metres long.

For safety reasons he always places the foot of the ladder at least 1.5 metres from the wall. He wants to know how far up the wall he can make his ladder reach.

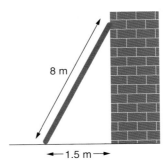

So far you have only been asked to find the length of the hypotenuse in a right-angled triangle. To solve Terry's problem, you need to be able to find one of the two shorter sides. You can use Pythagoras' rule to solve this kind of problem as well.

This is a simplified diagram of Terry's ladder.

$y$ stands for the height up the wall that the ladder reaches.

Pythagoras' rule is:

$$a^2 + b^2 = c^2$$

In this case:

$$1.5^2 + y^2 = 8^2$$
$$2.25 + y^2 = 64$$

To find $y^2$, you need to subtract 2.25 from both sides of the equation.

$$y^2 = 64 - 2.25$$
$$y^2 = 61.75$$

Now you can find $y$ by taking the square root of 61.75.

$$y = \sqrt{61.75} = 7.86$$

**The ladder reaches 7.86 metres up the wall.**

*Why must the value of y be less than 8 metres?*

Remember!

To find the hypotenuse, you have to **add**.

To find one of the shorter sides, you have to **subtract**.

In the calculation above you obtained $y = \sqrt{61.75}$. Your calculator gives the value of $y$ as 7.8581 ... and you round this to 7.86. Numbers like $\sqrt{61.75}$ which do not work out exactly are called **surds**. Sometimes lengths are given in surd form.

## 39: Pythagoras' rule

**1** Find the lengths of the sides marked *x* in each of these triangles. In some of them you have to find the hypotenuse, in others you have to find one of the shorter sides.

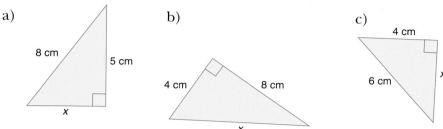

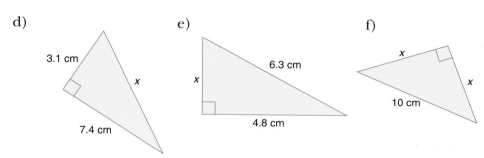

**2** A ladder 6.2 metres long is to be placed so that it just reaches a window 5.7 metres from the ground. How far from the wall is the foot of the ladder?

**3** The diagram shows an isosceles trangle split into two congruent right-angled triangles.

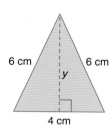

a) Use Pythagoras' rule to find the height, *y*, of the triangle.

b) Find the area of the triangle.

**4** The diagram shows a right-angled triangle ABC. The line BN has been drawn in, splitting the triangle into two smaller right-angled triangles ANB and CNB.

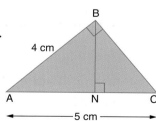

a) Work out the length of the side BC.

b) Using AB as the base of the triangle, work out the area of the triangle.

c) Using AC as the base of the triangle, use your answer to b) to work out the length of BN.

d) Work out the lengths of AN and CN.

**5** In triangle ABC, angle B = 90°, AB = 4 cm and AC = 7 cm. Work out the length BC giving your answer in surd form.

## 39: Pythagoras' rule

# Finishing off

**Now that you have finished this chapter you should be able to:**

★ use Pythagoras' rule to find the hypotenuse of a right-angled triangle

★ use Pythagoras' rule to find one of the shorter sides of a right-angled triangle

Use the questions in the next exercise to check that you understand everything.

## Mixed exercise

**1** Find the lengths of the sides marked with letters in these triangles.

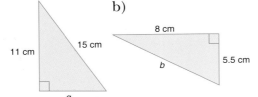

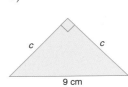

**2** Find the distance between each pair of points.
Give your answers in surd form where appropriate.

a) (1, 4) and (4, 0)

b) (−2, 3) and (2, −1)

c) (−3, −4) and (−1, 1)

**3** A ship leaves the port of Harwich (on the east coast) and sails 30 km. It is then 12 km north of Harwich.

How far east is it from Harwich?

**4** Jenny designs this ramp to provide access to a building.

Work out the length of the ramp giving your answer to the nearest centimetre.

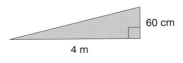

**5** A doorway is 1.96 m high and 0.65 m wide. Harry is outside with a thin panel of a self-assembly bedroom unit. The panel is 2.05 m high. Harry tilts the panel in an attempt to get it through the door. Will it go through or not?

# 39: Pythagoras' rule

**6** Mark and Imogen are putting up Christmas decorations (streamers) in their office.

This is a plan of their office.

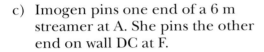

a) How long must a streamer be to go from A to C?

b) Mark pins one end of a 6 m streamer at A. He pins the other end on wall BC at E.

How far is E from C?

c) Imogen pins one end of a 6 m streamer at A. She pins the other end on wall DC at F.

How far is F from C?

d) What length of streamer is needed to go from E to F?

**7** This drawing shows Debbie's house.

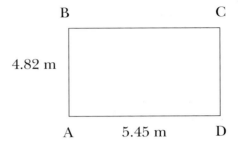

She is building it herself and wants to work out how many roof tiles she will need.

A row of tiles is needed approximately every 25 cm.

a) How many rows of tiles are needed to cover the roof?

b) A tile is 30 cm wide. How many tiles are there in a row?

c) Work out the total number of tiles needed to cover the roof.

**8** A thin rod is 50 cm long. It is straight and does not bend. Will the rod fit into a box 40 cm long, 25 cm wide and 20 cm high? Explain your answer.

## Investigation

This triangle is a right-angled triangle.

$3^2 + 4^2 = 9 + 16 = 25 = 5^2$

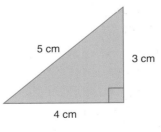

The numbers 3, 4, 5 are called a **Pythagorean triple** because they obey Pythagoras' rule.

Find as many different Pythagorean triples as you can.

(Don't count triples which are just a multiple of one you have already found, like 6, 8, 10 or 9, 12, 15, which are both multiples of 3, 4, 5.)

# Index

## A

acute angles, 2, 58
acute-angled triangle, 78
adding
    decimals, 66–67, 296–297
    fractions, 42–45
    negative numbers,
        116–117, 138–139
    powers, 254
    probabilities, 242–243
algebra, in setting up
    equations, 234–235
alternate angles, 282–283
angles
    acute, 2
    alternate, 282–283
    bisecting, 412–413
    corresponding, 282–283
    exterior, 284–285
    interior, 284–285, 288–289
    on a line, 36–37
    measuring, 58, 262, 264–265
    obtuse, 2
    opposite, 282–283
    properties, 2
    of a quadrilateral, 286–287
    reflex, 2
    right, 2, 36, 58
    round a point, 36–37
    and shapes, 282–295
    of a triangle, 284–285
    types, 58–59
approximations, 218–219
arc of a circle, 84–85
area
    calculating, 184–185
    of a circle, 3, 192–193
    formula, 350–351
    of a parallelogram,
        382–383
    and radius, 192–193
    of a rectangle, 188,
        350–351
    of shapes, 382–391
    of a trapezium, 382–383
    of a triangle, 2, 186–187
    *see also* surface area

arrow diagrams, 122, 130
averages, 104–105, 334–335
axes, 8

## B

back-substitution, 236
bank loan, 162–163
bank/building society
    accounts, interest,
    212–213
bar charts, 98–99
base, of a triangle, 186–187
base year, for social statistics,
    150
bearings, 266–267
best fit, line, 342–343
brackets
    expanding, 236, 346–347,
        354–355
    using, 136–137, 258–259,
        348–349
building societies *see* banks

## C

calculator
    brackets, 137
    checking results by
        estimating, 225–226
    index notation, 256–257
    order of operations, 258
    using, 2, 190, 253
cancelling, 270
capacity
    measuring, 14
    *see also* volume
centimetres, 3, 12
chord of a circle, 84–85
circles, properties, 3, 84–85,
    189–190, 192–193
circumference of a circle, 3,
    84, 190
clock times, 12 and 24 hour,
    16, 17
co-ordinates, 4–11, 320–329
    mathematical, 8–10
    positive and negative,
        118–119

and reflections, 396–397
    three-dimensional, 320–321
collecting like terms,
    134–135
column vectors, 394–395
commission, 210–211
common factors, 22, 249
comparisons, 110–111
    using fractions or
        percentages, 278–279
compound interest, 212–213
congruent *see* shapes,
    congruent; triangles,
    congruent
construction *see* drawing
conversion graphs, 126–127
converting
    currency, 376–377
    units, 3, 12, 14, 70–71, 73
corners of a cuboid, 200–201
correlation, 338–339, 342
corresponding angles,
    282–283
costs, estimating, 224–225
cross section of a prism, 204
cube number, 248
cube root, 248
cubes (shape), 26–27
    drawing, 198
    net, 204
cuboids
    drawing, 198
    volume, 3, 202–203,
        384–385
currency, converting, 376–377
curves, drawing, 326–327
cylinder, properties, 204,
    352–353, 386–387,
    388–389

## D

data
    discrete and continuous,
        106
    displaying, 98–99, 108–109
    grouped, 106–109,
        336–337, 340–341
    recording, 142–143, 147
    types, 142–143

# Index

decimal places, 220–221
decimal point, moving, 68, 298, 312
decimals
  adding and subtracting, 66–67, 296–297
  multiplying and dividing, 68–71, 298–299
  and percentages, 88–89, 312–313
  recurring, 275
  using, 62–75, 296–297
degrees of angle, 58
denominator, 40
density, 384
diagonal
  of a parallelogram, 286–287
  reflection, 396–397
diameter of a circle, 3, 84–85, 190–191
directed numbers, 114–121
discount, percentage, 314–315
discussion points, 2
distance, 378–379
distance–time graphs, 174–179
dividing
  decimals, 72–73, 298–299
  fractions, 272–273
divisibility, 22
division, long, 3
drawing
  angle bisector, 412–413
  perpendicular bisector, 410–411
  scale, 54–61
  solid objects, 196–197

## E

edges, of a cuboid, 200–201
elevation, 386
enlargement, 400–401
equations, 230–239
  brackets in, 236
  doing the same to both sides, 232
  and graphs, 168, 324–325
  and inequalities, 358–371
  of the line, 324
  setting up, 234–235
  solving, 230–237, 360–361
  of a straight line, 170
  using, 234–235
equilateral triangles, 2, 78
Eratosthenes, 25
Escher, M.C., 292
estimation, 218–229
  of probabilities, 244–245
even numbers, 24
exchange rates, 376–377
expansion *see* brackets
expressions, simplifying, 134, 156–157
exterior angles, 284–285
  of polygons, 290–291

## F

faces of a cuboid, 200–201
factors, 22–23, 248
fairness, 242
feet, 3, 12
FOIL (Firsts Outsides Insides Lasts), 354
formulae
  changing the subject, 352–353
  using, 130–141, 346–357
fractions, 40–53
  adding and subtracting, 42–45
  and decimals, 62–65, 274–275
  dividing, 272–273
  equivalent, 40–41, 90, 94
  improper, 46–47, 48–49
  multiplying, 270–271
  and percentages, 88–89, 94–95, 275–276, 312–313
  of a quantity, 50–51
  simplest form, 64, 94
  simplifying, 40
  top heavy *see* fractions, improper
  using, 270–281
frequency, relative, 244
frequency polygons, 108–109, 336–337
frequency tables, 142–143, 336–337

## G

gallons, 3, 14
gradients, 168–169
  positive/negative, 170
  of a travel graph, 178–179
grams, 3, 14, 297
graphs, 168–181, 320–329
  curved, 326–327
  distance–time (travel), 174–179
  and equations, 324–325
  intercept, 168–169, 170–171, 172
  obtaining information from, 172–173
grid references, 7, 56
grids
  co-ordinates, 4–5, 322–323
  on a map or plan, 4–8
grouped data, 106–109
  displaying, 108–109

## H

halves, 64–65
  as percentage, 88
hexagon, 82
highest common factor (HCF), 250–251
hire purchase, 162–163
horizontal line, reflection, 396–397
household bills, 160–161
hundredths, 62–63
  as decimals, 296
hypotenuse, of a right-angled triangle, 416–417

## I

imperial units, 3, 12, 14
inches, 3, 12
index/indices
  on a calculator, 256–257
  form, 254
  notation, 252–253, 256–257
  rules, 254–255

# Index

using, 152, 248–261
inequalities, 358–371
   combining, 366–367
   solving, 368–369
   symbols, 364
information, obtaining from a graph, 172–173
intercepts, 168–169, 170–171, 172
interest, simple and compound, 212–213
interior angles
   of polygons, 288–289
   of triangles, 284–285
isometric paper, 198–199
isosceles triangles, 2, 78

## K
kilograms, 3, 14
kilometres, 3, 12, 71, 73
kite (shape), 2, 80
knot (unit), 126

## L
length
   estimating, 12
   measuring, 12–13
like terms, collecting, 134–135
line graphs, 102–103
lines
   of best fit, 342–343
   equation, 170–171, 324
   gradient, 170–171
   parallel, 170–171, 282–283
   perpendicular, 36, 170–171
   of reflection symmetry, 30–33
litres, 3, 14
locus/loci, 408–415
   simple, 404–409
long division, examples, 3
long multiplication, examples, 3
lowest common multiple (LCM), 250–251

## M
maps, 54–61
   co-ordinates, 4–7
   Ordnance Survey, 7, 56
   scale, 54–57
mass, measuring, 14
mean, 104–105, 334–335
   of grouped data, 340–341
measuring
   angles, 58, 262, 264–265
   and drawing, 262–269
   mass/weight, 14
   time, 16–17
   volume/capacity, 14
median, 104–105, 334–335
   of grouped data, 340–341
meter readings, 160–161
metres, 3, 12, 297
metric system, units, 3, 12, 14, 297
miles, 3, 12, 71, 73
milligrams, 297
millilitres, 3, 14
millimetres, 3, 12, 297
mirror line *see* lines of reflection symmetry; reflections
mixed numbers, 46–47
   multiplying, 270
modal class, 108, 336–337
modal group, 144
mode, 104–105, 334–335
   of grouped data, 340–341
money
   calculations, 18–19
   changing, 376–377
   decimal notation, 66
   earning, 208–217
   spending, 160–167
multiples, 22–23, 248
   of 10, 68–69
multiplication, long, 3
multiplying
   decimals, 68–71
   fractions, 270–271
   mixed numbers, 270
   negative and positive numbers, 156
multiplying out brackets, 346–347
musical notation, 45
Mystic Rose pattern, 85

## N
negative correlation, 338–339
negative numbers
   adding and subtracting, 116–117, 138–139
   calculations, 156–157
   rules for signs, 154–157
   substituting into expressions, 154
   using, 114–115
nets, 197, 200–201, 204
notation
   on a calculator, 256–257
   decimal, for money, 66
   index form, 254
   music, 45
   *see also* symbols
$n$th term, of a sequence, 306–307
number lines
   for decimals, 62–65, 220, 296–297
   for directed numbers, 116–117, 138–139, 154, 156
   in estimation, 218–219
   for inequalities, 366–367, 368
number patterns, 248–249
numerator, 40

## O
obtuse angles, 2, 58
obtuse-angled triangle, 78
octagons, 82
   exterior angles, 290
odd numbers, 24
opposite angles, 282–283
order of operations, 258
order of rotational symmetry, 34–35
Ordnance Survey maps, 7, 56
origin, co-ordinates, 8
ounces, 14
outcomes, equally likely, 240, 244
overtime payments, 208

# Index

**P**

parallel lines, 170–171, 282–283
parallelograms, 2, 80, 286–287, 322–323, 382–383
patterns, 302–303
pentagons, 82, 288–289
percentages
    calculations, 92–93, 314–315
    converting to decimals or fractions, 94–95, 312–313
    of a quantity, 90–91
    using, 88–97, 312–319
perimeter, 182–183, 348–349
perpendicular bisector, drawing, 410–411
perpendicular height of a triangle, 186–187
perpendicular lines, 36, 170–171
pi ($\pi$), 190, 192
pictograms, 98–99
pie charts, 100–101, 332–333
pints, 3, 14
point
    finding co-ordinates, 6–7, 8
    locus, 410–411, 412–413
polygons, 82–83
    angle properties, 288–289, 290–291
    irregular, 82
    regular, 82, 288–289
positive correlation, 338–339
pounds (unit), 3, 14
powers, 152
    adding and subtracting, 254
    negative or zero, 254
    *see also* index/indices
prime factorisation, 250–251
prime numbers, 24–25, 248
prisms
    properties, 204–205, 386–389
    triangular, 196, 388
probability, 240–247
    adds up to 1, 242–243
    between 0 and 1, 240
    calculating, 240–241
    estimating, 244–245
profit, percentage, 314–315
proportion, 122–129, 372–381
    as percentage, 316–317
    simple, 122–123
protractor, 58, 262, 264–265
Pythagoras' rule (theorem), 416–421
Pythagorean triples, 421

**Q**

quadrilaterals
    angle properties, 286–287
    definition, 80
    irregular, 80
    special types, 2, 80–81, 286
quarters, 64–65
    as percentage, 88

**R**

radius, 84–85, 190, 192–193
range, 104–105, 334–335
ratios, 122–129, 372–381
    simplifying, 372–373
    unitary method, 374–375
reciprocals, 46–47
rectangles, 2, 80, 348–349
reflection symmetry, 30–33, 81
reflections, 392–393, 396–397
    recognising, 398–399
reflex angles, 2, 58
reports, writing, 148–149
rhombus, 2, 80
right angles, 2, 36, 58
right-angled triangles, 2, 78, 416–419
rotational symmetry, 34–35
rotations, 392–393, 398–399
    recognising, 398–399
rounding, 222–223
rules, of sequences, 302–309

**S**

salaries, 210–211
    gross and net, 214–215
scale, of a map or drawing, 54–57
scale factors, 400–405
scalene triangles, 2, 78
scatter diagrams, 338–339, 342
sequences, 302–311
shapes
    and angles, 282–295
    area, 188–189
    congruent, 82–83, 292, 392–393
    perimeter, 348–349
    similar, 400, 404–405
    sorting, 76–77
Sieve of Eratosthenes, 25
significant figures, 219, 222–223
similar *see* shapes, similar; triangles, similar
simple interest, 212–213
social statistics, 150–151
solid figures
    drawing, 196–197
    properties, 3
speed, 378–379
sphere, drawing, 196
spread, 104–105
square numbers, 248, 299
square roots, 26–27, 72–73, 248, 299
    positive and negative, 156
square (shape), 2, 26–27, 80
standard form, 252
standing charge, 160
statistics, 98–113, 330–345
stem-and-leaf diagrams, 144–145
stones (unit), 14
straight line
    equation, 170
    graph, 168–169
subject of a formula, changing, 352–353
substitution *see* formula
subtracting
    decimals, 66–67, 296–297
    fractions, 42–45
    negative numbers, 116–117, 138–139
    powers, 254

# Index

surface area, 204–205
surveys, 142–151
symbols
   for inequalities, 364
   using, 152–159
   *see also* notation
symmetry, 30–39

## T
tally charts, 142–143
tangent to a circle, 84–85
tax, 214–215
temperature conversion, 352–353
tenths, 62–63
   as decimals, 296
terms
   of an expression, 134
   collecting, 134–135
   of a sequence, 302–307
   unlike, 134
tessellations, 292–293
three-dimensional geometry, 196–207
three-quarters
   as a decimal, 64
   as a percentage, 88
time, 378–379
   measuring, 16–17
time series, 150
timetables, 16, 17

tonnes, 3, 14
tons, 3, 14
transformations, 392–407
translations, 392–393, 394–395, 398–399
trapezia, 2, 80, 382–383
travel graphs, 174–179
trial and improvement, 362–363
triangles
   acute-angled, 78
   angles, 284–285
   area, 2, 186–187
   congruent, 263, 264
   drawing, 262–265
   equilateral, 2, 78
   hypotenuse, 416–417
   isosceles, 2, 78
   obtuse-angled, 78
   properties, 78–79, 186–187
   right-angled, 2, 78, 416–419
   similar, 405
   types, 2
triangular prism, 196, 388

## U
unitary method, 374–375
units, 2, 3, 12
   converting, 126–127
unknowns, in equations, 232

## V
value added tax (VAT), 164–165
variables, 130
vertical line, reflection, 396–397
vertical line charts, 102–103
vertices, 200–201
views, 197
volume, 202–203, 382–391
   of a cuboid, 3, 202–203, 384–385
   of a cylinder, 352–353, 386–387
   measuring, 14
   of a prism, 386–387

## W
wages, 208–209
weight, measuring, 14
whole number, nearest, 218

## X
$x$ and $y$ co-ordinates, 8

## Y
yards, 12

## Z
zero power, 254